ASCE Manuals and Reports on Engineering Practice No. 130

Waterfront Facilities Inspection and Assessment

Second Edition

Edited by
Noah J. Elwood, P.E., BC.PE, BC.OE

Sponsored by
the Waterfront Facility Inspection Committee of the
Technical Committee on Ports and Harbors of the Coasts, Oceans,
Ports, and Rivers Institute of the American Society of Civil Engi

ASCE

COASTS, OCEANS,
PORTS & RIVERS
INSTITUTE

Published by the American Society of Civil Engineers

Library of Congress Cataloging-in-Publication Data

Names: Elwood, Noah J., editor. | Coasts, Oceans, Ports and Rivers Institute (American Society of Civil Engineers). Waterfront Facility Inspection Committee, sponsoring body.
Title: Waterfront facilities inspection and assessment / edited by Noah J. Elwood, P.E., BC.PE, BC.OE ; sponsored by the Waterfront Facility Inspection Committee of the Technical Committee on Ports and Harbors of the Coasts, Oceans, Ports, and Rivers Institute of the American Society of Civil Engineers.
Description: Second edition. | Reston, Virginia : American Society of Civil Engineers, 2025. | Series: ASCE manuals and reports on engineering practice ; no. 130 | Includes bibliographical references and index.
Identifiers: LCCN 2025006928 | ISBN 9780784416358 (soft cover) | ISBN 9780784486160 (PDF)
Subjects: LCSH: Harbors--Inspection--United States. | Hydraulic structures--Inspection--United States.
Classification: LCC TC223 .W38 2025
LC record available at https://lccn.loc.gov/2025006928

Published by American Society of Civil Engineers
1801 Alexander Bell Drive
Reston, Virginia 20191-4382
www.asce.org/bookstore | ascelibrary.org

***Errata:* Errata, if any, can be found at https://dx.doi.org/10.1061/9780784486160.**

Disclaimer: The publisher has provided alt-text for the EPUB version of this book.

ISBN 978-0-7844-1635-8 (soft cover)
ISBN 978-0-7844-8616-0 (PDF)
ISBN 978-0-7844-8617-7 (ePub)

Manufactured in the United States of America.
29 28 27 26 25 1 2 3 4 5

Waterfront Facilities Inspection and Assessment

Other Titles of Interest

Design of Marine Facilities: Engineering for Port and Harbor Structures, by John W. Gaythwaite (ASCE Press 2016). Gaythwaite provides the essential reference for civil and structural engineers designing port and harbor structures for all types of marine facilities. (ISBN 978-0-7844-7993-3)

Planning and Design Guidelines for Small Craft Harbors, Third Edition, by the Task Committee on Marinas 2020 (ASCE/COPRI 2012). MOP 50 provides new, state-of-the-art guidelines for the planning, design, and development of small craft harbors. (ISBN 978-0-7844-7649-9)

Managing Water in River and Reservoir Systems: Water Resources, Institutional Practices, and Constructed Infrastructure in Texas, by Ralph A. Wurbs (ASCE Press 2024). Author Ralph A. Wurbs provides an overview of effective river and reservoir system water management, discussing the collaborative endeavors of international, federal, interstate, state, river basin, and local agencies, private sector enterprises, and the water management community, using the state of Texas as a case study. (ISBN 978-0-7844-8538-5)

Mooring of Ships to Piers and Wharves, by the Mooring Analysis Task Committee; edited by John W. Gaythwaite (ASCE/COPRI 2014). MOP 129 provides guidelines for the determination of safe mooring design practices for vessels at fixed piers and wharves in ports and harbors. (ISBN 978-0-7844-7842-4)

Sea Level Rise and Coastal Infrastructure: Predictions, Risks, and Solutions, edited by Bilal M. Ayyub and Michael S. Kearney (ASCE/COPRI 2012). CDRM 6 analyzes the challenges posed by rising sea levels and climate change. Using a risk analysis and management framework, this report focuses on various aspects of sea level rise, including associated risks and socioeconomic impacts. (ISBN 978-0-7844-7651-2)

MANUALS AND REPORTS ON ENGINEERING PRACTICE

(As developed by the ASCE Technical Procedures Committee, July 1930, and revised March 1935, February 1962, and April 1982)

A manual or report in this series consists of an orderly presentation of facts on a particular subject, supplemented by an analysis of limitations and applications of these facts. It contains information useful to the average engineer in his or her everyday work, rather than findings that may be useful only occasionally or rarely. It is not in any sense a "standard," however, nor is it so elementary or so conclusive as to provide a "rule of thumb" for nonengineers.

Furthermore, material in this series, in distinction from a paper (which expresses only one person's observations or opinions), is the work of a committee or group selected to assemble and express information on a specific topic. As often as practicable, the committee is under the direction of one or more of the Technical Divisions and Councils, and the product evolved has been subjected to review by the Executive Committee of the Division or Council. As a step in the process of this review, proposed manuscripts are often brought before the members of the Technical Divisions and Councils for comment, which may serve as a basis for improvement. When published, each manual shows the names of the committees by which it was compiled and indicates clearly the several processes through which it has passed in review, so that its merit may be definitely understood.

In February 1962 (and revised in April 1982), the Board of Direction voted to establish a series titled "Manuals and Reports on Engineering Practice," to include the manuals published and authorized to date, future Manuals of Professional Practice, and Reports on Engineering Practice. All such manual or report material of the Society would have been refereed in a manner approved by the Board Committee on Publications and would be bound, with applicable discussion, in books similar to past manuals. Numbering would be consecutive and would be a continuation of present manual numbers. In some cases of joint committee reports, bypassing of journal publications may be authorized.

A list of available Manuals of Practice can be found at https://ascelibrary.org/page/books/s-mop.

CONTENTS

CONTRIBUTORS

Noah J. Elwood, P.E., BC.PE, BC.OE, *Chair and Editor*
Appledore Marine Engineering, LLC.

Danielle M. Somma, P.E., BC.PE, BC.OE, *Vice Chair*
Appledore Marine Engineering, LLC.

Brian L. Phan, P.E., *Vice Chair*
Simpson Gumpertz & Heger

Maximo L. Argo, P.E.
Simpson Gumpertz & Heger

Ben Armitage
NDT Division

Ed Bailey
Northwestern Michigan College

Albert "Lee" Barco, P.E., BC.PE
APM Terminals

Jonathan P. Benvenuto, P.E.
United States Coast Guard

Sean Chapman, P.E.
Marine Solutions

Brian J. DeBaun, P.E.
Jacobs

William Dubbs, P.E.
Moffatt & Nichol

Steven J. Gardner, P.E.
S. T. Hudson Engineers, Inc.

Robert F. Garrity, P.E.
Jacobs

Casey Gomez
Versar Global Solutions, Inc.

Elizabeth Greer, P.E., S.E.
Alaska Railroad Corporation

Rune Iversen, P.E.
Simpson Gumpertz & Heger

Zach Jenkins, P.E., S.E.
Collins Engineers, Inc.

Bryan N. Jones, P.E., BC.PE, BC.OE
HDR

Brad A. Koch, P.E.
Marine Solutions

Derek R. Lobedan, P.E.
Jacobs

Douglas W. Mann, P.E., BC.CE
Aptim Environmental & Infrastructure, LLC

David L. Marcotte, P.E.
Appledore Marine Engineering, LLC.

Konrad Mech, P.Eng.
Clear Horizon Consultants

Todd Mitchell
Ventura Port District

Aziah North
Kongsberg

Kirk F. Riden, P.E., BC.PE, BC.OE
Appledore Marine Engineering, LLC.

Amanda D. Schindhelm, P.E., BC.PE
Marine Solutions

Chris Skorny
Versar Global Solutions, Inc.

Shelley Sommerfeld, P.E.
Echelon Engineering, Inc.

Thomas Spencer, P.E., S.E., BC.PE
Moffatt & Nichol

Bradley A. Syler, P.E., S.E.
Marine Solutions

Matthew L. Teeden, P.E., S.E, BC.PE, BC.OE
Appledore Marine Engineering, LLC.

BLUE-RIBBON PANEL

We thank the following members of the Blue Ribbon Panel for their careful review of this publication and their excellent recommendations for improvements.

Gregor S. Bo, P.E.
US Navy

Joseph Bodzewski, PMP
US Coast Guard

William M. Bruin, P.E.
Simpson Gumpertz & Heger

Ronald E. Heffron, P.E., BC.PE
GHD

Matthew N. Martinez, S.E., BC.PE
Moffatt & Nichol

Henk Voogt
Port of Rotterdam

ACKNOWLEDGMENTS

Special thanks to the support from COPRI, the Ports & Harbors Committee, our esteemed Blue-Ribbon Panel and ASCE Publications and Standards Division. We would also like to acknowledge the Waterfront Facility Inspection Committee comprising more than 50 individual members.

PREFACE

I am pleased to introduce you to the second edition of MOP 130, *Waterfront Facilities Inspection and Assessment*. This manual is the result of a collaboration of industry professionals who share their expertise and experience. It guides the industry in the specialized field of inspecting and assessing waterfront facilities. Ten years after the original publication, this second edition is a complete update of all sections, including new content on polymer composite materials and the latest evolution of inspection tools and technologies.

Waterfront facilities support various uses, from transporting materials and cargo to supporting critical shoreside infrastructure. Too often, deterioration of these facilities below deck or underwater goes unchecked, leading to the loss of operations and damage to the environment or safety of personnel. This manual provides the needed guidance to engineers, inspectors, owners, and operators to effectively manage their waterfront facilities and maximize the return on the infrastructure investment.

This manual provides guidance from the initial scoping of a waterfront inspection project, details of conducting the inspection and documentation and reporting. A wealth of information is contained in the appendixes including a library of structure types, causes of defects and deterioration, and specialized inspection techniques.

I thank all our contributors and the Blue-Ribbon Panel for investing their time to make this latest edition a reality.

Noah J. Elwood, P.E., BC.OE, BC.PE

CHAPTER 1
INTRODUCTION

1.1 INTENT OF THIS MANUAL

This Manual of Practice intends to provide guidelines and methodology for conducting engineering inspections and assessments of existing waterfront facilities exposed to freshwater, brackish water, and seawater environments. The manual covers near-shore, waterfront, and inland facilities constructed of concrete, masonry, metal, polymer composite, and wood materials including, but not limited to, the following:

- Piers (jetties) and wharves (quays);
- Mooring, breasting, and turning dolphins;
- Bulkheads, seawalls, quay walls, and gravity block walls;
- Relieving platforms;
- Caissons and cofferdams;
- Wave screens/attenuators;
- Marina docks and boat ramps;
- Marine railways;
- Floating structures;
- Single-point and multibuoy moorings; and
- Slope protection systems.

For convenience, the manual will use the term *waterfront facilities* to refer to any of the types of facilities listed previously, where appropriate and not expressly limited.

The manual includes guidelines for conducting engineering inspections and assessments of waterfront facilities including structures, fendering, moorings, pavements, drainage, utilities, equipment, and appurtenances.

This manual does not address specialty items such as cranes, material offloading and conveyance equipment, and cargo handling equipment.

The manual is not applicable to bridges, dams, hydraulic structures, (deep water) offshore structures, nuclear facilities, or dry docks. The inspection and assessment of these facility types are more appropriately covered in other industry standards, guidelines, and practice manuals.

This manual is intended as a guide to be used by engineering professionals as part of an engineering inspection and assessment. As engineering inspections and assessments involve numerous factors and require engineering judgment, this manual should be used only by properly trained, qualified, and experienced engineering professionals.

The scope of this manual is limited to the engineering and technical aspects of conducting waterfront facility inspections and assessments. Training in the techniques for safely accessing waterfront facilities to identify and assess deficiencies is beyond the scope of this manual.

1.2 IMPORTANCE OF INSPECTIONS

Inspections are a necessary part of effective waterfront facility maintenance management programs. They play an important role in protecting the public, providing reliable service, protecting the environment, and reducing maintenance and repair costs. Structural conditions above water that could lead to failure, loss of life, or property damage are often observed by engineers, maintenance workers, operation personnel, and the public. These groups almost never observe similar structural conditions located below the deck or underwater until the distress has progressed to the point that damage is evident above water. A history of failures caused by deterioration has led to routine inspection mandates in the United States. Many waterfront facility owners have also adopted similar routine inspection policies for their facilities, recognizing the importance of these assets.

Inspections are also important for waterfront facility maintenance programs. Waterfront facilities are subject to environmental and external physical loads, and their capacity to support these loads diminishes as they deteriorate over time. An engineer cannot fully define the extent of the deterioration or distress nor design an appropriate repair without the benefit of an inspection (both above and below water).

1.3 HOW TO USE THIS MANUAL

Eight unique inspection types are defined in this manual and each has a different scope or purpose. A knowledgeable owner or specifying entity may specify the necessary type(s) when initiating an inspection effort. Less-knowledgeable owners or specifying entities may seek professional engineering guidance on what inspection type(s) to perform. This manual provides guidance in Chapter 2, "Standards of Practice," on various inspection types and how to match specific inspection types to a project's needs.

In many instances, combining inspection types and conducting them simultaneously is desirable. For example, when a single waterfront facility is being inspected because of obvious significant deterioration and the need for repairs is obvious and imminent, combining the routine inspection and repair or upgrade design inspection into one effort may be beneficial.

Once the inspection type(s) are identified, the owner or specifying entity can simply request that the inspection be performed in accordance with this manual, or the professional engineer can state in their proposal that the inspection will be performed in accordance with this manual.

Chapter 2, "Standards of Practice," points out instances in which going beyond the basic scope of work and including additional scope items such as a structural analysis, service life extension analysis, or repair cost study may be desirable.

The recommended interval between inspections can often differ for different portions of a facility (e.g., the above deck portion versus the below deck portion). In most cases, it is ideal to conduct the inspection of the entire facility at the same frequency using the same engineering team. Inspecting the entire facility at the same time using the same team is generally more cost-effective, and it produces a more complete data set, which is better for fully understanding the condition of the entire facility.

Unless otherwise specified, a typical inspection effort should include the waterfront structures, fendering, moorings, pavements, drainage, utilities, equipment, appurtenances, slope protection, cathodic protection, and surrounding mudline.

1.4 ASSESSMENT LIMITATIONS

Waterfront facilities are subject to deterioration over time. A waterfront facility assessment should be considered a snapshot of the condition at a moment in time over the history of the asset. The validity of the assessment is limited in light of this continuum of ongoing deterioration. The manual provides guidance as to the frequency of routine inspections during the facility's life cycle, with the frequency dependent on the condition assessment rating from the previous inspection, the type of construction materials involved, and the nature of the service environment. The manual also provides guidance for quantitatively predicting the remaining service life of waterfront facilities.

Chapter 2, Section 2.6.2, "Condition Assessment Ratings," explains that routine inspections are intended to compare the facility's current condition to the original as-constructed condition. Whether the facility was designed for current design standard loading should have no influence on the condition assessment ratings. Similarly, functional obsolescence and operational loading should not be considered when inspecting or rating a facility, as such information is generally not readily available to the inspection team.

1.5 SIGNIFICANT CHANGES AND OWNER'S RESPONSIBILITIES

An owner may make changes or modifications to a waterfront facility over its lifetime. Changes may include reconfiguring, adding or removing topside facilities, increasing or decreasing topside or vessel loading, or modifying mooring/berthing configurations and/or loading distributions. In these situations, the owner may have certain responsibilities to ensure safe operation of the facility beyond those intended when the facility was originally constructed. Similarly, environmental laws now commonly require owners to protect the environment from adverse impacts more than they likely did when the facility was originally constructed.

When a waterfront facility is to be repaired or rehabilitated to restore it to its original design capacity, consideration should be given to the residual strength of the structural components. Waterfront facilities generally have some reserve load capacity and can typically tolerate some deterioration without impacting the overall structural capacity. Unless the authority having jurisdiction or the engineer of record dictates otherwise, if no significant deterioration or damage is observed, the repair or rehabilitation can typically proceed without performing a structural analysis. The definition of *significant deterioration or damage* in this context should be based on the judgment of a professional engineer using recent inspection data and, when available, a review of the original design documents. This professional engineering judgment should consider the severity of the deterioration or damage as well as whether the deterioration or damage is localized or widespread.

A condition assessment rating of "poor" or below as defined in Chapter 2, Section 2.6, "Overall System Ratings," or deterioration or damage that reduces the original design capacity of a primary structural member by 20% or more is generally considered to be potentially significant. If the deterioration or damage is considered to be potentially significant, a professional engineer should determine the appropriate level of additional testing, evaluation, and analysis. These additional activities may be localized to an individual component and a group of components or generalized to the global structural system.

If a facility is to be upgraded—defined as significantly increasing loads or changing the load path(s)—ensuring the adequacy of the upgraded structural system is the responsibility of the owner. The appropriate level of structural inspection, assessment, testing, evaluation, and analysis required to upgrade a facility should be determined by a professional engineer. In this context, *significantly increasing loads* is when the demand–capacity ratio with the additional loads exceeds 10% of the demand–capacity ratio without the additional loads.

If loads are significantly increased or load paths are significantly modified as part of an upgrade project, an upgrade to comply with current code standards may also be required. The requirements of the authority having jurisdiction often dictate this, but it may also be a matter of prudence, depending on the scope of the upgrades being implemented.

1.6 LIMITS OF INSPECTION

Routine inspections are typically limited to accessible components. Buried components such as tie backs, deadman anchors, or below-grade portions of piles are typically excluded from the scope of work in a routine inspection.

When inspecting a waterfront facility with the intent to upgrade, which involves increasing loads or changing load paths, exposing representative samples of buried or inaccessible components is typically necessary to verify their integrity. Exposure and inspection of inaccessible components may also be required when evidence of potential or likely loss of integrity of the inaccessible components is observed.

Underwater inspection should be performed on all portions of facilities that are below the waterline at the time of inspection. When possible, the inspection should maximize the inspection of tidal areas above water at low tide because of the increased efficiency and improved data collection.

In some special circumstances, performing some limited underwater excavation along the structure/mudline interface may be necessary. Such excavation should be clarified in the scope of work and may require special environmental permits. Underdeck inspections are an essential task to understand deterioration fully. Often, underdecks may require special access requirements from vessels or divers in low freeboard instances.

1.7 TERMINOLOGY

The following terms are used throughout this manual, and defining their meaning in the context of waterfront facilities is important:

- Preservation: The facility is not being repaired but is rather being treated, such as with jackets, cathodic protection, coatings, sealants, or preservatives, to prevent or slow further deterioration.

- Sustainment: The facility has not exceeded its design life and is being repaired or maintained.
- Rehabilitation: The facility is being repaired or restored to its original, as-built condition. This is also referred to as *restoration* when the original design life is exceeded and is now being extended.
- Upgrade: The facility will have new or larger loads applied or the load paths will be significantly modified, thus necessitating an evaluation and analysis of the global structural system. This is also referred to as *retrofit* or *strengthening* when structural modifications are required.
- Inspection: The visual, tactile, and testing associated with determining the physical condition of the facility, compared with the original, as-built condition. This is also commonly referred to as *investigation* or *condition survey*.
- Assessment: The evaluation of the inspection results to determine the significance of observed damage and deterioration on the design capacity of the facility.

1.8 ORGANIZATION

This manual of practice has been organized into chapters and appendixes to provide both general and specific guidance for waterfront facility inspection and assessment projects.

Chapter 2, "Standards of Practice," provides an overview of the types of inspections, guidance on how to choose the appropriate inspection type, general requirements for conducting waterfront facility inspections, and guidelines for inspection frequencies. In addition, the recommended qualifications of inspectors are presented along with assessment rating guidelines that are applicable to or can be readily adapted to most waterfront facilities.

Chapter 3, "Scope of Inspection Work," defines the objective and detailed scope of work for each of the eight inspection types. Guidance is also provided regarding methodology and how to make appropriate recommendations for any follow-on actions required.

Chapter 4, "Service Life Estimation," provides guidance on quantitatively assessing the remaining useful service life of a waterfront facility. Field inspection and testing requirements, laboratory testing requirements, and numerical modeling techniques are provided.

Chapter 5, "Documentation and Reporting," provides guidelines for the preparation of a routine inspection report, often referred to as a condition assessment inspection report.

Chapter 6, "Administrative Considerations and Planning," provides general guidelines for developing agreements between consultants and facility owners. It also presents an overview of the special requirements for insurance requirements related to performing waterfront facility inspections.

Appendix A, "Special Considerations for Specific Structure Types and Systems," provides descriptions of common problem areas associated with specific structure types and systems.

Appendix B, "Types and Causes of Defects," provides general information on the various mechanisms of damage and deterioration that are applicable to the types of materials commonly used to construct waterfront facilities.

Appendix C, "Overview of Specialized Techniques," provides an overview of specialized inspection and testing techniques that can be used to investigate a waterfront facility more thoroughly.

Appendix D, "Inspection Nomenclature," provides an overview of waterfront facility nomenclature commonly used to record and report configuration and defect information.

The Glossary provides a list of generally accepted standard terms related to waterfront facilities.

CHAPTER 2
STANDARDS OF PRACTICE

This chapter presents requirements for the above-water and underwater inspection of waterfront facilities. It describes recommended practices intended to ensure that facilities are adequately maintained to protect life, the environment, and property and maximize the structure's service life. These requirements include

- Type and frequency of inspections,
- Selection of the proper inspection method,
- Service life modeling or estimation,
- Minimum qualifications of inspection personnel,
- Element-level ratings,
- System ratings, and
- Recommended action guidelines.

2.1 TYPE AND FREQUENCY OF INSPECTIONS

2.1.1 General

This manual covers the following eight inspection types. The inspection types cover a range of specific purposes to inform an owner in the maintenance and management of a waterfront facility.

- Routine inspection,
- Structural repair or upgrade design inspection,
- New construction inspection,
- Baseline inspection,
- Due diligence inspection,
- Special-purpose inspection,

- Repair construction inspection, and
- Post event inspection.

Routine inspections, structural repair or upgrade design inspections, special-purpose inspections, and repair construction inspections define routine maintenance activities. The routine inspection is the most common type and completed on a cyclical basis to assess a facility's fit for purpose and to support a maintenance and repair program. The routine inspection should be used as a screening mechanism to determine when other inspections should be conducted. Due diligence inspections are conducted when considering a change in ownership, an investment, insurance valuation, or negotiation of a lease. Postevent inspections are conducted only in response to a significant loading or environmental event.

New construction inspections are conducted only in association with a newly constructed structure or component. Baseline inspections are typically conducted near the completion of new construction, prior to owner acceptance, but they may also be conducted on existing structures coinciding with the first routine inspection.

The typical flow and context of inspection activities associated with the eight inspection types are shown in Figure 2-1. The figure represents a typical model of how inspection activities may flow. In many cases, combining inspection types may be necessary or advantageous to avoid duplication of effort or minimize mobilization costs. Deviating from the typical flow of activities may also be necessary to tailor the inspection scope of work to the global project requirements. These inspection types include both above-water and underwater efforts.

Routine inspections should be conducted with sufficient detail to evaluate the structure's overall condition, determine whether further maintenance is necessary, and determine the priority of such activities. Documentation of inspection results should be limited to the collection of data necessary to support these objectives to minimize the expenditure of maintenance resources.

Structural repair or upgrade design inspections are conducted only when repairs must be performed, as determined by the routine inspection. Structural repair or upgrade design inspections may take more time to execute than routine inspections because they require detailed documentation of existing conditions and all defects to be repaired. By using this two-tiered approach to the inspection process, inspection resources are utilized efficiently. For example, completing a repair design inspection on an entire facility is expensive and a waste of resources on portions of the structures or elements not requiring repairs.

A routine inspection is not always necessary prior to conducting a structural repair or upgrade design inspection. In situations where the need for repairs is known or obvious, or for small facilities, conducting the

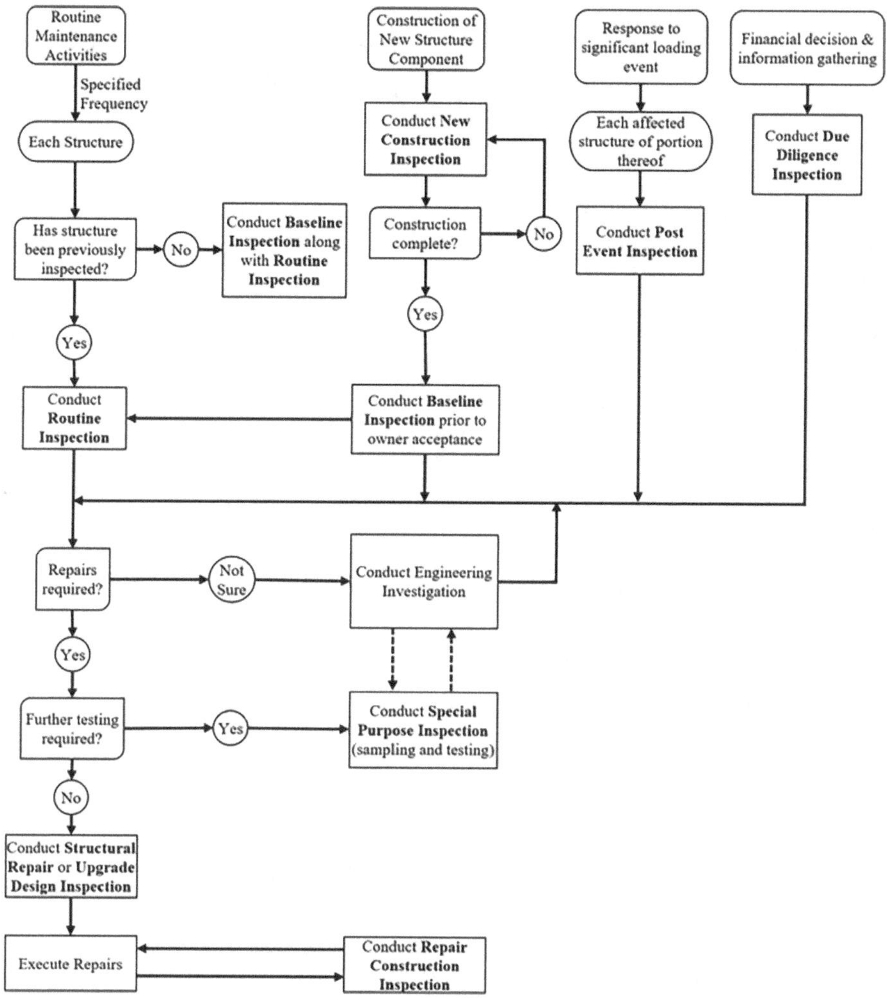

Figure 2-1. Flow and context of inspection activities.
Source: Courtesy of Moffatt & Nichol Inc., reproduced with permission.

routine inspection and the structural repair or upgrade design inspection simultaneously may be advantageous.

During the construction of new structures, above-water and underwater inspections should be conducted to ensure proper quality control and conformance with the contract documents. This type of inspection is called a new construction inspection.

To document the as built conditions of a facility including member sizes and details of the original construction a baseline inspection is conducted.

A limited inspection of a structure may establish the general condition and valuation; order-of-magnitude repair costs, repair methodology, and interval; and inspection costs and inspection intervals. This type of inspection is called a due diligence inspection.

In some cases, a more in-depth investigation involving various types of in situ and laboratory testing may be required. This type of inspection is called a special-purpose inspection. Special-purpose inspections are conducted when the cause of deterioration is unknown, or where multiple degradation mechanisms contribute to the deterioration.

While implementing repairs, above-water and underwater inspections should be conducted to ensure proper quality control and documentation as well as verify compliance with contract documents. This type of inspection is called a repair construction inspection.

In the event of an earthquake, vessel impact, tsunami, fire, or flood, conducting a rapid inspection to verify damage and ensure the safety of personnel and equipment may be necessary. This type of inspection is called a postevent inspection. Table 2-1 summarizes the purpose and frequency of inspection types. A description of each inspection type follows. Chapter 3, "Scope of Inspection Work," presents details of the scope of work for each inspection type.

2.1.2 Routine Inspections

2.1.2.1 Purpose and Frequency. A routine inspection aims to assess the structure's general condition, assign a condition assessment rating, and make recommendations. Routine inspections should be performed on a cyclical basis and represent a proactive approach to maintenance.

The frequency with which routine inspections should be conducted is a function of several variables. The most important variables are material type, service environment, and condition rating. The recommended frequency between inspections varies and is presented in Table 2-2. The frequencies represent maximum intervals between inspections and should be reduced as appropriate on the basis of the extent of deterioration observed in a structure, the rate of further anticipated deterioration, and the importance of the structure.

Conducting routine inspections on a cyclical basis allows for the detection of deteriorated elements and provides the opportunity to repair them before structural integrity is threatened. In addition, significant damage caused by impact from vessels or floating debris will be detected.

2.1.2.2 Scope of Work Overview. The above-water inspection includes a visual observation of the topside and above-water components. The above-water components are typically inspected from a boat; however, depending on circumstances, using divers may be necessary for low-clearance facilities

Table 2-1. Summary of Inspection Types.

Inspection type	Purpose	Frequency
Routine inspection	To assess general condition, assign condition rating, and provide recommendations for future maintenance	As indicated in Table 2-2
Structural repair or upgrade design inspection	To record relevant attributes of each defect to be repaired such that repair bid documents may be generated	Conducted only when designing repairs or upgrades and only on the portion of the structure to be repaired
New construction inspection	To ensure proper ongoing quality of new construction in accordance with plans and specifications	During the construction of new structures or components
Baseline inspection	For new construction, to verify that construction plans have been followed and to ensure that construction is free of significant defects prior to owner acceptance. For existing structures, to verify dimensions and construction configuration details	Prior to owner acceptance of newly constructed structures or at the time of the first routine inspection on existing structures
Due diligence inspection	To form an engineering opinion of the general condition of a structure and estimate order-of-magnitude replacement costs and repair costs	Conducted as part of a financial, investment, or insurance valuation process
Special-purpose inspection	To perform detailed testing or investigation of a structure required to understand the nature and extent of the deterioration prior to determining the need for and type of repairs required	Conducted only when deemed necessary as a result of a routine or structural repair or upgrade design inspection
Repair construction inspection	To ensure proper quality of repairs, resolve field problems, and ensure proper documentation of payment quantities	During repair projects involving structures or components
Postevent inspection	To perform a rapid evaluation of a structure following an earthquake, storm, vessel impact, fire, tsunami, or similar event to determine whether further attention to the structure is necessary as a result of the event	Following a significant potentially damage-causing event

Table 2-2. Recommended Maximum Interval between Routine Inspections (Years).[a]

Condition rating from previous inspection		Construction material: Unwrapped timber or unprotected steel (no coating or cathodic protection)		Concrete, masonry, wrapped wood, protected steel, or polymer composite materials		Mudline scour[b] (soundings)	
		Benign[c] environment	Aggressive[d] environment	Benign[c] environment	Aggressive[d] environment	Benign[c] environment	Aggressive[d] environment
6	Good	6	4	6	5	6	2
5	Satisfactory	6	4	6	5	6	2
4	Fair	5	3	5	4	6	2
3	Poor	4	3	5	4	6	2
2	Serious	2	1	2	2	2	1
1	Critical	Use and occupancy of facilities in critical condition is highly discouraged until repairs can be made.					

[a]The maximum interval between routine inspections may be reduced on the basis of the extent of deterioration, anticipated deterioration, and importance of the structure. Intervals may be increased for atypical cases where special construction materials are used. Regulations may dictate a maximum inspection interval.

[b]Two maximum intervals are shown, one for the assessment of construction material (wood, concrete, steel, etc.) and one for scour (last two columns). The frequency may dictate more frequent intervals for soundings compared to the routine inspection.

[c]*Benign environments* include freshwater with low-to-moderate currents (current < 0.75 knots).

[d]*Aggressive environments* include brackish water, seawater, polluted water, or waters with currents > 0.75 knots. Facilities that handle chemicals containing elements detrimental to the structure's durability, such as chlorides, sulfates, or alkalis, are aggressive environments.

or a snooper vehicle for high-clearance or open-ocean structures. The topside is typically inspected from the deck.

Underwater inspection is limited by poor visibility and the coverage of components by marine growth. A visual inspection of all submerged component surfaces during a routine inspection is impractical. For this reason, routine inspections focus on three levels of effort: visual and tactile inspection (Level I), partial marine growth removal of a representative sample (Level II), and non-destructive testing or partially destructive testing (Level III). The three levels are further defined in Chapter 3, "Scope of Inspection Work."

During the inspection, damage ratings are assigned to each element inspected to characterize the existing level of damage or deterioration. Upon completion of the inspection, a condition assessment rating is assigned to each element group or structure inspected considering each of the damage ratings assigned. Recommendations for additional follow-up activities are provided as appropriate, as also the recommended interval to the next routine inspection. If significant damage or deterioration is observed, a quantitative engineering assessment of the damage's effect on the structure's capacity should be recommended.

The assessment is typically limited to an evaluation of the capacity of typical components relative to their original, as-built condition and does not consider the actual or anticipated loading (structural demand), because such information is typically not readily available to the inspectors at the time of the routine inspection and beyond the scope of work. The results of such structural assessments should be used to assign a condition assessment rating. Should conditions warrant, an engineering evaluation should be recommended to evaluate the actual or anticipated loading against the reduced capacity determined as a result of the routine inspection.

Further details of the scope of work involved in conducting routine inspections are presented in Chapter 3, "Scope of Inspection Work."

2.1.3 Structural Repair or Upgrade Design Inspections

2.1.3.1 Purpose and Frequency. The purpose of the structural repair or upgrade design inspection is to collect existing conditions around the repairs and record all relevant defect attributes, such that repair bid documents may be generated. Structural repair or upgrade design inspection should be conducted only when repairs are to be performed and should be conducted with as little interval as practicable between the time of the structural repair or upgrade design inspection and the execution of repairs. Minimizing the interval between repair and the design inspection will not only help restore the facility but will minimize the number of change orders due to unforeseen conditions. Typically, one year has been accepted as a practical interval yielding sufficient time to develop the repair documents, bid and award the construction contract, and initiate construction without the design level inspection data changing sufficiently to cause significant change orders.

Typically, structural repair or upgrade design inspection results from a recommendation made following a routine inspection; these results are not sufficient for a repair design and require additional engineering resources to prepare a sufficient bid document. However, when the need for repairs is obvious, and the priority is clear, a structural repair or upgrade design inspection may be conducted without being preceded by a routine inspection or may be combined with a routine inspection.

2.1.3.2 Scope of Work Overview. Prior to commencing a structural repair or upgrade design inspection, the criteria for what defects should be repaired should be established. In addition, determining the method(s) of repair for all typical situations on the structure may be beneficial. The structural repair or upgrade design inspection is then performed to document the location and size of defects to be repaired and to assign a method of repair that is based on the preestablished criteria. Careful attention should be given to potential construction challenges for access to the work or interferences from utilities or other appurtenances.

Estimating the size of each defect to be repaired is important to prepare reasonably accurate quantity and cost estimates for the repair project. Ideally, the structural repair or upgrade design inspection should be conducted only on the components identified as needing repair during the routine inspection. Therefore, if the components requiring repair are limited to a certain area of the structure or specific component types, the scope of the structural repair or upgrade design inspection may be tailored to these specific areas to optimize resources and minimize costs.

The method of investigation used in a structural repair or upgrade design inspection must be tailored to the type and extent of deterioration observed. Details on the scope of structural repair or upgrade design inspection are presented in Chapter 3, "Scope of Inspection Work."

2.1.4 New Construction Inspections

2.1.4.1 Purpose and Frequency. New construction inspections should be performed during the construction of new structures, or the construction of components, to ensure proper construction quality in accordance with the design plans and specifications.

2.1.4.2 Scope of Work Overview. The scope of the new construction inspections should include quality control inspection of structures under construction for compliance with the construction documents and resolution of field problems. Details of the scope of new construction inspections are presented in Chapter 3, "Scope of Inspection Work."

2.1.5 Baseline Inspections

2.1.5.1 Purpose and Frequency. A baseline inspection aims to verify that the structure is built according to the design drawings and to ensure no significant defects exist before owner acceptance. Baseline inspections are typically conducted on newly constructed facilities prior to owner acceptance or final payment to the contractor. If the baseline inspection is not conducted at the time of original construction, it may be conducted simultaneously with the first routine inspection.

The purpose of conducting a baseline inspection on an existing structure is to verify that the structure is built according to the design drawings. Where no drawings exist, the purpose of the baseline inspection is to gather sufficient information to develop plan and section drawings of the structure.

2.1.5.2 Scope of Work Overview. The scope of a baseline inspection typically includes confirmation of overall dimensions, pile plan, and other physical features. For new construction, the baseline inspection may also include confirmation of water depths or dredging, ensuring that construction is free of significant defects and that construction debris has been removed. For existing structures, the scope may include detailed measuring and testing to develop drawings of the structure where none exist. Details of the scope of a baseline inspection are presented in Chapter 3, "Scope of Inspection Work."

2.1.6 Due Diligence Inspections

2.1.6.1 Purpose and Frequency. The purpose of the due diligence inspection is to provide the necessary information regarding the general condition of an existing facility; recommend repairs and provide an opinion on potential costs; and provide a projection of future repair, inspection, and maintenance costs. The inspection gathers only the limited information necessary for an engineer to form an opinion about the general condition of a structure; order-of-magnitude valuation; order-of-magnitude repair costs, repair methodology, and repair interval; and inspection costs and inspection interval recommendations.

2.1.6.2 Scope of Work Overview. The scope of the due diligence inspection is a visual/tactile inspection to determine the general condition of the facility, identify significant damage requiring repairs, identify any maintenance repairs, and determine the ability of the facility to support the current or anticipated operations. Financial or insurance considerations often dictate the need and scope of a due diligence inspection. Examples include the purchase of a marine terminal, negotiating or renegotiating a lease, or considering a new insurance carrier. Acknowledging the financial importance of a due diligence inspection, attention should be given to recovering all available records of the facility, including past inspection reports.

The inspection and ratings should follow the guidelines of a routine inspection as described in Section 2.1.2. Documentation resulting from a due diligence inspection may not be as detailed as a routine inspection, depending on the time required to complete the inspection and report. Because due diligence inspections are performed as part of a financial transaction, the results are often required sooner than performing a routine inspection would allow. Should this be the case, the observations' limited inspection effort and generalization should be stated explicitly, and the inspection results should be qualified as appropriate. Details of the scope of a due diligence inspection are presented in Chapter 3, "Scope of Inspection Work."

2.1.7 Special-Purpose Inspections

2.1.7.1 Purpose and Frequency. Special-purpose inspections are conducted to collect more detailed information than normally collected during a routine or structural repair or upgrade design inspection. Such information may be necessary to understand the nature and extent of deterioration prior to determining the need and type of repairs; special testing, analysis, monitoring, or investigation using nonstandard equipment/techniques is typically required. Special-purpose inspections may also be performed to estimate the remaining useful service life of the structure.

An example of the need for a special-purpose inspection is a concrete structure with piles that are soft below the waterline. In such a case, the special-purpose inspection would include coring, testing, and analysis to determine whether the cause of the softness is related to sulfate attack, alkali–silica reaction (ASR), or delayed ettringite formation. Just proceeding with installing jackets on piles with expansive reactions prior to determining the root cause can result in premature failure of the repairs. Another example of the need for a special-purpose inspection would include the use of ground-penetrating radar to determine whether there are structures, utilities, voids, and so on, within a facility causing a particular issue. Determining the root cause of a defect or deterioration is the fundamental step in assuring a lasting repair.

Special-purpose inspections are typically performed on a case-by-case basis as a result of a recommendation made following a routine inspection. However, a special-purpose inspection may also be performed concurrently with a routine inspection, structural repair, or upgrade design inspection where appropriate.

2.1.7.2 Scope of Work Overview. The scope of a special-purpose inspection may vary widely depending on the objectives of the inspection and the nature of the deterioration. Examples of common special-purpose inspections used along waterfront facilities are listed subsequently:

- Concrete coring for physical testing and laboratory analysis to include strength testing, composition analysis, dynamic modulus of elasticity testing, static modulus of elasticity testing, specific gravity and absorption testing, petrographic analysis, scanning electron microscopy, differential thermal analysis, and so on;
- Concrete chloride content evaluation;
- Half-cell potential measurements;
- Ultrasonic remaining thickness measurements;
- Rebound hammer;
- Penetration resistance testing;
- Pulse velocity and pulse echo testing;
- Crack monitoring;
- Settlement monitoring;
- Ground-penetrating radar investigations;
- Sub-bottom profiling;
- Side scan or multibeam sonar investigations;
- Wood preservative retention;
- Timber component removal and dissection;
- Timber resistance drilling test;
- Fluorometer dye/leak detection testing;
- Scour analysis or soil particle size analysis;
- Magnetic particle inspection;
- Coating thickness or continuity inspection;
- Coupon sampling and metallurgical analysis; and
- Postmarine growth removal inspection.

The requester of the services must specify the type of inspection or testing technique, which is typically based on a recommendation made following a routine inspection. Details on the scope of a special-purpose inspection are presented in Chapter 3, "Scope of Inspection Work."

2.1.8 Repair Construction Inspections

2.1.8.1 Purpose and Frequency. Repair construction inspections should be performed during the execution of repair projects to ensure proper quality of repairs, resolve field problems, and ensure impartial documentation of payment quantities. While evaluating work against the contract specifications, the on-site inspector may also evaluate contractor claims for progress payments or additional work. Repair quantities and specified repair methods are estimates based on the best judgment of the inspection engineer during the structural repair or upgrade design inspection. Widespread removal of unsound materials may reveal differing conditions. The need for attentive inspection during the execution of repairs is important to protect the interests of the owner. This is particularly true where payment for repair work will be based on unit pricing.

2.1.8.2 Scope of Work Overview. The scope of the repair construction inspection should include quality control inspection of repairs for compliance with the specifications and may also include resolution of field problems, evaluation of contractor claims, and determination that repair quantities are properly recorded where necessary. Details of the scope of repair construction inspection are presented in Chapter 3, "Scope of Inspection Work."

2.1.9 Postevent Inspections

2.1.9.1 Purpose and Frequency. Postevent inspections should be conducted following a significant, potentially damage-causing event such as a flood, earthquake, storm, vessel impact, or tsunami. The primary purpose of a postevent inspection is to rapidly assess the structural stability of the structure, document and quantify the damage, and determine whether further action is necessary.

2.1.9.2 Scope of Work Overview. Postevent inspections are intended to be somewhat rapid visual/tactile inspections to determine whether the event resulted in any significant damage requiring repairs or load restrictions. The need for and scope of a postevent inspection is generally dictated by the type and severity of the event. For example, a major flood may result in scour conditions that necessitate an underwater inspection. An earthquake or vessel impact often results in damage above the waterline as well as below; an underwater inspection may be triggered only where above-water damage is visible.

If a postevent inspection is required, the amount of marine growth removal required for the inspection should be based on the type of damage that may have occurred, whereas gross breakages or mudline evaluations may require limited marine growth removal. However, overstressing cracks on concrete piles may dictate higher levels of growth removal.

Documentation resulting from a postevent inspection may be minimal. A simple rating system is used to indicate whether further action is required and the level of urgency. The rating system used for a postevent inspection is different from the rating system used during a routine inspection because the postevent rating should focus on event-related damage only. However, general observations of significant damage not related to the event may be cause for the owner to complete a full routine inspection. Details of the scope of a postevent inspection are presented in Chapter 3, "Scope of Inspection Work."

Table 2-3. Matching Inspection Objectives with Eight Inspection Types.

Objective	Inspection type	Included in standard scope of work	Addition to the standard scope of work
Ensure quality control during new construction	New construction inspection	✓	
Verify installed quantities for contractor payment	New construction inspection	✓	
Respond to field questions and problems during new construction	New construction inspection	✓	
Verify that the structure is built in general compliance with the design drawings, if available	Baseline inspection	✓	
Ensure that the new structure has no significant defects prior to the owner's acceptance	Baseline inspection	✓	
Generate design drawings where no drawings exist	Baseline inspection		✓
Assess and rate the overall condition of an existing structure	Routine inspection	✓	
Determine what future maintenance activities are necessary for a structure	Routine inspection	✓	

(Continued)

Table 2-3. (*Continued*).

Objective	Inspection type	Included in standard scope of work	Addition to the standard scope of work
Quantitatively evaluate the local loss of structural capacity of typical components as a result of damage or deterioration	Routine inspection	✓	
Quantitatively evaluate the global structural integrity relative to actual loads on the structure, considering observed damage or deterioration	Routine inspection		✓ (Engineering evaluation)
Estimate the remaining useful service life of the structure	Routine inspection		✓ (May require special-purpose inspection)
Develop order-of-magnitude estimates of probable costs for rehabilitation work	Routine inspection		✓
Assess the general condition of an existing facility	Due diligence inspection	✓	
Develop order-of-magnitude valuation of the facility	Due diligence inspection	✓	
Develop order-of-magnitude rehabilitation costs, including future maintenance and inspections	Due diligence inspection	✓	

(*Continued*)

Table 2-3. (*Continued*).

Objective	Inspection type	Included in standard scope of work	Addition to the standard scope of work
Provide an opinion of the facility's ability to support present and anticipated operations	Due diligence inspection	✓	
Document details of defects and components to be repaired	Structural Repair or Upgrade Design Inspection	✓	
Develop detailed quantity estimates for rehabilitation work	Structural Repair or Upgrade Design Inspection	✓	
Develop detailed repair plans (bid documents), including drawings and specifications	Structural Repair or Upgrade Design Inspection		✓
Determine the cause of observed deterioration to fix or prevent it in the future, where such cause is not readily apparent	Special-Purpose Inspection*		✓
Quantify the extent of observed deterioration	Special-Purpose Inspection*		✓
Determine the structural significance of observed damage	Special-Purpose Inspection*		✓

(*Continued*)

Table 2-3. (*Continued*).

Objective	Inspection type	Included in standard scope of work	Addition to the standard scope of work
Determine the significance of observed damage on future durability	Special-purpose inspection*		✓
Ensure quality control of repairs during construction	Repair construction inspection	✓	
Respond to field questions and problems during construction of repairs	Repair construction inspection	✓	
Ensure impartial documentation of repair quantities during the construction of repairs	Repair construction inspection	✓	
Assess and rate structural integrity following a significant loading or environmental event	Postevent inspection	✓	
Determine whether additional remedial action is necessary on a structure as a result of a significant loading or environmental event	Postevent inspection	✓	

*Special-purpose inspections have no standard scope of work. Each special-purpose inspection should be conducted for a predefined purpose, and such purposes may vary considerably.

2.2 SELECTION OF THE PROPER INSPECTION TYPE

Each of the eight inspection types defined herein has a distinct purpose; these inspection types are not necessarily exclusive. Inspection types may be combined freely to meet the global objectives of a project.

Table 2-3 lists the most common inspection objectives and provides guidance on choosing the inspection type or types that meet the needs of the project. Guidance is also provided in Table 2-3 to indicate whether the inspection objective is included in the standard scope of work for an inspection type or whether the objective is nonstandard. Nonstandard objectives must be specifically stated when defining the scope of work for an inspection or repair project.

2.3 SERVICE LIFE MODELING: PURPOSE AND VALUE

The use of inspection data, engineering assessments, and cost data for planning, budgeting, and management of facilities often necessitates estimating the predicted remaining service life. Typically, this estimate of remaining service life is based on the assessing engineer's experience and judgment. In most cases, this is reasonable and acceptable. In cases where quantitative information is needed, such as critical facilities, or in cases that may be adversarial or controversial, a predictive tool to estimate the service life of materials can be valuable. Service life modeling uses site-specific materials and boundary conditions to predict the service life (defined in more detail in Chapter 4, "Service Life Estimation"). Situations in which such service life modeling would be valuable include:

- **Performance Specifications**: When specifying concrete service life, a quantitative tool is desirable to measure the predicted service life of the various alternatives under consideration. For example, when considering various cements, pozzolans, admixtures, aggregate, sand, water/cementitious ratios, concrete cover, and reinforcing types, optimizing the concrete mixture to meet the specified service life may be beneficial to a contractor.
- **Existing Structures**: When assessing an existing structure and its various elements that may have not reached the corrosion threshold (average critical chloride content at the initiation of corrosion for a specific material), modeling the remaining service life can be beneficial for the planning and future management of a facility. Being able to quantitatively measure the impact of various alternatives or options on the remaining service life of the various elements of a structure can aid in managing and planning the maintenance of a facility. For example, when considering when or how to repair a facility, knowing

the impact on the remaining service of the elements is essential in determining the most economical solution to reach the desired service life goals. When assessing steel structures, corrosion rates can be modeled for the remaining service life, with and without treatment.

- **Construction Variances**: When assessing a variance's impact on the service life of a structure or an element, having a tool to quantitatively measure the remaining service life is valuable in modeling the difference in service life between the as-designed condition and the as-built condition. For example, if concrete is placed with a lower water/cementitious ratio but with less cover than designed, does it have the same, more, or less service life than that designed? If it is less, various alternatives could be modeled to determine which alternative would achieve the desired service life.

2.4 MINIMUM QUALIFICATIONS OF INSPECTION PERSONNEL

2.4.1 General

The inspection of waterfront structures requires a specialized understanding of the engineering applications and technologies unique to waterfront structures and the complex marine environment. Professionals with the appropriate education, experience, and expertise must conduct the inspections. All personnel should have the qualifications, including education and practical experience, required to perform the inspection or specific technical task assigned if working as part of a team. The entire process, from scoping to reporting, and all inspection types should be performed under the direction of a registered professional engineer qualified in the appropriate discipline.

A properly executed inspection goes well beyond the mere logging of observed defects. The nature of on-site inspection work necessitates that engineering judgment be applied to decisions made throughout the inspection process. The minimum qualifications for on-site personnel are to ensure that they have the necessary training, education, and experience to observe, assess, and exercise sound judgment. Specific reasons for such judgment include

- Assigning condition assessment ratings to a structure requires understanding load paths and the structural significance of observed damage. For example, assessing the significance of a deteriorated wood cross brace versus a deteriorated timber pile requires understanding structural redundancies and alternate load paths and understanding where the section loss occurs on the member relative to the point of maximum bending moment or shear.

- Quantifying and evaluating the structural significance of damage requires a first-hand knowledge of the deterioration and the judgment to know what specific data should be collected to support the structural analysis. For example, corroded steel piles may require corrosion profiling of representative members to evaluate section loss against axial forces and bending moments at various points along the piles.
- Estimating the remaining useful life of a structure requires a detailed understanding of the deterioration mechanism(s) and rates. For example, the rate of chloride intrusion into a particular concrete structure may be determined and compared to the corrosion threshold to estimate the remaining useful life of the structure.
- Determining the most appropriate method of conducting a structural repair or upgrade design inspection requires a detailed understanding of which repair methods will be cost-effective and economical. For example, piles that will be jacketed may not require the same level of inspection effort as piles on which each defect will be repaired individually.
- Determining the proper repair method for each defect requires understanding the deterioration process. For example, distinguishing an overstressing crack from a corrosion crack is important because, whereas an overstressing crack may be repaired by epoxy injection, such a repair on a corrosion crack would be inappropriate as a long-term repair solution.

This section defines the recommended minimum qualifications of the project manager, team leader, and team members. The number of members of the inspection team and the qualifications and expertise of the individual members should be evaluated in consideration of the engineering requirements of the task and the requirements of the facility owner and regulatory agencies involved. Although safety considerations are not covered in this manual, the safety-related qualifications should be evaluated separately, and all inspection personnel should be trained or certified and experienced in the applicable safety practices necessary to conduct the inspections.

The minimum qualifications for personnel performing underwater inspections should be equivalent to those performing above-water inspections. All underwater inspections and diving operations should be conducted in accordance with applicable regulations. In the United States, diving is usually performed in accordance with the requirements of the federal Commercial Diving Standards of the Occupational Safety and Health Administration (OSHA) and state OSHA requirements, as applicable.

2.4.2 Project Manager

The inspection team should be under the direction and supervision of a project manager who understands the overall goals and objectives of the inspection. The project manager should be a registered professional engineer specializing in civil, structural, or ocean engineering with at least 10 years of experience in a responsible capacity for the inspection, design, and construction of waterfront structures.

2.4.3 Team Leader

The inspection team should be led by and be under the direct on-site supervision of a team leader. The team leader should be a registered professional engineer and should have a minimum of 5 years of experience performing similar waterfront inspections. The team leader should be at the site for the duration of the inspection and direct the inspection team to ensure that each element is inspected and its condition properly documented.

When unusual conditions, significant structural deficiencies, or unusual construction are encountered, the team leader should personally observe and evaluate the condition. The team leader should periodically communicate with the project manager to report the inspection findings and receive instructions. For underwater inspections, the team leader should also be a trained commercial diver and actively participate by personally conducting the underwater inspection of a minimum of 25% of the structure.

2.4.4 Team Members

Team members involved in inspection and note-taking or documentation work shall be trained inspectors with relevant certifications, such as having completed the ASCE Waterfront Facilities Inspectors Certificate training course or are graduates of a 4-year engineering curriculum and certified as an engineer-in-training (EIT). Team members trained under the prior standards before the publication of this edition would be considered grandfathered.

Other personnel performing manual tasks, such as removing marine growth, or supporting diving operations, but not conducting or reporting inspections, may have lesser qualifications. In addition, other technicians and divers with special knowledge, skills, or experience may be part of the team as required to support the objective. Team members involved in underwater inspections should also be certified commercial divers.

2.5 ELEMENT-LEVEL DAMAGE RATINGS

A damage rating is assigned to each element inspected during an investigation. The rating reflects the condition of the individual element only and is independent of the element's structural importance and the type of inspection being conducted.

Element-level damage ratings are standardized to provide a qualitative description of an element's condition on the basis of a quantified level of damage. By using a quantified scale, objectivity is maintained throughout the inspection.

The following sections present damage ratings typically used for timber, steel, reinforced concrete, and prestressed concrete elements.

2.5.1 Timber Elements

Typical damage ratings used for timber elements are described in Table 2-4 and depicted in Figure 2-2.

2.5.2 Steel Elements

Typical damage ratings used for steel elements are described in Table 2-5 and depicted in Figure 2-3.

2.5.3 Reinforced Concrete Elements

Typical damage ratings for reinforced concrete elements are described in Table 2-6 and depicted in Figure 2-4.

2.5.4 Prestressed Concrete Elements

Typical damage ratings for prestressed concrete elements are described in Table 2-7 and depicted in Figure 2-5.

2.5.5 Mooring System Elements

Typical damage ratings for mooring systems are described in Tables 2-8 and 2-9 and depicted in Figures 2-6 through 2-8.

2.5.6 Fender System Elements

Typical damage ratings for fender system elements are described in Tables 2-10 through 2-13 and depicted in Figures 2-9 through 2-13.

2.5.7 Fiber-Reinforced Polymer Composite Systems

Failure modes for polymer composites consist of the following: material degradation, mechanical damage, overstressing, deformation, and breakage/cracking.

Typical damage ratings for polymer composite systems are described in Tables 2-10, and 2-14 and depicted in Figures 2-9, 2-14 and 2-15.

Table 2-4. Damage Ratings for Timber Elements.

Damage rating	Existing damage[a]	Exclusions [defects requiring elevation to the next higher damage rating(s)]
NI Not inspected	• Not inspected or inaccessible, or passed by[b]	
ND No defects	• Sound surface material	
MN Minor	• Checks, splits, or gouges less than 1/2 in. wide • Minor corroded hardware • Evidence of marine borer or fungal decay without section loss	Minor damage rating not appropriate if • Loss of cross-sectional area • Marine borer and fungal decay with loss of section • Displacements, loss of bearing, or connections
MD Moderate	• Checks, splits, and gouges wider than 1/2 in. • Diameter loss of up to 15% • Cross-sectional area loss up to 25% • Moderately corroded hardware • Evidence of marine borers or fungal decay with loss of section	Moderate damage rating not appropriate if • Displacements, loss of bearing, or connections
MJ Major	• Checks or splits extending through cross section • Loss of diameter between 15% and 30% • Cross-sectional area loss between 25% and 50% • Heavily corroded hardware with section loss • Displacement and misalignments at connections	Major damage rating not appropriate if • Partial or complete breakage
SV Severe	• Diameter loss of more than 30% • Cross-sectional area loss of more than 50% • Loss of connections and/or fully nonbearing condition • Breakage	

Note: Diameter loss is assumed from outside toward the center and does not account for internal loss of cross-sectional area.

[a]At least one defect listed within a damage rating is sufficient to assign the corresponding damage grade, but it is not necessary; the assessing professional engineer shall make final determination on the basis of the overall system.

[b]If not inspected because of inaccessibility or passed by, note as such.

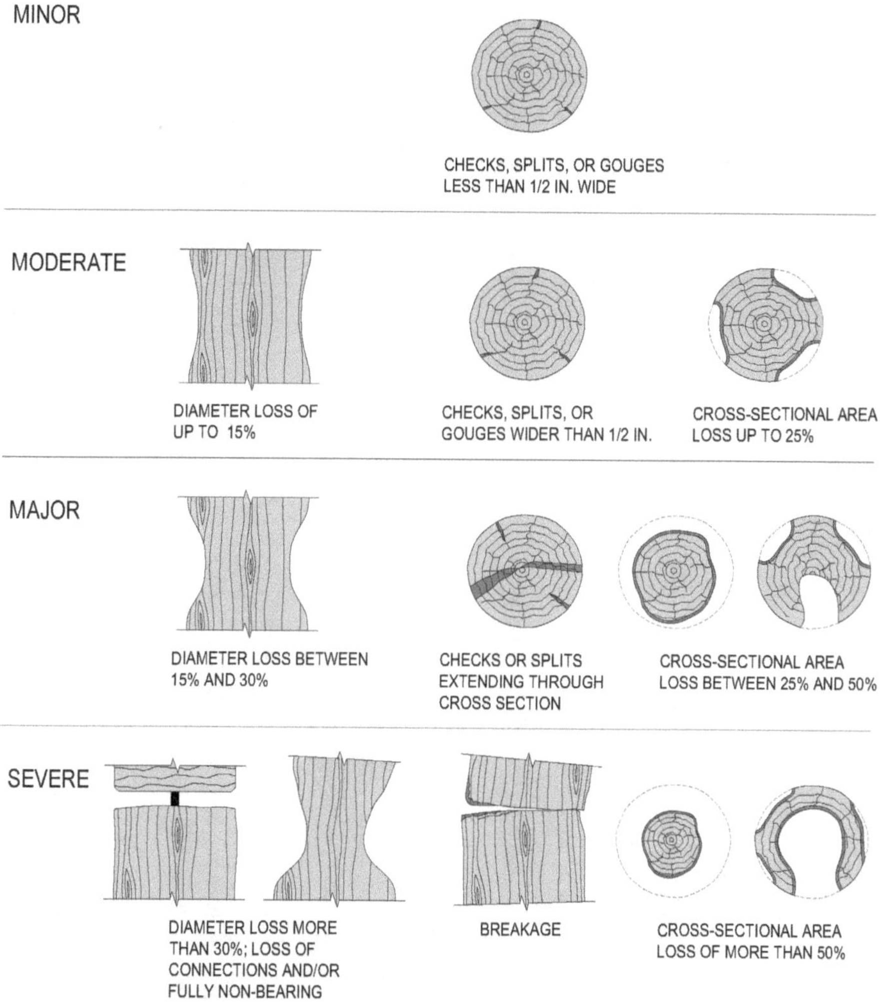

Figure 2-2. Damage ratings for timber elements.
Source: Courtesy of Appledore Marine Engineering, LLC, reproduced with permission.

Table 2-5. Damage Ratings for Steel Elements.

Damage rating	Existing damage[a]	Exclusions [defects requiring elevation to the next higher damage rating(s)]
NI Not inspected	• Not inspected or inaccessible, or passed by[b]	
ND No defects	• Protective coating or wrap intact • Light surface corrosion for uncoated elements • No apparent loss of material or visible damage	No defects damage rating not appropriate if • Protective coating or wrap is damaged • Changes in straight line configuration or local buckling
MN Minor	• Protective coating damaged on up to 25% of the surface area • Loss of nominal thickness of up to 15% at any location • Minor corrosion at connections. • Less than 50% of the surface area affected by corrosion at any elevation or cross section	Minor damage rating not appropriate if • Changes in straight line configuration or local buckling • Corrosion loss exceeding fabrication tolerances (at any location)
MD Moderate	• Protective coating damaged on greater than 25% of the surface area • Loss of nominal thickness between 15% and 30% at any location • Corrosion at connections with minor section loss • More than 50% of the surface area affected by corrosion at any elevation or cross section	Moderate damage rating not appropriate if • Changes in straight line configuration or local buckling • Loss of nominal thickness exceeding 30% at any location
MJ Major	• Partial loss of flange edges or visible reduction of wall thickness on pipe piles • Loss of nominal thickness between 30% and 50% at any location • Heavily corroded connections with significant section loss	Major damage rating not appropriate if • Changes in straight line configuration or local buckling • Perforations or loss of nominal thickness exceeding 50% at any location
SV Severe	• Loss of nominal thickness exceeding 50% at any location • Structural bends or buckling • Breakage and displacement at supports • Loose or Loss of any connections • Cracks through cross section	

[a]At least one defect listed within a damage rating is sufficient to assign the corresponding damage grade.

[b]If not inspected because of inaccessibility or passed by, note as such.

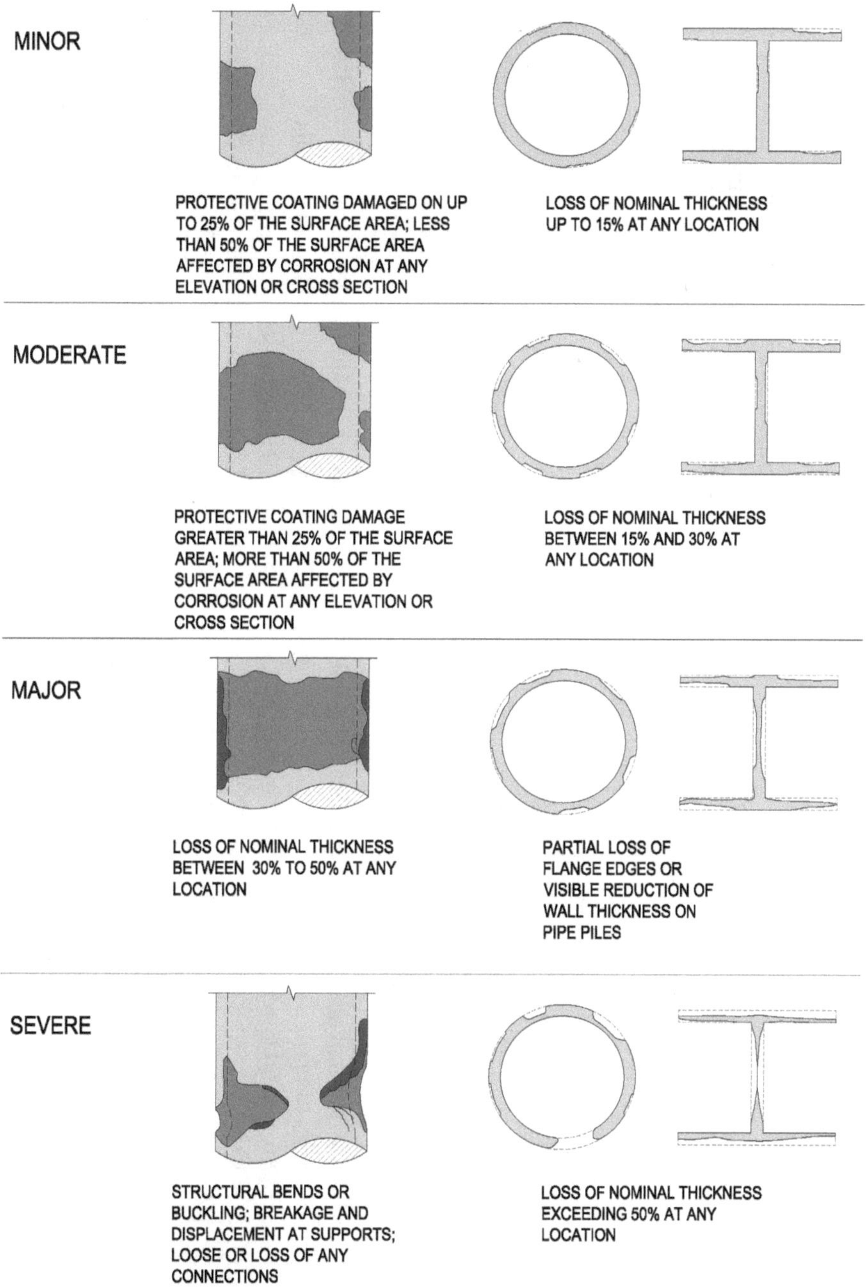

Figure 2-3. Damage ratings for steel elements.
Source: Courtesy of Appledore Marine Engineering, LLC, reproduced with permission.

Table 2-6. Damage Ratings for Reinforced Concrete Elements.

Damage rating	Existing damage[a]	Exclusions [defects requiring elevation to the next higher damage rating(s)]
NI Not inspected	• Not inspected or inaccessible, or passed by[b]	
ND No defects	• Original hard and sound material • Very small shrinkage cracks from curing process.	
MN Minor	• Mechanical abrasion or impact spalls up to 1 in. deep • Occasional corrosion stains or small pop-out corrosion spalls • General cracks up to 1/16 in. wide	Minor damage rating not appropriate if • Structural damage • Corrosion cracks or spalling • Chemical deterioration[c]
MD Moderate	• Mechanical abrasion or impact spalls greater than 1 in. deep • Structural cracks up to 1/16 in. wide • Corrosion cracks up to 1/4 in. wide • Closed corrosion spalling up to 10% of the total surface area of the element • Chemical deterioration: Random cracks up to 1/16 in. wide; Soft concrete and rounding of corners up to 1 in. deep	Moderate damage rating not appropriate if • Structural breakage and spalls • Exposed reinforcement • Loss of cross section due to chemical deterioration beyond rounding of corner edges • Spalling greater than 10%

(*Continued*)

Table 2-6. (*Continued*).

Damage rating	Existing damage[a]	Exclusions [defects requiring elevation to the next higher damage rating(s)]
MJ Major	• Mechanical abrasion or impact spalls exposing the reinforcing • Structural cracks between 1/16 and 1/4 in. wide with breakage internal beyond reinforcing steel • Corrosion cracks wider than 1/4 in. • Open corrosion spalls between 10% and 30% of the total surface area of the element • Multiple cracks and disintegration of the surface layer due to chemical deterioration • Less than 30% reinforcing section loss	Major damage rating not appropriate if • Loss of cross section exceeding 30% due to any cause • Corrosion spalls greater than 30% of the total surface area • More than 30% reinforcing section loss
SV Severe	• Structural cracks wider than 1/4 in. • Breakage • Complete loss of concrete cover due to corrosion of reinforcing steel with more than 30% section loss to any main reinforcing bar • Loss of bearing and displacement at connections • Loss of concrete cover (exposed steel) due to chemical deterioration • Open corrosion spalls greater than 30% of total surface area of the element • Loss of more than 30% of the cross section due to any cause	

[a]At least one defect listed within a damage rating is sufficient to assign the corresponding damage grade.

[b]If not inspected because of inaccessibility or passed by, note as such.

[c]Chemical deterioration: sulfate attack, alkali–silica reaction, alkali–aggregate reaction, alkali–carbonate reaction ettringite distress, or other chemical/concrete deterioration.

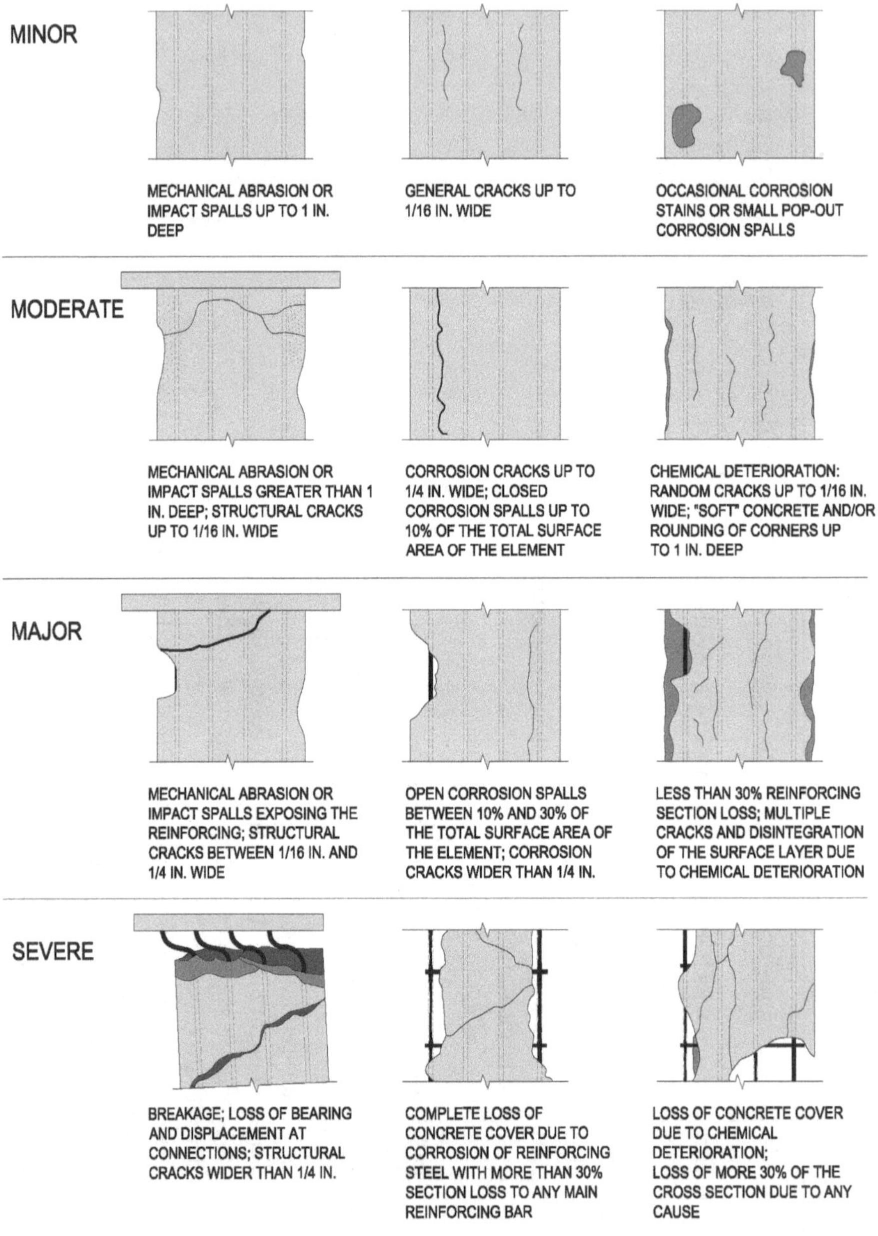

Figure 2-4. Damage ratings for reinforced concrete elements.
Source: Courtesy of Appledore Marine Engineering, LLC, reproduced with permission.

Table 2-7. Damage Ratings for Prestressed Concrete Elements.

Damage rating	Existing damage[a]	Exclusions [defects requiring elevation to the next higher damage rating(s)]
NI Not inspected	• Not inspected or inaccessible[b]	
ND No defects	• Original hard and sound surface and material	
MN Minor	• Mechanical abrasion or impact spalls up to 1/2 in. deep that do not expose reinforcing	Minor damage rating not appropriate if • Structural damage • Corrosion damage • Chemical deterioration[c] • Cracks of any type or size • Mechanical or impact spalling greater than 1/2 in. deep
MD Moderate	• Mechanical abrasion or impact spalls greater than 1/2 in. deep that do not expose reinforcing • Structural cracks up to 1/32 in. wide • Chemical deterioration or general cracking up to 1/32 in. wide	Moderate damage rating not appropriate if • Structural breakage • Corrosion cracks • Loss of cross section in any form • Softening of concrete • Any abrasion or impact spalls that exposes reinforcing material

(Continued)

Table 2-7. (*Continued*).

Damage rating	Existing damage[a]	Exclusions [defects requiring elevation to the next higher damage rating(s)]
MJ Major	• Structural cracks between 1/32 and 1/8 in. wide • Any corrosion cracks generated by strands or cables • Chemical deterioration yielding cracks between 1/32 and 1/8 in. wide • Softening of concrete up to 1 in. deep	Major damage rating not appropriate if • Exposed prestressing steel • Cracks greater than 1/8 in. wide • Breakage or loss of bearing
SV Severe	• Structural cracks wider than 1/8 in. • Breakage or loss of bearing • Partial spalling and loss of cross section due to chemical deterioration • Corrosion spalls exposing any prestressing steel • Any exposed prestressing steel	

[a]At least one defect listed within a damage rating is sufficient to assign the corresponding damage grade.

[b]If not inspected because of inaccessibility or passed by, note as such.

[c]Chemical deterioration: sulfate attack, alkali–silica reaction, alkali–aggregate reaction, alkali–carbonate reaction ettringite distress, or other chemical/concrete deterioration.

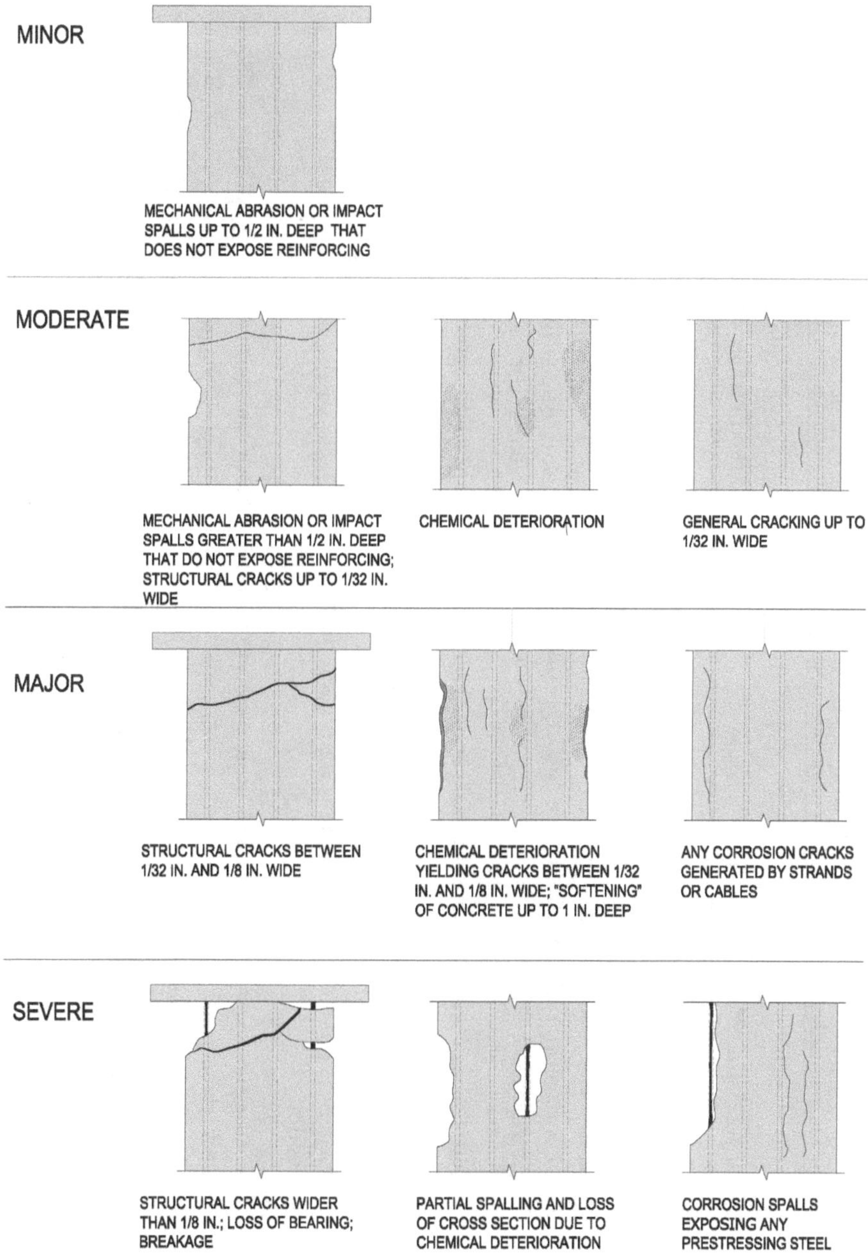

Figure 2-5. Damage ratings for prestressed concrete elements.
Source: Courtesy of Appledore Marine Engineering, LLC, reproduced with permission.

Table 2-8. Damage Ratings for Mooring Hardware.

Damage rating	Existing damage[a]	Exclusions [defects requiring elevation to the next higher damage rating(s)]
NI Not inspected	• Not inspected or inaccessible, or passed by[b]	
ND No defects	• Material sound, surfaces smooth without indications of corrosion, fully intact surface coating, sound connections	No defect damage rating not appropriate if • Surface coatings worn or damaged • Visible corrosion on fasteners
MN Minor	• Fitting has corrosion between 10% and 25% of its surface area • Wear marks or pitting on the surface of the fitting are less than 1/8 in. deep • Fasteners have corrosion with no significant loss of section	Minor damage rating not appropriate if • Deep pits, gouges, or wear on fitting surfaces • Any noticeable loss of section on fastener threads, if visible
MD Moderate	• Fitting has corrosion with loose scale on up to 50% of its surface area • Significant surface wear marks or pitting on fitting between 1/8 and 1/4 in. deep • Fasteners have corrosion with up to 25% section loss	Moderate damage rating not appropriate if • Loose scale on fasteners • Inability to remove fasteners because of heavy corrosion, if accessible

(*Continued*)

Table 2-8. (*Continued*).

Damage rating	Existing damage[a]	Exclusions [defects requiring elevation to the next higher damage rating(s)]
MJ Major	• Fitting has corrosion with loose scale over 50% of its surface area and less than 25% section loss • Significant surface wear marks or pitting on fitting greater than 1/4 in. deep • Fasteners have corrosion with loose scale or section loss greater than 25%	Major Rating not appropriate if • Displaced, damaged, or broken fitting components • Loose or missing fasteners
SV Severe	• Fitting has heavy corrosion and loose scale with greater than 25% section loss at critical areas of the fitting • Structural displacement, deformation, or rotation of the fitting are present; fitting components are broken, cracked, or delaminated • Loose, broken, or missing fasteners	

[a]At least one defect listed within a damage rating is sufficient to assign the corresponding damage grade.

[b]If not inspected because of inaccessibility or passed by, note as such.

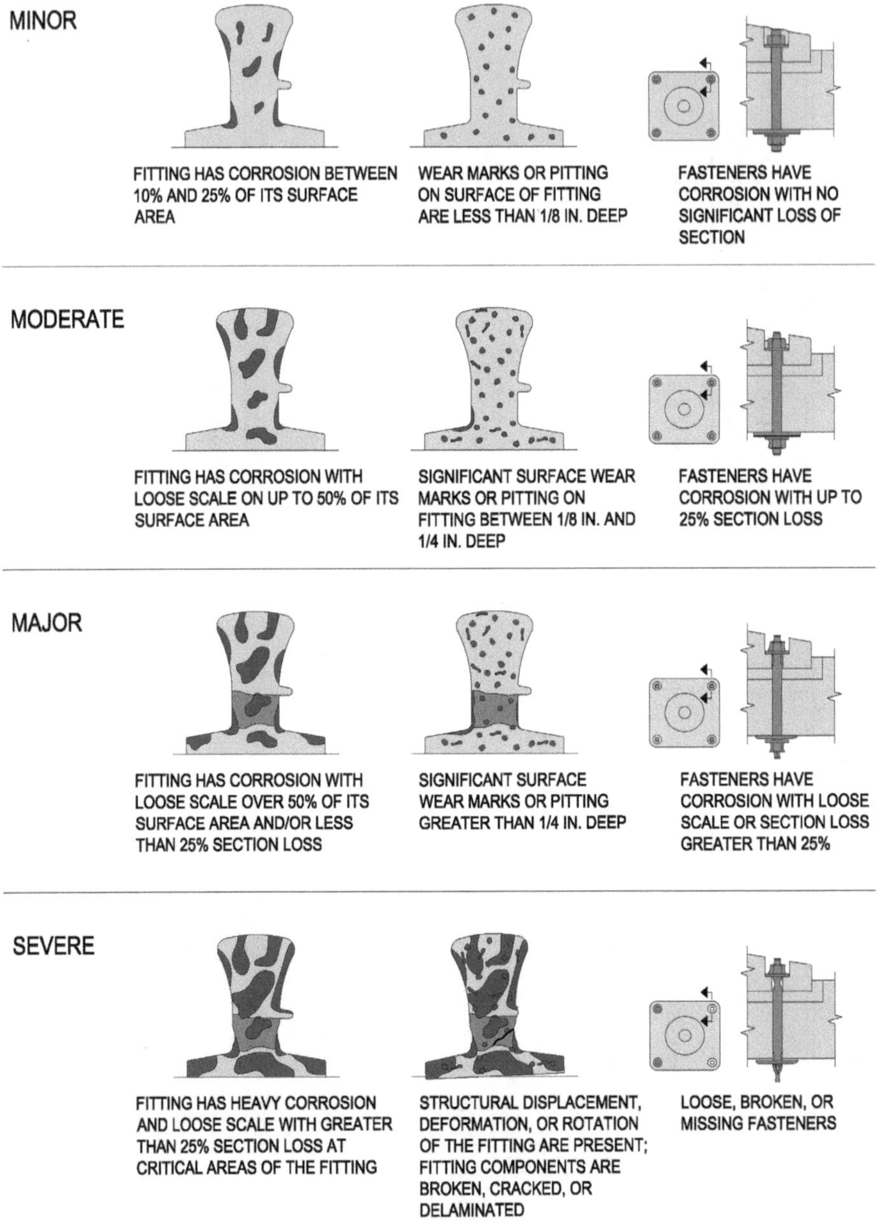

Figure 2-6. Damage ratings for mooring hardware elements.
Source: Courtesy of Appledore Marine Engineering, LLC, reproduced with permission.

Table 2-9. Damage Ratings for Mooring Foundations.

Damage rating	Existing damage[a]	Exclusions [defects requiring elevation to the next higher damage rating(s)]
NI Not inspected	• Not inspected or inaccessible, or passed by[b]	
ND No defects	• Original hard and sound surface and material	No defect damage rating not appropriate if • Weathering on timber, steel, or polymer composite elements • Hairline cracks in concrete elements
MN Minor	• Timber: Weathered timber; evidence of fungal decay; checks, splits, and gouges up to 1/4 in. wide • Steel: Weathering of steel coating, light surface corrosion • Concrete: No significant section loss to load-bearing areas, hairline cracking of the concrete due to corrosion of the mooring hardware • Polymer composite: Weathered surfaces	Minor damage rating not appropriate if • Load-bearing areas around mooring hardware are not sound • Displacements, loss of bearing, or connections • Timber: Fungal decay, insect infestation within or adjacent to the bearing area • Steel: Corrosion loss exceeding fabrication tolerances (at any location) • Concrete: Structural damage or corrosion cracking • Polymer composite: Deformed or cracking

(Continued)

Table 2-9. (*Continued*).

Damage rating	Existing damage[a]	Exclusions [defects requiring elevation to the next higher damage rating(s)]
MD Moderate	• Timber: Splits and checks up to 1/2 in. wide; weathered surfaces; fungal decay under or adjacent to the mooring hardware with a maximum depth of 1 in. • Steel: Corrosion with section loss between 10% and 25% at any location • Concrete: Noticeable cracking of concrete larger than hairline but with no loss of interlock • Polymer composite: deformation or cracking that does not extend through the cross section	Moderate rating not appropriate if • Displacements, loss of bearing, or connections • Changes in straight-line configuration or local buckling • Steel: Loss of nominal thickness exceeding 30% at any location • Concrete: Structural breakage, spalls, or corrosion cracks; Chemical deterioration[c] or softening • Polymer composite: Cracking through cross section
MJ Major	• Timber: Checks and splits greater than 1/2 in. wide; fungal decay with a depth between 1 and 3 in.; up to 25% loss of bearing • Steel: Corrosion with section loss between 25% and 50% at any location • Concrete: Noticeable cracking of concrete resulting in loss of interlock • Polymer composite: Cracking through cross section, split elements, or yielding	Major damage rating not appropriate if • Breakage or displacement of any element • Concrete: Exposed steel strands in prestressed concrete elements • Steel: Perforations or loss of section exceeding 50%

(*Continued*)

Table 2-9. (*Continued*).

Damage rating	Existing damage[a]	Exclusions [defects requiring elevation to the next higher damage rating(s)]
SV Severe	• Displacement/yielding of any support members • Loss of full bearing of fitting under hardware • Timber: Fungal decay with a depth greater than 3 in. • Steel: Significant corrosion with greater than 50% section loss at any location • Concrete: Cracking or spalling of concrete base under hardware • Polymer composite: Broken or damaged elements	

[a]At least one defect listed within a damage rating is sufficient to assign the corresponding damage grade.

[b]If not inspected because of inaccessibility or passed by, note as such.

[c]Chemical deterioration: sulfate attack, alkali–silica reaction, alkali–aggregate reaction, alkali–carbonate reaction ettringite distress, or other chemical/concrete deterioration.

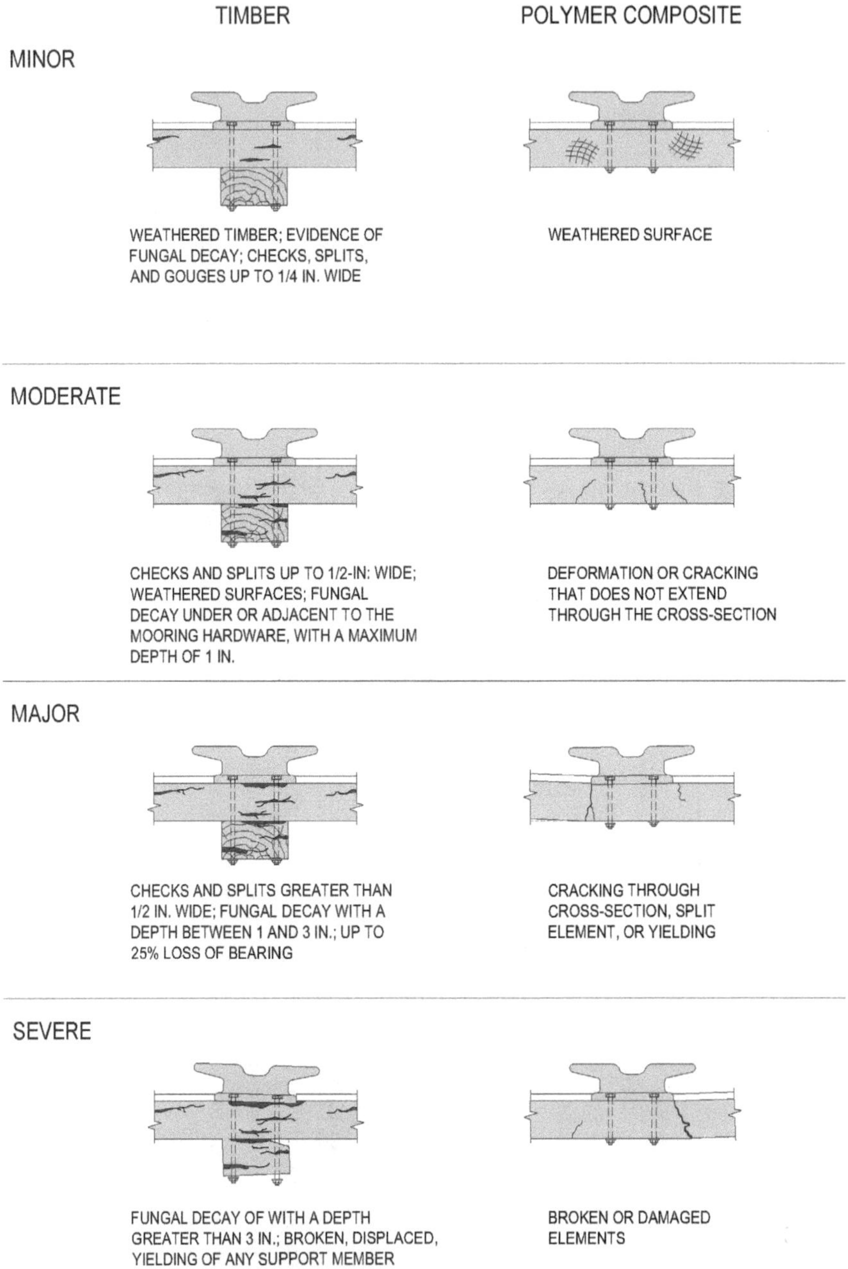

Figure 2-7. Damage ratings for timber and polymer composite mooring foundations. Source: Courtesy of Appledore Marine Engineering, LLC, reproduced with permission.

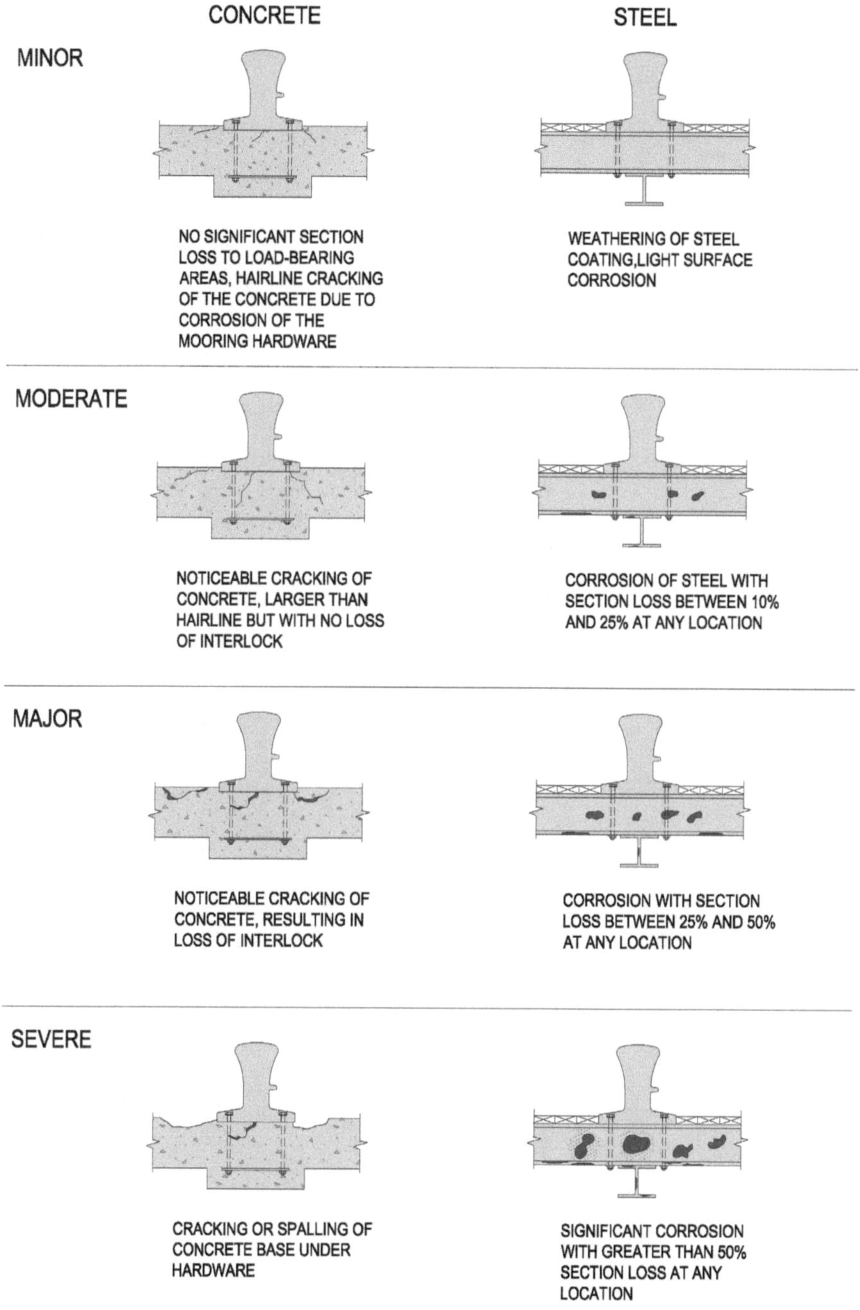

Figure 2-8. Damage ratings for steel and concrete mooring foundations.
Source: Courtesy of Appledore Marine Engineering, LLC, reproduced with permission.

Table 2-10. Damage Ratings for (Timber/Steel/Concrete/Polymer Composite) Fender Piles.

Damage rating	Existing damage[a]	Exclusions [defects requiring elevation to the next higher damage rating(s)]
NI Not inspected	• Not inspected or inaccessible, or passed by[b]	
ND No defects	• Original sound surface with no defects observed	No defect damage rating not appropriate if • Surface coatings worn or damaged
MN Minor	• Timber: Light abrasion and fungal decay with a maximum depth of 1/2 in.; minimal marine borer activity observed with less than 5% section loss • Steel: Weathering of steel coating; surface corrosion with no significant pitting • Concrete: Hairline cracking • Polymer composites: Weathered elements	Minor damage rating not appropriate if • Timber: Abrasion and fungal decay to timber element exceeding 1/2 in. depth • Steel: Any section loss to element due to corrosion • Concrete: Softening of concrete • Polymer composites: Any abrasions or delamination
MD Moderate	• Timber: Checks and splits up to 1/2 in. wide; fungal decay with a depth between 1/2 and 1 in.; abrasion up to 2 in. deep; between 5% and 10% section loss due to marine borers	Moderate damage rating not appropriate if • Timber: Checks or splits greater than 1/2 in. wide; fungal decay with a depth greater than 1 in. • Steel: Corrosion with greater than 25% section loss • Concrete: Softening of the concrete greater than 1 in. deep

(Continued)

Table 2-10. (*Continued*).

Damage rating	Existing damage[a]	Exclusions [defects requiring elevation to the next higher damage rating(s)]
	• Steel: Corrosion with up to 25% localized section loss • Concrete: Noticeable cracking of concrete but with no loss of interlock; softening of concrete up to 1 in. deep • Polymer composites: Abrasions and delamination of the top layer	• Prestressed concrete fender piles (with a low effective prestress) are expected to crack under load; therefore, they should be rated minor if no corrosion is present and the cracks are closed • Polymer composites: Cracked or split elements; delamination through multiple layers of element
MJ Major	• Timber: Checks and splits greater than 1/2 in. wide; fungal decay with a depth between 1 and 3 in.; abrasion damage greater than 2 in. deep; between 10% and 25% section loss due to marine borers • Steel: Corrosion with localized section loss between 25% and 50%; localized buckling of a flange • Concrete: Noticeable cracking of concrete with loss of interlock; softening of the concrete between 1 and 3 in. deep • Polymer composites: Elements cracked or split; delamination through multiple layers of element; direct exposure to reinforcing	Major damage rating not appropriate if • Broken elements • Timber: Fungal decay with a depth greater than 3 in.; greater than 25% section loss due to marine borers • Steel: Corrosion with greater than 50% section loss • Concrete: Exposed reinforcing steel or prestressing steel strands; softening of concrete greater than 3 in. deep • Polymer composites: Major part of the cross section is delaminated

(*Continued*)

Table 2-10. (*Continued*).

Damage rating	Existing damage[a]	Exclusions [defects requiring elevation to the next higher damage rating(s)]
SV Severe	• Broken • Timber: Fungal decay greater than 3 in. deep; greater than 25% section loss due to marine borers • Steel: Corrosion with localized section loss greater than 50%; yielded member • Concrete: Exposed reinforcing steel or prestressing steel strands; spalling of the concrete; softening of the concrete greater than 3 in. deep • Polymer composites: A major part of the cross section is delaminated	

[a]At least one defect listed within a damage rating is sufficient to assign the corresponding damage grade.

[b]If not inspected because of inaccessibility or passed by, note as such.

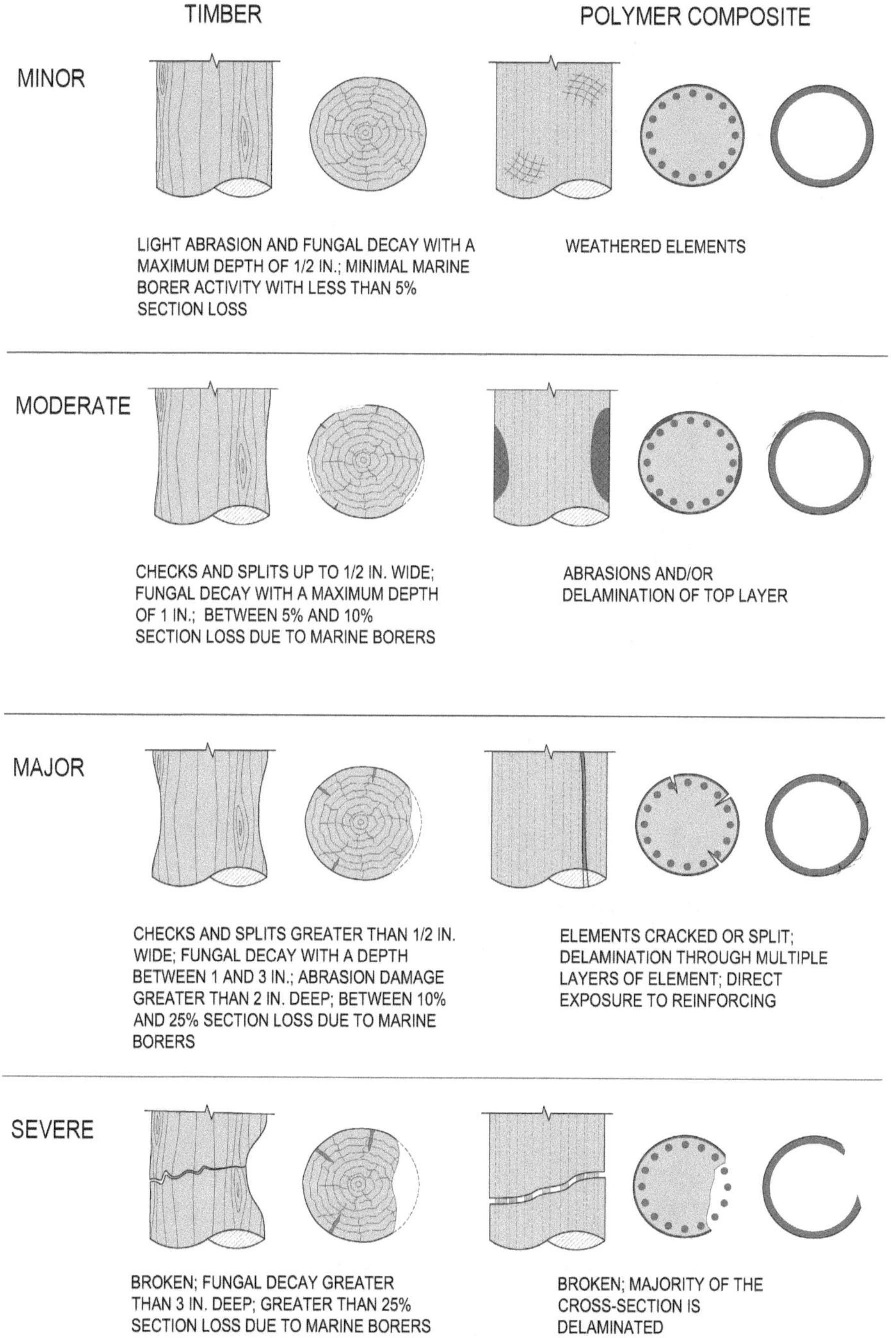

Figure 2-9. Damage ratings for timber and polymer composite fender piles. Source: Courtesy of Appledore Marine Engineering, LLC, reproduced with permission.

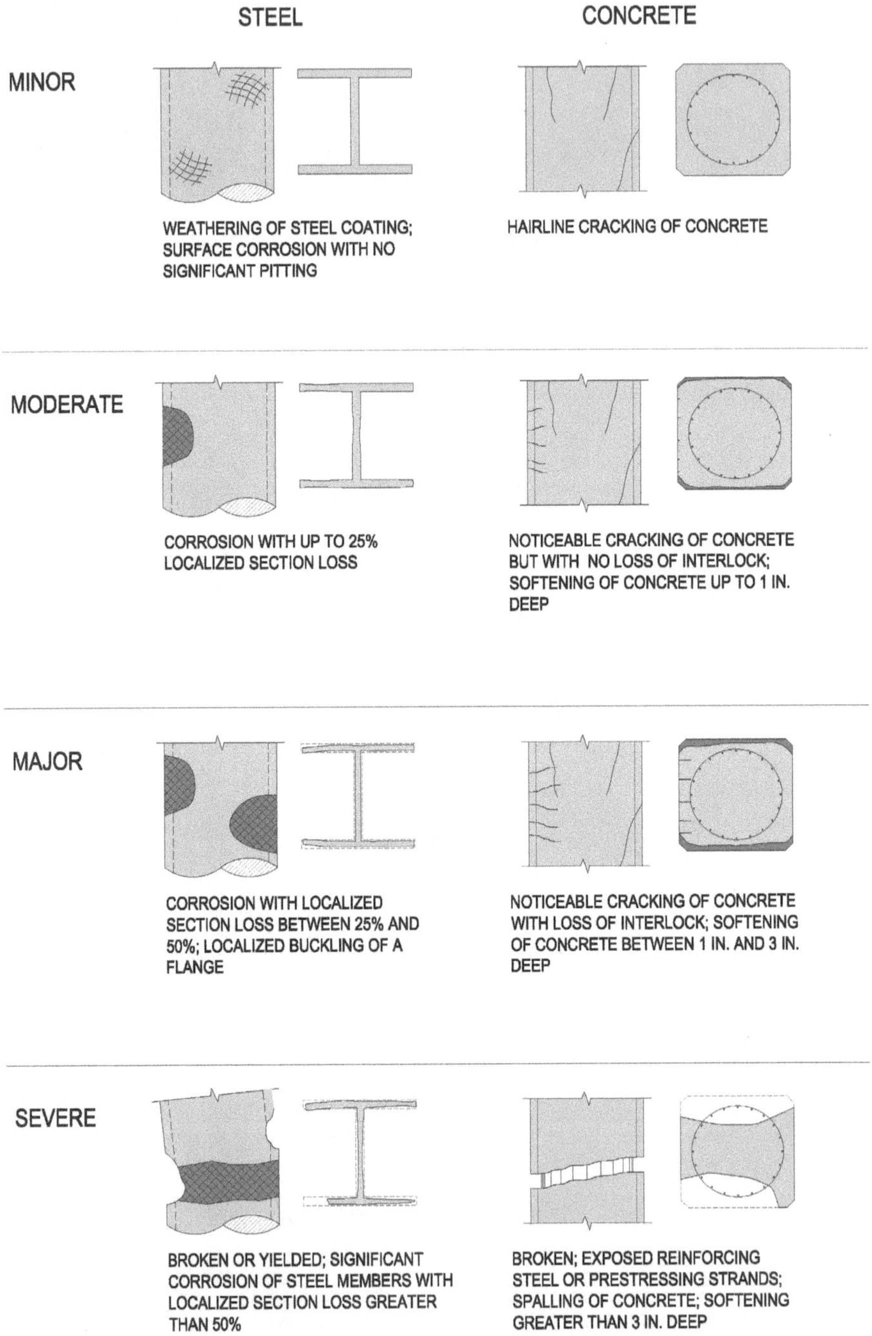

Figure 2-10. Damage ratings for steel and concrete fender piles.
Source: Courtesy of Appledore Marine Engineering, LLC, reproduced with permission.

Table 2-11. Damage Ratings for Pneumatic, Foam-Filled, and Hydropneumatic Fenders.

Damage rating	Existing damage[a]	Exclusions [defects requiring elevation to the next higher damage rating(s)]
NI Not inspected	• Not inspected or inaccessible, or passed by[b]	
ND No defects	• Original hard surfaces • Components sound • All hardware intact and operable	No defect damage rating not appropriate if • Components are weathered, worn, or torn
MN Minor	• Wear on the fender unit with no visible belting • Hardware intact with light surface corrosion with less than 10% section loss • Swivel operable but binding	Minor damage rating not appropriate if • Fender unit with visible belting • Hardware with greater than 10% section loss
MD Moderate	• Wear on the fender with belting visible to a maximum depth of 1 in. • Hardware is intact with section loss between 10% and 25% • Swivel heavily corroded and or bound.	Moderate damage rating not appropriate if • Fender unit permanently set or deformed

(*Continued*)

Table 2-11. (*Continued*).

Damage rating	Existing damage[a]	Exclusions [defects requiring elevation to the next higher damage rating(s)]
MJ Major	• Wear on the fender • Belting visible to a maximum depth of 2 in. • Permanent deformation of the unit • Hardware is loose or heavily corroded with section loss between 25% and 50% • Swivel heavily corroded and or bound, or with 25% to 50% section loss • Air pressure inflation and valves do not appear operable[c]	Major Rating not appropriate if • Components missing or broken
SV Severe	• Considerable wear on the fender with belting visible to a depth greater than 2 in. • Punctures, tears, or holes in fender; foam exposed • Hardware heavily corroded with greater than 50% section loss; missing or broken hardware • Swivel heavily corroded and or bound, or with greater than 50% section loss or is broken • Air pressure inflation and valves are broken or damaged[c]	

[a]At least one defect listed within a damage rating is sufficient to assign the corresponding damage grade.

[b]If not inspected because of inaccessibility or passed by, note as such.

[c]For pneumatic and hydropneumatic fenders, an assessment of the air pressure and inflation/pressurization system should be confirmed.

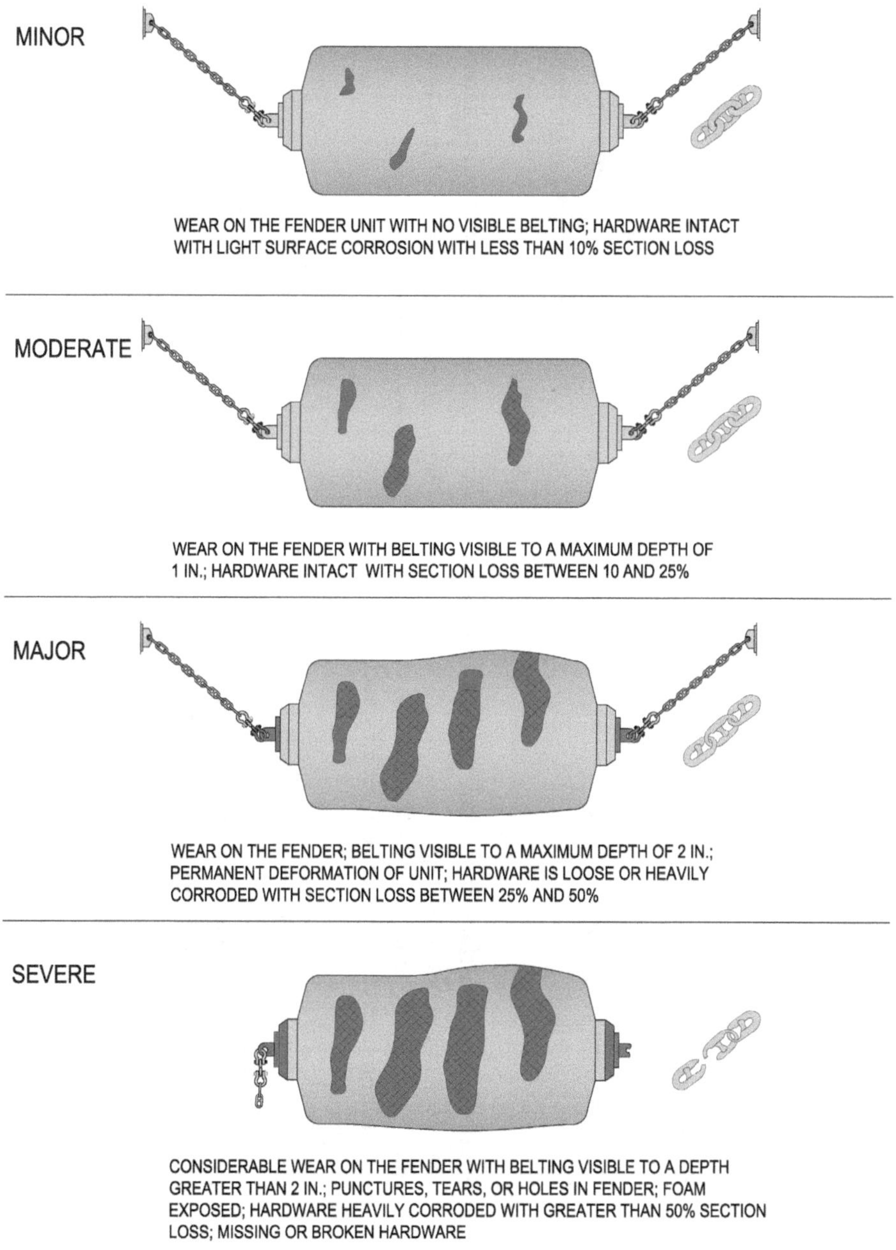

Figure 2-11. Damage ratings for pneumatic, foam-filled, and hydropneumatic fender elements.
Source: Courtesy of Appledore Marine Engineering, LLC, reproduced with permission.

Table 2-12. Damage Ratings for Rubber Fender Elements.

Damage rating	Existing damage[a]	Exclusions [defects requiring elevation to the next higher damage rating(s)]
NI Not inspected	• Not inspected or inaccessible, or passed by[b]	
ND No defects	• Original hard and sound surface • Connections intact and tight	No defect damage rating not appropriate if • Noticeable abrasion or wear of rubber surfaces
MN Minor	• Small gouges or surface defects are present and are less than 10% of the nominal depth • Intact and tight connection; light corrosion of connection hardware with less than 10% section loss at any location	Minor damage rating not appropriate if • Surface cracking or degradation of rubber components
MD Moderate	• Gouges, wear, or tears between 10% and 25% of the nominal depth • Rubber damaged at the connectors or connection plates • Connections are loose; corrosion of connection hardware with section loss between 10% and 25% at any location	Moderate damage rating not appropriate if • Permanent deformation or misalignment of rubber elements • Missing connection hardware

(Continued)

Table 2-12. (*Continued*).

Damage rating	Existing damage[a]	Exclusions [defects requiring elevation to the next higher damage rating(s)]
MJ Major	• Cracks, gouges, or tears between 25% and 50% of the nominal depth • Rubber is torn at the connectors or connection plates • Connections are loose; corrosion of connection hardware with section loss between 25% and 50% at any location	Major damage rating not appropriate if • Rubber element is split or torn through • Missing connection hardware
SV Severe	• Cracks, gouges, or tears greater than 50% of the nominal depth • Rubber is torn through at the connectors or connection plates • Missing connection hardware; corrosion of connection hardware with greater than 50% section loss at any location	

[a]At least one defect listed within a damage rating is sufficient to assign the corresponding damage grade.

[b]If not inspected because of inaccessibility or passed by, note as such.

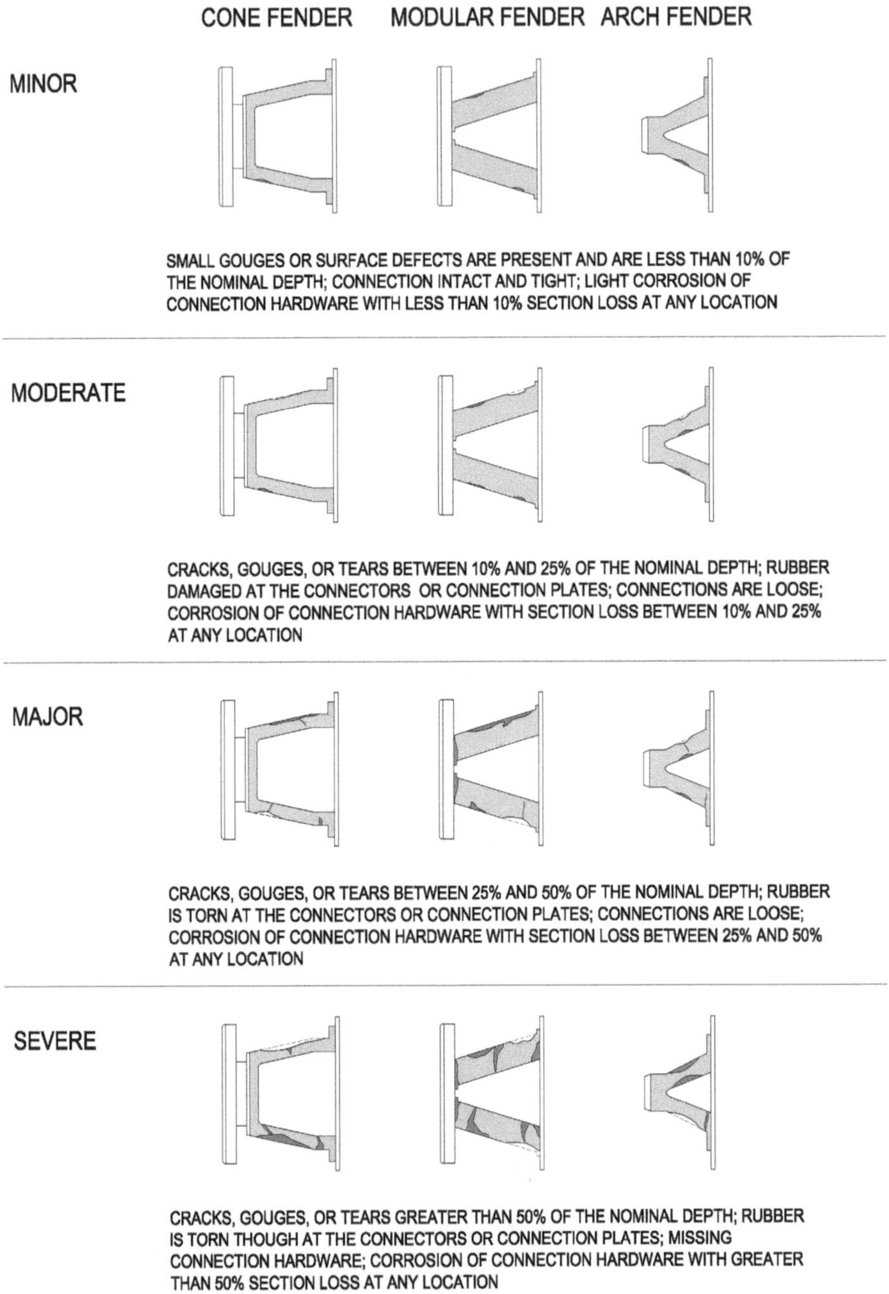

Figure 2-12. Damage ratings for rubber fender elements.
Source: Courtesy of Appledore Marine Engineering, LLC, reproduced with permission.

Table 2-13. Damage Ratings for Fender Panels.

Damage rating	Existing damage[a]	Exclusions [defects requiring elevation to the next higher damage rating(s)]
NI Not inspected	• Not inspected or inaccessible, or passed by[b]	
ND No defects	• Original hard surfaces • All connections intact • Backing frame sound • Support chains intact with no damage	No defect damage rating not appropriate if • Coating is damaged • Visible surface corrosion
MN Minor	• Facing elements have small cracks or gouges that are less than 10% of the nominal thickness • 90% of facing connections are intact • Backing frame has surface corrosion with no significant section loss • Support chains are intact with light surface corrosion	Minor damage rating not appropriate if • Facing elements are worn or damaged with cracks or gouges greater than 10% of nominal thickness
MD Moderate	• Facing elements have cracks or gouges between 10% and 25% of the nominal thickness • 75% of facing connections are intact • Facing elments are displaced from the backing frame • Corrosion of backing frame with up to 25% localized section loss • Support chains are intact with less than 25% section loss	Moderate damage rating not appropriate if • Facing elements are displaced or misaligned • Any loose or missing hardware

(*Continued*)

Table 2-13. (*Continued*).

Damage rating	Existing damage[a]	Exclusions [defects requiring elevation to the next higher damage rating(s)]
MJ Major	• Facing elements have cracks or gouges between 25% and 50% of the nominal thickness • 50% of the facing connections are intact or multiple facing elements are displaced from the backing frame • Corrosion of backing frame with section loss between 25% and 50% • Support chains are heavily corroded with more than 25% section loss	Major damage rating not appropriate if • Panel system is sagging, misaligned, or has limited bearing
SV Severe	• Facing elements have cracks or gouges that are greater than 50% of the nominal thickness • Less than 50% of the facing connections are intact or multiple facing elements are displaced from the backing frame • Backing frame heavily corroded with section loss greater than 50% • Sagging/displacement of the panel system • Support chains are heavily corroded with loose scale and are missing or broken	

[a]At least one defect listed within a damage rating is sufficient to assign the corresponding damage grade.

[b]If not inspected because of inaccessibility or passed by, note as such.

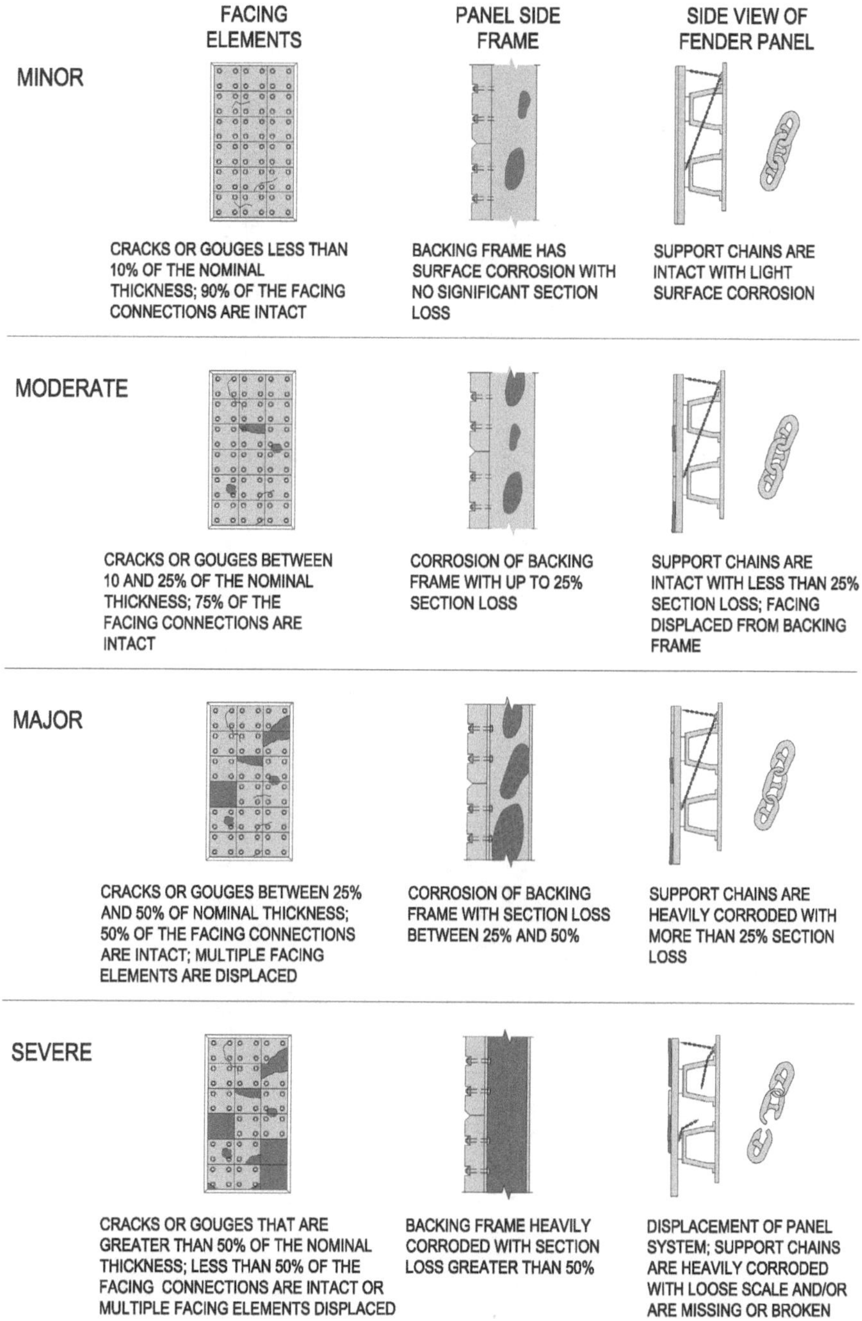

Figure 2-13. Damage ratings for fender panel elements.
Source: Courtesy of Appledore Marine Engineering, LLC, reproduced with permission.

Table 2-14. Damage Ratings for FRP Polymer Composites.

Damage rating	Existing damage[a]	Exclusions [defects requiring elevation to the next higher damage rating(s)]
NI Not inspected	• Not inspected or inaccessible, or passed by[b]	
ND No defects	• No signs of delamination • No signs of ultraviolet (UV) or chemical degradation • Coating (if originally equipped) intact • No signs of cracking, deformation, or damage	No defect damage rating not appropriate if • Any abrasion or fiber exposure other than at the top of a pile or end of the element
MN Minor	• Small abrasions or damage that may expose the top fiber layer up to 5% of the surface area and no deeper than the top fiber layer • Slight delamination at the ends of the element and around any damaged areas • Signs of UV fading to no more than 25% of the surface area • Slight damage to small portions of any element that are non-load-bearing	Minor damage rating not appropriate if • More than 5% of the surface area has damage to the fiber core or is deeper than the top fiber layer • Beyond slight damage non-load-bearing portions of the element

(Continued)

Table 2-14. (*Continued*).

Damage rating	Existing damage[a]	Exclusions [defects requiring elevation to the next higher damage rating(s)]
MD Moderate	• Abrasions or damage that may expose the top fiber layer up to 15% of the surface area and no deeper than the top layer of fiber; abrasions or damage that go below the top fiber layer up to 5% of the element • Delamination to layered fiber polymer composites, no more than 10% of the cross-sectional area • Signs of UV cracking or chemical damage between 1/32 and 1/16 in. • Cracking, deformation, overstressing, and other damage to large portions of the element that does not impact its load capacity	Moderate damage rating not appropriate if • More than 15% of the surface area has damage to the top fiber layer or more than 5% of damage extends to deeper fiber layers • Rate of degradation on load-bearing connections would cause further damage after report is published or one year, whichever is longer

(*Continued*)

Table 2-14. (*Continued*).

Damage rating	Existing damage[a]	Exclusions [defects requiring elevation to the next higher damage rating(s)]
MJ Major	• Abrasions or damage that may expose the top fiber layer up to 50% of the surface area and no deeper than the top fiber layer; abrasions or damage that go below the top layer of fiber up to 15% of the cross-sectional area • Exposed or strained internal fiber reinforcing on solid fiber-reinforced elements • Delamination to layered fiber polymer composites between 10% and 25% of the cross-sectional area • Signs of UV cracking or chemical damage between 1/16 and 1/8 in. • Cracking, deformation, overstressing that impact load capacity and would cause reduction in the load rating • Exposure of fiber-reinforced strands at the center of the element	Major damage rating not appropriate if • More than 50% of the surface area has damage to the top fiber layer or more than 15% of damage extends to deeper fiber layers • Any abrasion or crack through full depth of the cross section
SV Severe	• Abrasions or damage that expose the top fiber layer to more than 50% of the surface area or more than 15% of damage extends to deeper fiber layers • Broken internal fiber reinforcing on solid fiber-reinforced elements • Delamination to layered fiber polymer composites, more than 25% of the cross-sectional area	

[a]At least one defect listed within a damage rating is sufficient to assign the corresponding damage grade.

[b]If not inspected because of inaccessibility or passed by, note as such.

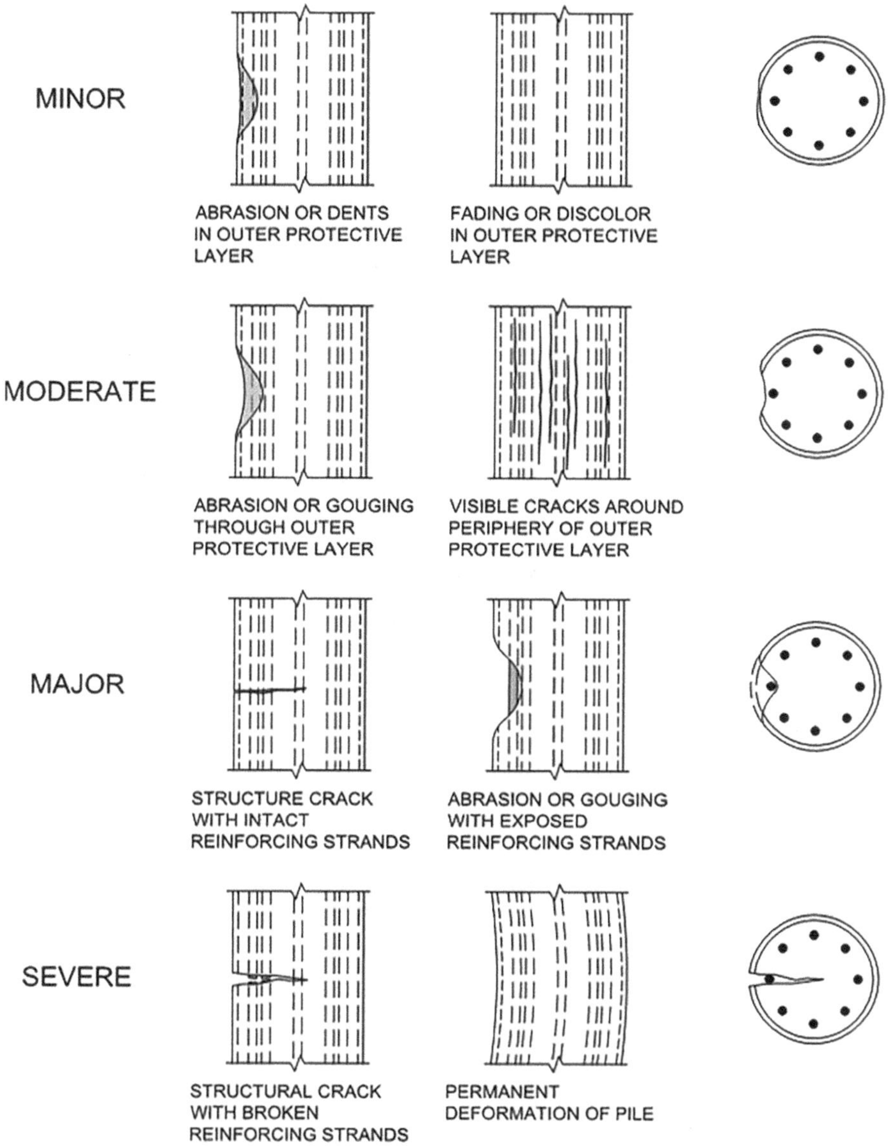

Figure 2-14. Damage ratings for solid polymer composite piles.
Source: Courtesy of Simpson Gumpertz & Heger Inc., reproduced with permission.

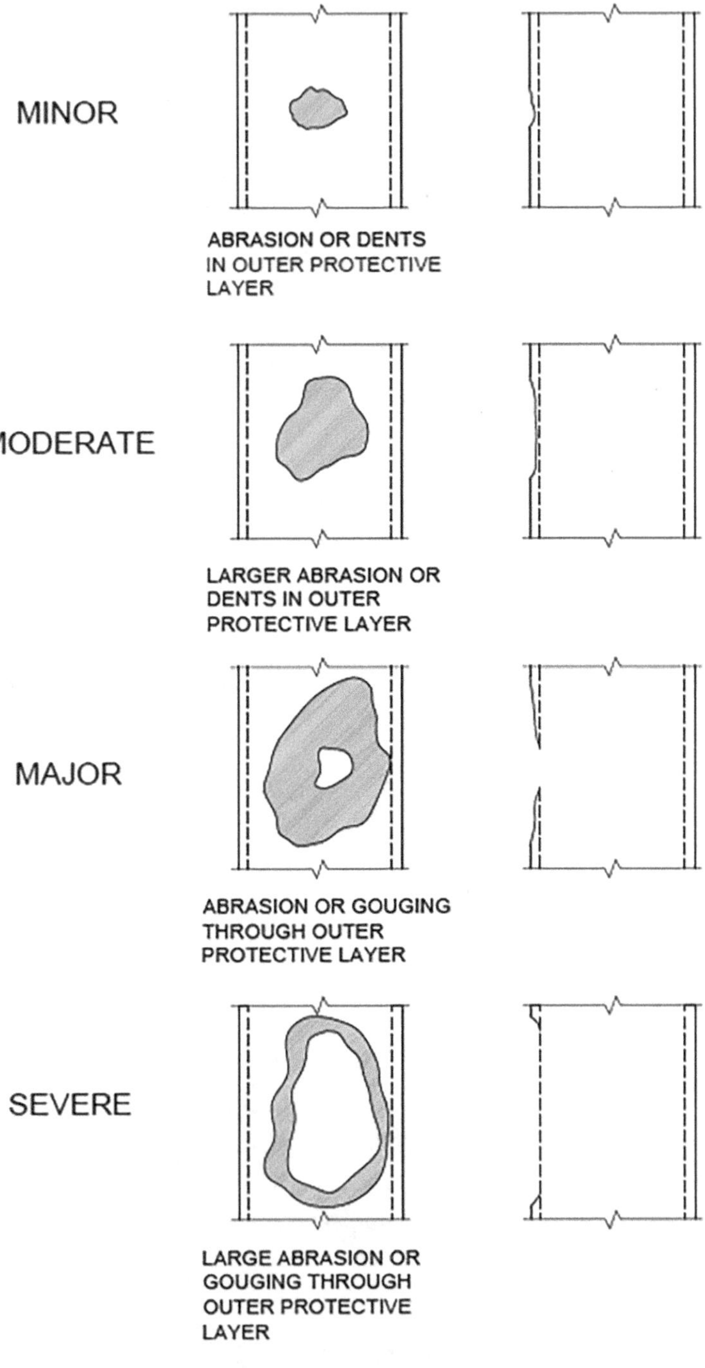

Figure 2-15. Damage ratings for tubular polymer composite piles.
Source: Courtesy of Simpson Gumpertz & Heger Inc., reproduced with permission.

2.5.8 Utility Systems

Damage ratings of waterfront utility systems provide a means of identifying the importance and severity of deficiencies observed during inspection. An inspection of the utilities is limited to the observed condition of the utility lines, risers, hangars, brackets, and accessories. The intent of the inspection is to ensure, to the extent possible, that the utility line and supports are structurally sound.

The remaining thickness and interior condition of the utility piping, valves, and other system elements are typically not part of a waterfront inspection. The typical rating system described in Section 2.6.2 is used as a basis for the damage ratings used for these systems. Utility lines, brackets, and accessories are rated on the basis of the degree of observed damage or deterioration and the corresponding urgency of repair. The severity of the rating indicates a condition that requires replacement, repair, or a more detailed inspection to determine the remaining material thickness or internal condition.

The results of the condition inspection may be represented on a plan using a minimum scale of 1 in. = 30 ft (1:300). The utility lines are color-coded to indicate observed conditions, the degree of damage, and the corresponding urgency of repair. For below-deck utilities, a reflected soffit plan is useful, with the utilities depicted as solid lines. For utilities located in a utility trench, an above-deck plan with utility lines shown as dashed lines is useful to indicate that the lines are enclosed in a trench. The utility plans should also indicate the location and damage rating of valves, risers, cleanouts, utility support brackets, pipe reducers, and expansion joints.

2.6 OVERALL SYSTEM RATINGS

2.6.1 General

Ratings are assigned to the inspected portions of each structure upon completion of routine inspections, as shown in Table 2-15. The ratings are important in establishing the priority of follow-up actions to be taken. This is particularly true when many structures are included in an inspection program, and follow-up activities must be ranked or prioritized due to limited resources.

The rating system used for postevent inspections differs from that used for routine inspections because postevent inspection ratings must focus on event-induced damage only, excluding long-term defects such as corrosion deterioration. An alphabetical scale is used for postevent inspections to distinguish from the numerical condition assessment scale used for routine inspections, as shown in Table 2-16.

2.6.2 Condition Assessment Ratings

The condition assessment rating should be assigned upon completion of the routine inspection and remain associated with the structural unit (as defined in Section 3.1.1) until the structure is rerated following a quantitative engineering evaluation and repairs or upon completion of the next scheduled routine inspection. The ratings should be assigned via a tiered approach by first utilizing distinct structural unit ratings to help define a group of unit rating that will translate into a rating for the overall facility.

A scale of 1 to 6 is used for the rating system, as shown in Table 2-15. A rating of 6 represents a structure in good condition, whereas a rating of 1 represents a structure in critical condition. Other suitable rating systems may be substituted for a particular owner's purpose as appropriate.

It is important to understand that ratings are used to describe the existing in-place structure relative to its condition when newly built. The fact that the structure is designed for loads that are lower than the current standards for design or higher than the current usage shall not influence the ratings. This may result in a structure being given a condition assessment rating that does not necessarily reflect its fit-for-use capacity. At a minimum, the engineer shall provide a rating on the basis of the existing condition and original design criteria. If there is an alternative use of the structure, it may be analyzed for the specific intended use. In such cases, the engineer must perform an engineering assessment and analysis to determine whether the structure can support the actual loading (structural demand).

Equally important is understanding that the correct assignment of ratings requires experience and knowledge of the structural system. Judgment must be applied in considering

- Scope of damage (total number of defects),
- Severity of damage (type and size of defects),
- Distribution of damage (local versus general),
- Types of components affected (their structural sensitivity),
- Location of defect on component (relative to point of maximum moment/shear), and
- Serviceability.

The qualifications of individuals assigning ratings are important in ensuring that the ratings are assigned consistently and uniformly in accordance with sound engineering principles and the guidelines provided herein. With oversight from the project manager, the team leader should verify that the assigned ratings are appropriate.

Table 2-15. Condition Assessment Ratings.

Rating	Description
6 Good	No visible damage or only minor damage noted. Structural elements may show very minor deterioration, but no overstressing observed. No repairs are required
5 Satisfactory	Limited minor to moderate defects or deterioration observed but no overstressing observed. No immediate repairs are required
4 Fair	All primary structural elements are sound but minor to moderate defects or deterioration observed. Localized areas of major to severe deterioration may be present but do not significantly reduce the load-bearing capacity of the structure. Repairs are recommended, but the priority of the recommended repairs is low
3 Poor	Major to severe deterioration or overstressing observed on widespread portions of the structure but does not significantly reduce the load-bearing capacity of the structure. Repairs may need to be carried out with moderate urgency
2 Serious	Major to severe deterioration, overstressing, or breakage may have significantly affected the load-bearing capacity of primary structural components. Local failures are possible, and loading restrictions may be necessary. Repairs may need to be carried out on a high-priority basis with urgency
1 Critical	Major to severe deterioration, overstressing, or breakage has resulted in localized failure(s) of primary structural components. More widespread failures are possible or likely to occur, and load restrictions should be implemented as necessary. Repairs may need to be carried out on a very high-priority basis with strong urgency

2.6.3 Postevent Damage Ratings

The postevent damage rating should be assigned upon completion of the postevent inspection, preferably prior to leaving the site. The rating should be used to reflect whether additional attention is necessary and, if so, at what priority level. Table 2-16 shows the four postevent damage ratings. A rating of "A" indicates no further action is required, whereas a rating of "D" indicates major structural damage requiring urgent attention.

The following guiding principles should be followed when assigning postevent damage ratings:

- Ratings should reflect only damage that is likely caused by the event. Long-term or pre-existing deterioration, such as corrosion damage, should be ignored unless the structure's structural integrity is immediately threatened.
- Ratings are used to compare the existing in-place structure to the original structure. The fact that the structure is designed for loads that are lower than the current standards for design should have no influence on the ratings.
- Assignment of ratings should reflect an overall characterization of the entire structure being rated. Correct assignment of a rating should consider both the severity of the deterioration and the extent to which it is widespread throughout the structure.
- Assignment of rating codes will require judgment. Standard rating guidelines are intended to make assignment of these ratings uniform among inspection personnel.

Table 2-16. Postevent Damage Ratings.

Rating	Description
A	No significant event-induced damage observed; no further action is required
B	Minor to moderate event-induced damage observed, but all primary structural elements are sound. Repairs may be required, but the priority of repairs is low
C	Moderate to major event-induced damage observed that may have significantly affected the load-bearing capacity of primary structural elements. Repairs are necessary on a priority basis
D	Major event-induced damage has resulted in localized or widespread failure of primary structural components. Additional failures are possible or likely to occur. Urgent remedial attention is necessary

2.7 RECOMMENDED ACTION GUIDELINES

Condition assessment and postevent damage ratings describe the urgency with which or when follow-up action should be taken, whereas recommended actions describe what specific actions should be taken. Recommended actions are assigned upon completion of each inspection type described in Section 2.1. New construction inspections and repair construction inspections are in-process activities that typically require immediate follow-up action in the event of nonconformance. Tables 2-17 and 2-18 and Figure 2-16 serve as a tool to assist clients/owners in systematically evaluating and prioritizing repair and maintenance activities on the basis of the assigned condition assessment rating from Section 2.6.

A description of each recommended action choice is provided in Table 2-16. Typical recommended action options for each inspection type are depicted in Figure 2-16. Multiple recommended actions may be assigned upon completion of each inspection; however, guidance should be provided to indicate the order in which the recommended actions should be carried out.

For example, a structure that has received a routine inspection may be assigned recommended actions for additional inspections due to broken or deteriorated piles. This may include repair or upgrade design inspection to collect the required design-level data to design repairs. In addition, a special-purpose inspection might be necessary if the cause of the deteriorated piles is not known, and coring, testing, and analysis are required. The first action is to advise the owner on the need for erecting barricades or closing a portion of the structure. Next, the special-purpose inspection should be executed to determine the cause of the deterioration. Finally, the repair or upgrade design inspection should follow.

2.7.1 Repair Priority

The repair priority matrix serves as a strategic tool to assist waterfront facility owners in systematically evaluating and prioritizing repair and maintenance activities. By considering key criteria such as operations, potential loss of property, and other relevant factors, owners can make informed decisions to mitigate risks and enhance the resilience of their waterfront assets. The matrix tool is used in conjunction with owner/client input to determine risk on the basis of intended use, goals to create the criteria, and a target for the matrix.

These criteria can include operations impact, potential loss of property, or other relevant factors (e.g., environmental impact, regulatory compliance). Four levels of risk are presented within the matrix. On the basis of the risk tolerance and condition assessment rating, a low to very high priority designation is assigned to the structure. This repair priority

Table 2-17. Description of Recommended Action Options.

Recommended action	Description
Emergency action	Recommended whenever an unsafe condition is observed. If the situation is life-threatening, significant property damage may occur, or significant environmental damage may occur, and appropriate owner representatives should be contacted immediately. Emergency actions may consist of barricading or closing all or portions of the structure, placing load restrictions, or unloading portions of the structure
Engineering evaluation	Recommended whenever significant damage or defects are encountered that require a structural investigation or evaluation to quantify the structural capacity, determine whether repairs are required, or determine what method of repair is appropriate. Although the scope of the routine inspections should include the structural assessment on the capacity of typical structural components relative to their new condition, the engineering evaluation should consider the actual/anticipated loads that are or will be imposed on the structure
Structural repair or upgrade design inspection	Recommended whenever repairs are required, typically as a result of a routine inspection, but may also result from a special-purpose inspection or postevent inspection
Special-purpose inspection	Typically recommended to determine the cause or significance of nontypical deterioration, usually prior to designing repairs. Special testing, analysis, monitoring, or investigation using nonstandard equipment/techniques is typically required
Repair plans development	Recommended when the structural repair or upgrade design inspection has been completed and any special-purpose inspections recommended have been completed. Indicates that the field data have been collected, and the structure is ready to have repair documents prepared
No action	Recommended when no further action is necessary on the structure until the next scheduled routine inspection

matrix provides a structured approach and effective tool for owners to assess and prioritize repair and maintenance activities for their waterfront facilities.

Table 2-18. Repair Priority Matrix.

Condition assessment rating	Risk tolerance 1 Acceptable	 2 Tolerable	 3 Unacceptable	 4 Intolerable
6—Good	Low	Low	Low	Low
5—Satisfactory	Low	Low	Medium	Medium
4—Fair	Low	Medium	Medium	High
3—Poor	Medium	Medium	High	High
2—Serious	High	High	Very High	Very High
1—Critical	High	High	Very High	Very High

Interpretation:

- **Very High:** Issues categorized as very high priority demand immediate and urgent attention because of their severe impact on operations, potential loss of property, or other critical factors. Immediate intervention is imperative to prevent catastrophic consequences.
- **High:** High-priority items require prompt action to address significant impacts on operations, potential loss of property, or other crucial factors. Timely intervention is necessary to prevent further deterioration and mitigate risks.
- **Medium:** Medium-priority items may not demand immediate action but should be closely monitored. Consider scheduling repairs or implementing preventive measures to avoid escalation to high-priority status.
- **Low:** Low-priority items have minimal immediate impact and can be addressed over a longer timeframe. Regular monitoring and periodic assessment are recommended to prevent these issues from escalating over time.

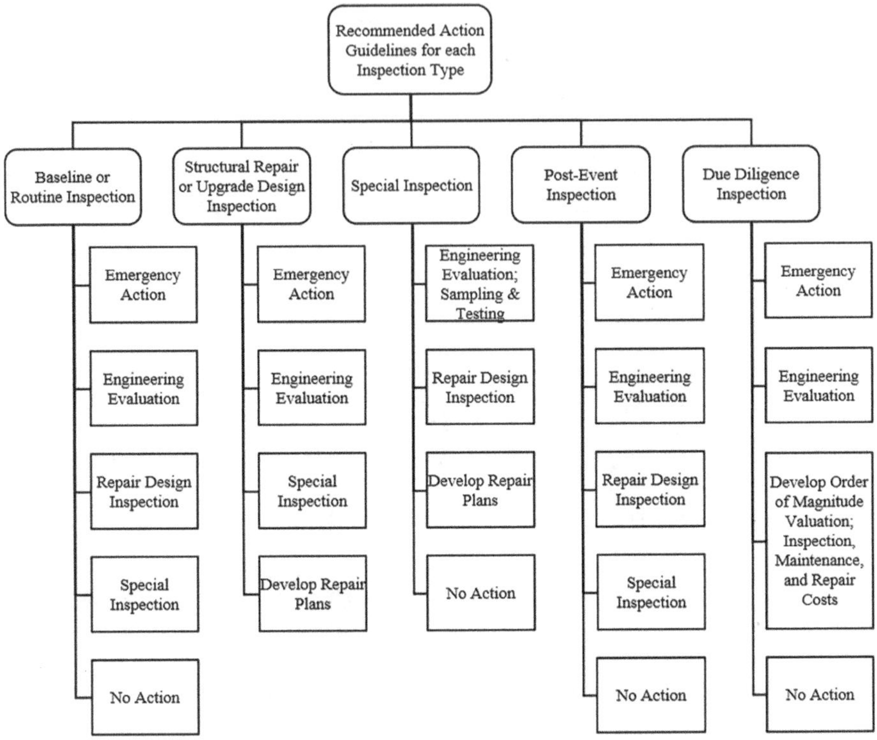

Figure 2-16. Recommended action guidelines.
Source: Courtesy of Moffatt & Nichol Inc., reproduced with permission.

CHAPTER 3
SCOPE OF INSPECTION WORK

3.1 GENERAL

The scope and methods for conducting the eight inspection types are presented in this chapter. These methods are general and apply to all structure types and construction materials. This manual does not present detailed *how to* techniques but provides methodology guidance. The provider of inspection services is assumed to possess the required expertise to competently and professionally implement the methodologies described herein. Confirming and documenting the specific scope and limits of the work with the owner at the outset are imperative.

Appendix A, "Special Considerations for Specific Structure Types and Systems," addresses unique aspects of inspection scopes of work applicable to specific structure types and systems. Appendix B, "Types and Causes of Defects," presents definitions of defect types and deterioration mechanisms for each material and component type.

3.1.1 Structural Boundaries

Inspections should be conducted and ratings assigned to distinct structural units. For example, a timber pier projecting from a steel sheet pile bulkhead should be divided into at least two distinct structures to inspect and assign condition ratings. Structural units should typically be of uniform construction type and material and, in the case of pile-supported structures, should be in a continuous pile bent numbering sequence.

The boundaries of structures must be clearly defined at the outset of the work. Breaking other structures into separate sections may be advantageous for large structures such as piers or wharves. Common boundaries include property lines, expansion joints, configuration changes, changes in the age of construction, changes in direction, or changes in bent numbering sequence.

3.1.2 Limits of Inspection

Inspections should be conducted on all accessible components as agreed to with the owner as indicated in the scope of work for the specific project. The upper limit may be defined as the elevation of the highest element to be inspected. The lower limit is typically defined as the channel bottom, mudline, or sea floor. It is important for the inspector and owner to agree upon the reference elevation datum [mean lower low water (MLLW), North American Vertical Datum of 1988 (NAVD88), local state plane, etc.] prior to the start of the inspection. Inspections may be conducted on generally inaccessible components as agreed with the owner, if means of access are arranged (e.g., excavation). Typically, the owner will arrange for the inaccessible items to be uncovered. An example of such is the excavation and exposure of sheet pile wall tie-rods located behind the wall.

Accessible components are defined as those components that are readily accessible without the need for excavation or extensive removal of materials that may impair visual inspection. The inspection report should identify inaccessible components that are not included in the scope of the inspection. Specialty equipment and/or procedures used to complete the inspection should be noted in the report for future inspections.

3.1.3 Definition of Underwater Inspection Levels of Effort

Defining the level of effort in the underwater inspection scope of work is important as the relative cost of performing this underwater portion is generally much higher than the above-water inspection. This manual recommends the minimum levels of inspection effort. These levels are based on industry experience, which has demonstrated that the sample size is appropriate for most situations.

Because of limited visibility, the inherent access restrictions of the underwater environment, and the presence of marine growth, certain inspection types, such as routine and baseline inspections, focus on investigating a statistically representative sample of underwater components. Three levels of underwater inspection effort are defined as follows:

- Level I Effort: Includes a close visual examination above and underwater or a tactile examination using large sweeping motions of the hands where visibility is limited underwater. Although the Level I effort is often referred to as a *swim by* inspection, it must be detailed enough to detect obvious major damage or deterioration due to overstress or other severe deterioration. It should confirm the continuity of the full length of all members and system components and detect undermining or exposure of normally buried elements. A Level I effort may also include limited probing of the substructure and sounding of the adjacent mudline.

- Level II Effort: A detailed inspection above and underwater may require the removal of corrosion and/or marine growth from portions of the structure. Underwater marine growth removal is costly; hence, the inspection must be based on a representative sample of components. For piles, a 12 in. high band should be cleaned at designated locations, generally near the low waterline, at the mudline, and midway between the low waterline and the mudline. On a rectangular pile, the marine growth removal should include at least three sides; on an octagonal pile, at least six sides; and on a round pile, at least three-fourths of the perimeter. On large-diameter piles, 3 ft or greater, 1 ft × 1 ft areas should be cleaned at four locations approximately equally spaced around the perimeter, at each elevation. On large solid-faced elements, such as retaining structures, 1 ft × 1 ft areas should be cleaned at these three elevations. The Level II effort should also focus on typical areas of weakness such as connections, attachment points, and welds. The Level II effort is intended to detect and identify damaged and deteriorated areas that may be hidden by surface biofouling, coating, or corrosion, or not be readily accessible for a Level I inspection effort. Marine growth removal should be governed by what is necessary to discern the condition of the underlying material. Removal of all biofouling staining is generally not required. Means and methods for the removal of biofouling growth are not typically defined in the scope of work. It may be appropriate for owners to specify particular methods on the basis of environmental and site conditions or concerns about maintaining the integrity of coating materials. Methods may include hand scrapers or mechanical systems ranging from high-pressure water blasters to barnacle busters and pressurized air bubble devices.
- Level III Effort: A detailed inspection above and underwater typically involving nondestructive or partially destructive testing conducted to detect hidden or interior damage or to evaluate material homogeneity. Typical inspection and testing techniques include the use of ultrasonic, coring, boring or resistance drilling, physical material sampling, and in situ hardness testing. Level III testing is generally limited to key structurally suspect areas or areas representative of the structure or system.

Table 3-1 presents a summary of the information typically collected for these levels of effort.

Table 3-1. Summary of Inspection Levels.

		Detectable Defects			
Level	Purpose	Steel	Concrete	Timber	Polymer Composite
I	General visual/tactile inspection to confirm as-built condition and detect severe damage	• Extensive corrosion, holes • Severe mechanical damage • Broken piles	• Major spalling and cracking • Severe reinforcement corrosion • Broken piles	• Major loss of section • Broken decking, caps, stringers, piles, and bracings • Severe abrasion or marine borer damage	• Permanent deformation • Broken piles • Major cracking or mechanical damage
II	Visual/tactile inspection to detect surface defects normally obscured by marine growth, coating, corrosion, etc.	• Moderate mechanical damage • Corrosion pitting and loss of section	• Surface cracking, spalling, and erosion • Rust staining • Exposed reinforcing steel and/or prestressing strands	• External pile damage caused by marine borers • Splintered piles • Loss of bolts and fasteners • Rot or insect infestation	• Cracking • Delamination • Material degradation
III	Visual/tactile inspection to detect hidden or interior damage, evaluate loss of cross-sectional area, or evaluate material homogeneity	• Thickness of material • Electrical potentials for cathodic protection systems • Thickness of coatings	• N/A*	• Internal damage, voids, marine borer activity • Decrease in material strength	• N/A

*Concrete coring, sampling, and laboratory testing are typically completed as part of a special inspection.

3.2 ROUTINE INSPECTIONS

3.2.1 Objectives

The routine inspection is a basic function performed in support of normal maintenance of waterfront facilities. Decisions made as a result of the information gathered during routine inspections will dictate the course and priority of future maintenance activities. Objectives of the routine inspection include

- Assessment of the overall condition of underwater, intertidal, and above-water portions of the structure;
- Determination of element damage ratings for all elements of the structure;
- Assigning a condition assessment rating for the overall structural unit;
- Development of recommendation(s) for follow-up action; and
- Determination of the recommended interval for subsequent inspection.

3.2.2 Methods of Inspection and Documentation

3.2.2.1 General Considerations. A routine inspection should be limited to collecting sufficient information to address each key objective. In most cases, documenting the location and size of each observed defect is not necessary. Rather, rating components on a relative condition scale of no defects, minor, moderate, major, or severe damage is standard practice. These damage grades must be specifically defined, wherever they are used, in terms of cross-sectional loss, crack widths, causation, location of damage, and so on. Chapter 2, "Standards of Practice," provides guidance on damage ratings for various material types.

When observed damage could have a significant effect on the load-bearing capacity of a structural component, sufficient data should be collected in the field to allow the impact of the damage to be assessed by analysis. Depending on the severity of the defect, advanced or immediate notice to the owner and/or the appropriate responsible party may be necessary.

In some cases, the routine inspection may be expanded to include estimating the remaining useful life of the structure or developing order-of-magnitude estimates of probable costs for rehabilitation work. These items should be clearly identified in the project's scope of work. In such cases, expanding the scope of the inspection and documentation to include sufficient information to address these issues may be necessary. Conducting a special-purpose inspection along with the routine inspection may also be necessary to address such issues. The scope of work described in the remainder of this section does not include these additional efforts.

3.2.2.2 Routine Above-Water Inspections. A routine inspection of above-water elements should include the sampling and methods of inspection summarized in Table 3-2. A routine inspection of above-water structural elements should include a general overview inspection of all exposed portions of the structure, plus sufficient close-up scrutiny of specific structural elements to establish a condition rating of individual structural components, and to assign condition assessment rating for the whole structure.

Above-water inspections may involve considerations related to issues of access and safety. Limited below-deck (above-water) access may require the use of additional means such as scaffolding or mechanized inspection devices (snooper trucks or aerial work platforms). When using an apparatus that results in significant loading, the load-carrying capacity of the supporting structure should be evaluated. Appropriate measures should be taken to ensure applicable safety regulations are followed regarding the use of such equipment.

Destructive testing (DT) or nondestructive testing (NDT) techniques may be appropriate, depending on the structural material, to gather further condition data. For instance, ultrasonic thickness measurements of steel elements, material sampling and testing of concrete elements, and resistance drill of timber elements can yield valuable information. When such testing techniques require the removal of surficial material or protective covering, an appropriate repair material should be applied to maintain protection. Specific recommendations for DT/NDT related to above-water inspections are provided in subsequent sections of this manual.

3.2.2.3 Routine Underwater Inspections. A routine inspection of underwater elements should include the sampling and methods of inspection summarized in Table 3-2. At a minimum, routine underwater inspections should include a Level I inspection effort for all components within the defined scope. In addition, a Level II inspection effort should be conducted on at least 10% of the submerged structural components. Depending on the material being inspected, a Level III inspection effort should also be conducted. The Level III effort typically involves either partially destructive or nondestructive investigation of approximately 5% of the underwater components. The type of testing will depend on the material and the specific damage or deterioration mechanism to be quantified. Descriptions should include a description of biofouling (marine growth) and the approximate zone or elevation where the damage is present, such that the significance of the deterioration may be evaluated.

Alternative inspection techniques, such as scanning using multibeam devices, can provide additional information or inspection of elements that may not be possible with a standard underwater diving inspection.

Table 3-2. Levels of Inspection Effort.

		Level I		Level II		Level III	
Material	Element	Sample Size[a] (%)	Method	Sample Size[a]	Method	Sample Size[a]	Method
Topside and above Water							
Steel[b,c]	Piles (≤ 24 in. dia.) Large or solid-faced elements/piles (> 24 in.)[d]	100	Visual/ tactile	As necessary	Visual: Removal of corrosion to observe parent metal and pitting size	As necessary	Remaining thickness measurement
Concrete[c]	Piles (≤ 24 in. dia.) Large or solid-faced elements/piles (> 24 in.)[d]	100	Visual/ tactile	As necessary	Visual/auditory: Sounding	As necessary	Electrical potential measurements, corrosion mapping, coring
Timber[c]	Piles Large or solid-faced elements[d]	100	Visual/ tactile	As necessary	Visual/auditory: Sounding and probing	As necessary	Drilling, coring
Masonry	Pilasters (> 24 in. per side) Large or solid-faced elements (> 24 in. per side)[d]	100	Visual/ tactile	As necessary	Visual/probing: Sampling of joint material	As necessary	Coring

(Continued)

Table 3-2. (*Continued*).

Material	Element	Level I Sample Size[a] (%)	Level I Method	Level II Sample Size[a]	Level II Method	Level III Sample Size[a]	Level III Method
Polymer Composites	Piles	100	Visual/ tactile	As necessary	Visual	0%	N/A
	Large or solid-faced elements[d]						
Underwater							
Steel[b,c]	Piles (≤ 24 in. dia.)	100	Visual/ tactile	10%	Visual: Removal of marine growth in three elevation bands	5%	Remaining thickness measurement, electrical potential measurements, corrosion profiling as necessary
	Large or solid-faced elements/piles (> 24 in.)[d]			Every 100 LF/quarter points	Visual: Removal of marine growth in 1 SF area at three elevations	Every 200 LF/5%	

(*Continued*)

Table 3-2. (*Continued*).

		Level I		Level II		Level III	
Material	Element	Sample Size[a] (%)	Method	Sample Size[a]	Method	Sample Size[a]	Method
Concrete[c]	Piles (≤ 24 in. dia.)	100	Visual/ tactile	10%	Visual: Removal of marine growth in three elevation bands	0%	N/A
	Large or solid-faced elements/piles (> 24 in.)[d]			Every 100 LF/quarter points	Visual: Removal of marine growth in 1 SF areas at three elevations		
Timber[b]	Piles	100	Visual/ tactile	10%	Visual: Removal of marine growth on three bands Measurement: Remaining diameter	5%	Internal marine borer infestation evaluation
	Large or solid-faced elements[d]			Every 50 LF	Visual: Removal of marine growth in 1 SF area at three elevations	Every 100 LF	

(*Continued*)

Table 3-2. (*Continued*).

		Level I		Level II		Level III	
Material	Element	Sample Size[a] (%)	Method	Sample Size[a]	Method	Sample Size[a]	Method
Masonry	Pilasters (> 24 in. per side)	100	Visual/ tactile	10%	Visual: Removal of marine growth in three elevation bands	0%	N/A
	Large or solid-faced elements (> 24 in. per side)[d]			Every 100 LF/quarter points	Visual: Removal of marine growth in 1 SF area at three elevations		
Polymer Composites	Piles	100	Visual/ tactile	10%	Visual: Removal of marine growth in three elevation bands	0%	N/A
	Large or solid-faced elements[d]			Every 100 LF/quarter points	Visual: Removal of marine growth in 1 SF area at three elevations		

[a]The minimum inspection sample size for small structures shall include at least two components of each underwater component type.

[b]Coated elements: Inspect the element with the coating intact.

[c]Jacketed or encased elements: Visually inspect the jackets and encasements for deterioration. Inspect the base element to the extent possible.

[d]Large, solid-faced elements may include bulkheads, retaining walls, piers, gates, caissons, piles greater than 24 in. in diameter, and so on.

3.2.3 Methods of Inspection for Steel Components

3.2.3.1 Routine Above-Water Inspection. Above-water inspection efforts of steel structures should be conducted using visual/tactile means. Additional testing techniques, such as thickness and profile measurements, should be employed to supplement the visual findings.

Above-water/below-deck inspection efforts of steel structures should also be conducted using visual means. Additional testing techniques, such as thickness and profile measurements, should be employed to supplement the visual findings. Attention should be paid to waterline and splash zones as these areas often have higher corrosion than other above-water areas. It is important to complete a vertical profile of thickness measurements on each structure to identify elevated corrosion zones. For example, the low water and splash zones vary by location and site-specific factors. The profile should be completed by selecting one representative member and completing Ultrasonic Thickness (UT) measurements at increased intervals, typically at one foot along the full length.

3.2.3.2 Routine Underwater Inspection. Level I inspection efforts for steel structures are conducted using visual/tactile means. It is recommended that the Level I inspection includes 100% of the structure, and particular attention should be paid to areas of known elevated deterioration. This is particularly true on steel structures in the splash zone and slightly below low water. Corrosion rates vary significantly on the basis of environmental factors and site-specific characteristics; therefore, local knowledge should be applied.

Level II efforts are also visual upon removal of corrosion, marine growth, or other materials that prevent Level I inspection effort in areas to be inspected. Level II efforts are generally performed on 10% of the structure or system. The splash zone and immediately below the low water elevation should be closely evaluated to identify the magnitude and extent of corrosion losses. Attention should be given to identifying the low water zone as it can vary geographically and within harbors due to specific environmental factors.

Level III efforts for steel structures require the remaining thickness of the element to be measured in random locations that are representative of the structure. Such measurements may be taken by micrometer, pipe pit gauge, where feasible, or by ultrasonic thickness measuring devices or electrochemical impedance spectroscopy. At a minimum, Level III inspections should be conducted on 5% of the elements. Specific circumstances may warrant additional Level III efforts such as inspecting welds using magnetic particle testing, coupon samples, and voltage potential measurements; however, magnetic particle testing and coupon sampling would typically be completed as part of a special-purpose inspection. Testing at the mudline should be considered in structures with noted section loss.

If the inspected components exhibit significant corrosion that could affect the load-carrying capacity of the structure, corrosion profiling should be performed to establish the extent of corrosion as it varies along the height of the structure. Multiple profiles may be necessary to establish the uniformity or variability of the damage throughout the structure. The results of the corrosion profiling should be used to evaluate the significance of the corrosion. It is particularly important while inspecting fill-containing structures, such as sheet pile cells or bulkheads, to inspect closely for corrosion holes, as small holes may result in large voids behind the structure, potentially resulting in sinkholes and failures.

For steel members that are cathodically protected, an electrical potential survey should be conducted during the inspection. Potential measurements should be taken at points throughout the structure to determine the effectiveness of the cathodic protection system. An Ag–AgCl half-cell is used in the marine environment. A reading between −0.800 and −1.200 V indicates cathodic protection adequacy. This range can vary depending on factors such as water salinity and type of half-cell. Readings above –1.200 may provide adequate protection but could result in damage to coating systems.

For coated steel members, Level I and II efforts should focus on the evaluation of the integrity and effectiveness of the coating. For submerged elements, care should be taken to avoid damage to the coating during the removal of marine growth for the Level II effort. Level III efforts for coated steel members should include ultrasonic thickness measurements without removal of the coating, where feasible. Collecting coating samples for evaluation of hazardous materials, such as lead, may be prudent if a repair entails the removal of the coating system.

For steel members that have been wrapped, Level I and II efforts should focus on the evaluation of the integrity of the wrap. Care should be taken to avoid damage to the wrap during the removal of marine growth for the Level II effort. Because the effectiveness of a wrap may be compromised by removal and because the removal and reinstallation of wraps is time-consuming, such removals should not be routine. If evidence of significant corrosion exists, or if the effectiveness of the wrap may be in question, limited wrap removal may be considered to facilitate inspection and evaluation. The sample may be limited to zones or portions of the member if damage is suspected, for example, at the waterline. The sample size should be determined on the basis of the physical evidence of potential problems and the aggressiveness of the service environment. A minimum sample size of three members should be used. A 5% sample size, up to 30 total members, may be adequate as an upper limit on large structures.

For steel members that have been encased, Level I and II efforts should focus on the evaluation of the integrity of the encasement. Encasements are not typically removed for a routine inspection. If evidence of significant deterioration of the encasement is present, or if evidence of significant deterioration is present on the underlying member despite the encasement, the damage evaluation should consider whether the encasement was provided for protection, structural capacity, or both. For encasements where the formwork has been left in place, the inspection should focus on the integrity of the encasement and not on the formwork. Level I and II efforts in such cases should concentrate on the top and bottom of the encasement.

Encasements only installed to the low water elevation can elevate the rate of corrosion below the encasement, especially if a cathodic protection system is not active. This should be an area of focus.

3.2.4 Methods of Inspection for Concrete Components

3.2.4.1 Routine Above-Water Inspection. Above-water inspection efforts of concrete structures should be conducted using visual means, supplemented by audio methods, such as dragging a chain over the deck to find hollow-sounding areas, which indicate delamination and potential spalls. Additional testing techniques such as corrosion mapping and chloride profiling may be necessary to supplement the visual findings; however, this type of testing is not typically included in a routine inspection and should be included as a special-purpose inspection.

Above-water/below-deck inspection efforts of concrete structures should be conducted using visual means, supplemented by audio methods. A hammer, roller, or other metal probes may be used to find loose or delaminated concrete. Thermal cameras, either handheld or drone-mounted, are valuable devices for large structures. Additional testing techniques, such as corrosion mapping and chloride profiling, may be necessary to supplement the visual findings; however, this type of testing is not typically included in a routine inspection and should be included as a special-purpose inspection.

3.2.4.2 Routine Underwater Inspection. The underwater inspection effort is typically composed of Level I and Level II inspections. Level I inspections for concrete structures are conducted using visual/tactile means. Level II efforts are also visual, after removal of marine growth in areas to be inspected. Level III efforts are not typically required on concrete elements.

The inspection of encased concrete members should incorporate topside and above-water inspections. Level I and Level II underwater inspection efforts should focus on evaluating the integrity of the encasement.

Encasements are not typically removed for a routine inspection. If evidence of significant deterioration of the encasement is present, or if evidence of significant deterioration is present on the underlying member despite the encasement, then the evaluation of damage should consider whether the encasement was provided for protection, structural capacity, or both. Encasements where the formwork has been left in place should focus the inspection on the integrity of the encasement and not on the formwork. In such cases, Level I and II efforts should concentrate on the top and bottom of the encasements. If deterioration, debondment, or other significant problems with the encasement are suspected, conducting a special-purpose inspection may be necessary. The special-purpose inspection in such circumstances may include coring of the encasement and laboratory testing of the sample to evaluate the materials.

For wrapped concrete members, the above-water, Level I, and Level II underwater inspection efforts should focus on evaluating the integrity of the wrap. Care should be taken to avoid damage to the wrap during the removal of marine growth for the Level II effort. Because the effectiveness of a wrap may be compromised by removal and because the removal and reinstallation of wraps is time-consuming, such removals should not be done routinely. If evidence of significant damage exists, or if the effectiveness of the wrap is in question, then a sample of the wrap should be removed to facilitate the inspection and evaluation. The sample may be limited to a particular zone or portions of the member if damage is suspected, for example, at the waterline. The sample size should be determined on the basis of physical evidence of potential problems. A minimum sample size of three members should be used. A 5% sample size, up to 30 total members, may be adequate as an upper limit.

3.2.5 Methods of Inspection for Timber Components

3.2.5.1 Routine Above-Water Inspection. A routine inspection of above-water timber components primarily consists of visual assessment to verify that the elements are intact and functioning as intended. Specific focus should include noting evidence of section loss, rot, preservative treatment retention, insect damage, abrasion, weathering, wear, impact, or overstressing. Visual assessments should be supplemented with ice pick or awl penetration examination to check softness. Inspection efforts conducted on timber elements to detect internal deterioration should be implemented if timber elements have suspected deficiencies not apparent during visual inspections. Resistance drilling, boring, or coring to quantify internal deterioration is typical. Laboratory analyses of core samples are often conducted to determine the presence of marine borers and the density, specific gravity, and moisture content. Core samples extracted from creosote pressure-treated timber elements can be subject to composite creosote penetration and retention analysis. Bore and core holes should be filled with oversized treated hardwood dowels, epoxy, or nonshrink grout.

3.2.5.2 Routine Underwater Inspection. Level I inspection efforts for timber structures are conducted using visual/tactile means. An ice pick or awl should be used to probe for softness, except for wrapped members.

Level II efforts are also visual upon removal of marine growth or other materials that would preclude visual inspection in areas to be inspected. Level III efforts on timber elements have historically been conducted using several methods. The material's intrinsic nature causes deterioration after initiation to progress rapidly compared to other materials utilized in marine construction. Employing resistance drilling, boring, or coring is typical to quantify internal deterioration. Bore and core holes should be filled with oversized treated hardwood dowels, epoxy, or nonshrink grout.

For timber members that have been repaired by encasement, Level I and II efforts should focus on evaluating the integrity of the encasement. Such encasements should not be removed for a routine inspection. If evidence of significant deterioration of the encasement is present, or if evidence of significant deterioration is present on the underlying member despite the encasement, then the evaluation of damage should consider whether the encasement was provided for protection, structural capacity, or both. For encasements on which the formwork has been left in place, the inspection should focus on the integrity of the encasement and not the formwork. Level I and II efforts in such cases should concentrate on the top and the bottom of the encasement.

For wrapped timber members, the Level I and II efforts should focus on evaluating the integrity of the wrap. Care should be taken to avoid damage to the wrap during the removal of marine growth for the Level II effort. Level III efforts should consist of removal of the wraps from a representative sample of components to evaluate the condition beneath the wrap. The sample may be limited to a particular zone or portions of the member if damage is suspected, for example, at the mudline/bottom of the wrap or in the tidal zone. The sample size should be determined on the basis of the physical evidence of potential problems and the aggressiveness of the service environment. A minimum sample size of three members should be used. A 5% sample size, up to 30 total members, may be adequate as an upper limit. Upon removal of the wrap, the timber should be evaluated using visual/tactile means and boring or coring as described for nonwrapped timber elements. Wraps that are removed to facilitate such inspections should be restored or replaced in accordance with the wrap manufacturer's installation requirements. Cutting the wrap over half the perimeter of the member and then repairing it upon completion of the inspection may be more advantageous than removing the wrap entirely. The age and integrity of the wrap should be considered as to whether a partial removal may be prudent.

3.2.6 Methods of Inspection for Masonry Components

3.2.6.1 Routine Above-Water Inspection. Masonry structures, including grouted cyclopean walls, typically used for quay walls or abutments, rely on their overall geometry and gravity for stability and resistance to environmental and operational loads. While conducting the inspection, it is crucial to lay out a baseline to evaluate such factors as the location of sinkholes or depressions, alignment and batter changes, or displacement of discrete wall sections. Void networks located within/behind the masonry retaining structure may be evidence of sinkholes below paved surfaces and may warrant more detailed inspections such as ground-penetrating radar or excavation of test pits.

While conducting inspections, the following characteristics of the masonry structure should be noted:

- Type of stone and joint construction;
- Size and orientation of stones;
- Condition of stones and joint material;
- Size of voids, noting large voids by hand probing;
- Scour or erosion of the toe;
- Alignment and batter;
- Surface drainage or tidal water flow through the wall; and
- Attachments and embedded items.

The primary modes of deterioration of masonry structures are loss of joint material, displacement of stones, settlement of the toe, and sinkholes behind the structure. Sinkholes may not be outwardly apparent; therefore, collecting overall observations and probing for void networks to develop conclusions are important.

3.2.6.2 Routine Underwater Inspection. Level I inspection efforts for masonry components are conducted using visual/tactile techniques with a particular focus on overall alignment and attitude and position of individual blocks. Any missing masonry should be noted.

Level II efforts are also visual, after the removal of marine growth or other material that would prevent visual inspection in areas to be inspected. In general, marine growth removal underwater is based on the size of the structure, but, at a minimum, marine growth should be based on the frequency outlined in Table 3-2. This level of inspection should be more detailed and include an inspection of the joints and void networks that may be present.

Level III efforts are not typically required on masonry elements.

3.2.7 Methods of Inspection for Polymer Composite Components

3.2.7.1 Routine Above-Water Inspection. A routine inspection of above-water polymer composite components primarily consists of a visual assessment to verify that the elements are intact and functioning as intended. Specific focus should include noting evidence of ultraviolet degradation (often characterized by discoloration), creep, excessive deflection, deformation, or overstressing, particularly at hardware connections.

3.2.7.2 Routine Underwater Inspection. Level I inspection efforts for polymer composite components are conducted using visual/tactile means. Level II efforts are also visual upon removal of marine growth in areas to be inspected. Level III efforts are not typically required on polymer composite elements.

3.2.8 Methods of Inspection for Slope Protection

3.2.8.1 Routine Above-water Inspections. A routine inspection of exposed above-water portions of slope protection, revetments, and other similar structures primarily consists of rapid visual assessment to verify that the components are stable and remain somewhat unchanged when compared to the original cross section. This is best accomplished at low tide to take advantage of the ability to see as much exposed surface as possible.

Structures consisting primarily of various sizes of rock, such as graded revetments or rubble mound structures, are typically designed to tolerate a certain amount of in-place settlement and may have shifted accordingly. Weather-related phenomena, such as waves, ice, river runoff, and earthquakes, may also have resulted in significant deformation of these structures.

These structures may have hardened areas at or near the top of the structure, often as a means of retaining stone, or for use as a rat guard (concrete used to solidify the section to prevent the migration of rodents through the rock). If the overall slope of the section has undergone settlement, then a significant cross-section distortion may have occurred at the intersection between hardened and unhardened areas.

As with underwater inspections (see next section), remote sensing techniques can also be used in above-water inspections. A detailed survey of the features in question may be obtained by using laser scanning. Supplementing laser scanning with coincident photography for data interpretation is very useful. Multibeam and side scan sonar techniques, as well as light detection and ranging (LiDAR) scanning, may be conducted coincident with underwater inspections. Harmonizing captured data on both sides of the waterline can be highly valuable for assessing structures spanning the waterline. Remote sensing techniques such as these are only as accurate as their horizontal and vertical controls, and thus, a detailed and accurate baseline survey is needed with tie-ins to known benchmarks.

Besides visual means, an inclinometer may be used to measure the slope relative to original placement. Traditional survey methodology can also be implemented to verify profiles.

An above-water routine inspection should note the overall stability and performance of the structure, and any significant defects or discontinuities found as a result of exposure to weather-related phenomena or settlement.

3.2.8.2 Routine Underwater Inspection. Level I inspection efforts on slope protection, such as armor stone, riprap, gabions, concrete liners, and scour protection mattresses, are conducted using visual/tactile means. Level II efforts are not typically required, unless removal of marine growth from representative areas is necessary to judge the condition of the slope protection, as may be the case for concrete liners, or unless removal of silt is necessary to expose the structure.

For slope protection under pile-supported structures such as wharves, Level I and Level II inspections can be completed in conjunction with pile inspection. In this case, the elevation to top of rock or soil can be measured directly at each pile to give transections at each bent.

Performing Level I efforts may sometimes not be feasible or practical due to waves, currents, or restricted visibility. This is particularly true for the offshore face of breakwaters, which can pose higher risk. In such cases, performing the inspection using Level III remote-sensing techniques may be more cost-effective or technically advantageous. Common remote-sensing techniques for features below the water line are typically LiDAR, acoustic (sonar), side-scan sonar, and multibeam sonar systems. Laser scanning and photogrammetry may be conducted in parallel with above-water inspections. Coordinating captured data on both sides of the waterline can be highly valuable for assessing structures spanning the waterline.

The results of surveys may indicate areas of potential problems such as discontinuities, protrusions, or other inconsistencies, which should be further investigated by dive inspections. Remote-sensing techniques are only as accurate as their horizontal and vertical control; therefore, proper control is vital. Differential global navigation satellite systems, differential global positioning systems, inertial measuring unitss, and/or digital compasses are frequently used to interface with these sonar systems. In locations where obstructions prevent acquiring sufficient satellite signals, range-azimuth systems may also be employed. Differentially referencing to the appropriate geodetic or local coordinate control and reference to the correct datum is critical (e.g., North American Datum of 1983, NAVD88, tide gauges, MLLW).

3.2.9 Methods of Inspection for Mudline

Level I inspection efforts at the mudline near and around structural elements are conducted using lead-line soundings, and potentially visual/tactile means to evaluate scour or changes in the bottom conditions. Level II efforts are not typically required.

Performing Level I efforts may not be feasible or practical due to waves, currents, or restricted visibility. This is particularly true for the offshore face of breakwaters, which can pose high risk. In such cases, performing the inspection using Level III remote-sensing techniques may be more cost-effective or technically advantageous. A hydrographic survey plan that combines bathymetric water depth survey and subsurface topology features with surface topographic survey can be valuable for assessing structures spanning the waterline.

The results of hydrographic surveys may indicate areas of potential problems such as discontinuities, protrusions, or other inconsistencies, which should be further investigated by dive inspections. Remote-sensing techniques such as these are only as accurate as their horizontal and vertical controls; therefore, proper control is vital. Differential global positioning systems or range-azimuth systems may be used to interface with the sonar system, tidal depth gauge, and so on, to locate the anomaly for investigation.

Depending on the scope and use of an underwater or above-water survey, it may be appropriate to engage a licensed surveyor or hydrographer to oversee the work and certify the resulting documents.

3.2.10 Methods of Inspection for Mooring Hardware and Fender Systems

3.2.10.1 Mooring Hardware. Level I inspection efforts involve a walk-through inspection of the mooring hardware. Major defects and obvious damage should be noted. Level I inspections are typically done periodically after major storm events, or a mooring incident such as hardware breakage or lines parting.

Level II inspection efforts involve a detailed visual and tactile inspection of the mooring hardware components. The mooring hardware should be assessed for significant signs of wear or stress from the mooring lines. Exposed fasteners should be observed for signs of corrosion or whether displacement has occurred. The mooring fitting base should also be inspected for signs of cracking or other stress-related defects. In addition, the plumb and level of the mooring hardware should be measured to determine whether any movement has taken place. When mooring hardware is secured through the deck, the underside of the deck and visible fasteners should be inspected. Level II inspections are typically done periodically, or when results from a Level I inspection warrant it.

Level III inspection efforts involve a detailed inspection of the mooring hardware and are typically considered a special-purpose inspection. Level III efforts may include the removal of a fitting for inspection of the anchorage.

3.2.10.2 Fender Systems. Fender systems are made of different material types, and, depending on the type of material, the relevant subsection appears elsewhere in this section, such as the timber, polymer composite, rubber, or steel elements.

3.2.11 Methods of Inspection for Anchors and Chains

3.2.11.1 Routine Above-Water Inspections. A routine inspection of the exposed portions of the anchor and chain system primarily consists of assessing the chain and hardware connection that is connected to a structure. The chain should be inspected for wear, distortion, corrosion, and alignment with the attachment point. This is best accomplished at a low water elevation to take advantage of the ability to see as much of the chain above water and extended as possible. The connection hardware should be inspected for wear, distortion, corrosion, and a secure connection to the structure.

Some anchor system designs may involve dissimilar material types connected to each other, such as a galvanized anchor chain and connection hardware attached to an epoxy-coated, black steel mooring buoy. In this example, increased corrosion rates could occur at the intersection if the epoxy coating has deteriorated such that the two dissimilar metals are in contact. Another similar scenario involves the use of nongalvanized (or nonstainless) steel anchor bolts to secure a galvanized pad eye to a concrete surface; in this case, increased corrosion will occur on the anchor bolt. The remaining material thicknesses should be measured using calipers to determine the remaining capacity.

3.2.11.2 Routine Underwater Inspection. Level I inspections efforts for anchor and chain leg assemblies are conducted using visual/tactile means and consist of examining the overall alignment of the anchor and chain leg, the anchor connection components, the anchor position on the ground and surrounding ground material type, the orientation of the stabilizers and flukes, and any evidence of drag movement in the surrounding area. The anchor chain inspection should focus on the riser, catenary, dip, and ground sections.

Level II inspection efforts of an anchor consist of measuring the connection component's remaining thickness, and the chain's offset angle to the shank end. Level II efforts are visual upon the removal of marine growth in areas to be inspected.

Level III inspection efforts of the anchor chain legs consist of measuring the single-link diameter to determine the amount of remaining steel. A double-link measurement is used to determine the amount of wear at the adjoining two-link connection. A five-link measurement is used to check for chain stretch. A single-link measurement checks the individual chain link for stretching and determines whether it falls within the manufacturer's tolerance. In special mooring configurations, measuring the angle of the chain at the catenary and dip sections using an inclinometer device may be necessary. Connections, such as Kenter joining links, should be inspected for the connection stud, keeper pins, and lead pellets (plugs).

See Appendix A.9, "Mooring Buoy Systems," for additional information.

3.2.12 Methods of Inspection for Buoys

3.2.12.1 Routine Above-Water Inspection. A routine inspection of the exposed portions of buoys primarily consists of visually assessing the buoy and its topside hardware, fenders, and chafing strips. Also, the buoy should be assessed to verify that it has not been dragged from its intended location.

The buoy should be examined to detect physical damage such as holes, dents, distortion, or listing. For fiberglass or fiberglass-coated buoys, exterior noting cracks, wear, peeling, or rust staining must be assessed. The coating system for cracking, chipping, peeling, or debonding for painted buoys must be assessed. For all buoys, the freeboard must be measured and any penetrations and/or drains must be assessed. Fenders and chafing strips should be assessed for the remaining section, cracks, attachment, and fastener corrosion.

Mooring components such as chains, linkages, and jewelry should be assessed as indicated in Section 3.2.11.

3.2.12.2 Routine Underwater Inspection. Level I inspection efforts for submerged buoy hulls are conducted using visual/tactile means and consist of examining the overall exposed surface, identifying damage and/or deterioration as previously mentioned; the mooring leg connection components; and the integrity of the coating system. Level II efforts are also visual, upon removal of marine growth in areas to be inspected. Level III inspection efforts of a buoy consist of measuring the remaining thickness of the buoy hull.

3.2.13 Methods of Inspections for Utility Systems

Utility systems on marine structures may include steam, potable water, fire suppression, seawater, sanitary sewer, oily waste disposal, fuel, compressed air, electrical, data and communication services, and fire alarm systems. This section does not include specialty systems such as cargo-handling systems. For utilities, the inspection scope of this manual addresses the surface observations and associated support systems. Evaluating the utility's code compliance or ability to perform its function as designed is beyond the scope of this manual. In such cases, inspections require the inclusion of specialized engineers on the inspection team to evaluate the utility systems.

The scope of inspections should be performed in accordance with Table 3-2. Level I inspections efforts for utilities are conducted using visual/tactile means. Level II and III are not normally conducted, unless the utility extends underwater, and marine growth removal is needed to observe the condition. Caution must be exercised when removing marine growth or corrosion from deteriorated exposed utilities, because this may result in a rupture of the line, causing leakage (for wet utilities) and possibly environmental contamination. If the line is color-coded, ensure that it conforms with the American Public Works Association's (1999) "Uniform Color Code," is labeled, and matches the material in the line. Coordination and/or approval from the utility owner will be necessary.

Where utility systems are within sleeves, encasements, utility vaults, or underground, methods to determine the line condition are limited. Sleeves and encasements provided with drain holes should be checked for signs of leakage. Uncovering buried utility systems is not generally performed under the scope of a routine inspection. Equipment and valves should not be operated.

The surfaces of the utility lines, including coatings, should be inspected for their condition. Other common components requiring inspection include appurtenances such as hangars, mounts, and cradles.

3.2.14 Evaluation and Rating

A condition assessment rating should be assigned to each inspected element group and/or system in accordance with Section 2.5, "Element-Level Damage Ratings." Each structure should be assigned an overall rating in accordance with Section 2.6, "Overall System Ratings."

For structural elements, if significant damage or deterioration is observed that could affect the load-carrying capacity of the structure, then an engineering evaluation of the effect of the damage on the structural capacity should be conducted prior to assigning a condition assessment rating. Such a quantitative evaluation should typically be limited to the assessment of individual or typical components and should also be limited to an evaluation of the effect of the damage on the individual components' capacity, without considering the actual or anticipated loading (structural demand). A routine inspection should not include conducting a detailed structural analysis of the structure. Should the need for a more rigorous structural analysis become apparent as a result of a routine inspection, then an engineering evaluation should be recommended as a follow-up action.

Utility systems should be assigned a condition assessment rating. The purpose of a routine inspection is not to provide comprehensive functional testing or code compliance. Should the need for a more thorough inspection or function testing become apparent as a result of the routine inspection, then further investigation should be recommended as a follow-up action.

3.2.15 Recommendations

Upon completion of the routine inspection, recommended actions should be assigned to the structure in accordance with Section 2.7, "Recommended Action Guidelines." In addition, the recommended interval to the next regularly scheduled routine inspection should be assigned using the guidelines of Section 2.1.1, "General."

3.3 REPAIR OR UPGRADE DESIGN INSPECTIONS

3.3.1 Objectives

A repair or upgrade design inspection should be conducted once the decision to proceed with repairs has been made or if more accurate repair cost estimates are desired. A repair or upgrade design inspection is intended to be targeted to the specific areas or structural members needing repairs. It is generally not desired to complete a detailed repair or upgrade design inspection on the entire facility because of the cost.

Structural upgrades to the facility that improve operations or load-carrying capacity require a detailed inspection. Additional investigations such as geotechnical may also be required depending on the upgrade. Theoretically, the upgrade work could be accomplished in combination with or independent of repairs. By including the two together, upgrade work that involves removing and reconstructing deteriorated elements will eliminate the need to repair these items and thus reduce costs. The types of repairs or upgrades to be implemented should be made after the determination of the expected remaining service life of the facility. This is essential to ensure that repairs are targeted with a similar service life goal to maximize repair expenditures and to eliminate waste.

Examples of facility upgrades include improvement of deck structure capacity to resist concentrated loads from mobile truck cranes or other equipment; operational improvements such as additional space, new utility trenches, or larger gantry cranes; or improvement in the structural performance under seismic loading. Ideally, the decision to upgrade a structural component should be made prior to the commencement of the repair or upgrade design inspection. Preliminary results from the inspection may modify the extent of the initial upgrade plan to be more cost-effective. An example of this might be the extension of a deck reconstruction into an adjacent area that is highly deteriorated in lieu of repairs.

Data collected during a repair or upgrade design inspection will be used to prepare contract bid documents to address specific defects and/or upgrade of components. Information gained from previous inspections should be used as appropriate to supplement the repair or upgrade design inspection. Specific objectives of a repair or upgrade design inspection include

- Documentation of the location and size and selection of the proposed method of repair for each defect or component to be repaired,
- Thorough investigation of the areas of the upgrade reconstruction work where such upgrades tie back into or reuse the existing structure,
- Information leading to preparation of quantity estimates for the project, and
- Thorough documentation of all site-related conditions that could affect or impede straightforward execution of the work.

Upon completion of the repair or upgrade design inspection, bid documents can be prepared for the work. Bid documents typically include drawings and specifications covering each aspect of the work, along with bid forms, quantity estimates, and estimates of probable cost. Preparation of bid documents is beyond the scope of this manual. The owner should be aware of the differences in contract pricing methods for repair work (i.e., extensive use of unit-priced bid items with estimated quantities at bid time and final quantities determined in the field).

3.3.2 Methods of Inspection and Documentation

The repair or upgrade design inspection should focus primarily on the documentation of only those elements that are intended to be repaired or upgraded. For this reason, defining specific repair criteria prior to executing the inspection is important. Defects that do not meet these predefined criteria need not be recorded, thereby improving the efficiency of the field inspection work.

It is important to understand the owner's expectations for the inspection and potential repairs. Funding may limit planned repairs and/or upgrades and may require deferring less critical damage or deterioration for future repairs.

The information required to prepare contract bid documents from the repair or upgrade design inspection results typically includes the following:

- Types of defects;
- Locations and extent of demolition and removal (when required);
- Location of the defects on the structure (component ID);
- Position of the defect on each component;
- Size of the defects;
- Location of surrounding infrastructure, such as utilities, which could complicate the repairs, or be important for a contractor in developing staging and work platforms;

- Method of repair for the defects;
- Location of items to be upgraded; and
- Method of upgrade (supplement, strengthen, or replace).

Documentation of repair size should consider the increase in, or growth of, the defect size as a result of repair preparation, as in the case of preparation for repair of concrete spalls. In many cases, this can be a significant increase in the size of the visible or apparent repair area. It may be advantageous to prioritize the elements of work because some defects will require repairs more urgently than others and upgrade work may take precedence over repair work, or vice versa. The use of unit-price costing, bid alternatives, or bid options can be implemented to accomplish these goals. Such prioritization will allow the repair project to be readily altered to focus on the most important elements of the work if funding is restricted. The bid documents and pay clauses in the technical specifications should reflect these arrangements, if used.

The repair or upgrade design inspection should be conducted shortly after the decision to repair and/or upgrade the structure with minimal delay before soliciting bids for repair work. The repair or upgrade design inspection should focus only on those portions of the structure identified in previous inspections that require repair or upgrades. Should a lengthy time elapse between the original inspection and the implementation of the contract, additional investigation and effort may be required to inspect and document components that have sustained additional deterioration.

The methods used for repair or upgrade may significantly affect the amount of time necessary to conduct a repair or upgrade design inspection. If concrete piles supporting a wharf exhibit extensive defects such as cracks (above or below water), knowing the proposed repair method prior to the inspection is beneficial. The cause of the cracking and whether the piles will be repaired by jacketing, epoxy injection, sealing, or even replacement will influence the required efforts. If the cracks are to be repaired by epoxy injection or sealing, recording the location and size of each crack will be necessary to produce repair plans and properly estimate repair quantities. Jacketing may be sufficient knowledge that cracks exist on the pile that meet the predefined repair criteria. The specific attributes of each crack become irrelevant in this case.

The difference in inspection time required to support these two scenarios may be significant. The documentation of every defect could require the removal of all marine growth from each pile. By contrast, to establish only the presence of cracking, only minimal removal of marine growth may be required. It is important to tailor the methodology used for conducting the repair or upgrade design inspection to the repair method employed.

3.3.3 Inspection of Buried Elements

Prior to implementation of repairs or upgrades to a structure, verify the condition of buried or inaccessible elements that contribute to structural capacity. Such elements may include tiebacks, walers, dead men, soldier piles, cofferdams, relieving platforms, or similar items. Implementation of repair or upgrade projects without confirmation of the condition of buried structural members is not recommended because the integrity of the full structural load path has not been confirmed. An inspection of these elements requires a consideration of the following issues:

- Excavation pit size and methodology (i.e., laid-back soil or use of shoring),
- Identification of potential obstructions (buried or surficial),
- Verification of safety requirements,
- Investigation of dewatering requirements (if any) and appropriate disposal of dewatering effluent, and
- Preparation for the potential of encountering contaminated or hazardous materials.

3.3.4 Recommendations

Upon completion of the repair or upgrade design inspection, recommended actions should be assigned to the structure in accordance with Section 2.7, "Recommended Action Guidelines."

3.4 NEW CONSTRUCTION INSPECTIONS

3.4.1 Objectives

A new construction inspection should be performed on a new structure concurrent with the construction. The inspection should address the following objectives:

- Control quality to verify compliance with design documents.
- Verify quantities installed.
- Respond to field questions and resolve field problems.
- Develop a list of deficiencies for the contractor to correct.

3.4.2 Method of Inspection and Documentation

A new construction inspection is conducted during construction to ensure quality control for the waterfront project. The resolution of field questions and issues ensures that the design intent and construction documents are properly interpreted, implemented, and contribute to keeping the project within the established budget and schedule.

Documentation obtained during a new construction inspection should provide details about modifications to the design documents and the quantity of installed materials. Specific information should include confirmation of dimensions, verification of installed members (e.g., cotter pins and plates for anchor systems), locations, and other physical features. Field issues and associated resolutions should be documented for future reference, including deviations from the design plans, and documented in the record drawings.

The scope and level of inspection required may vary significantly depending on the structure type and construction methods used. At a minimum, a visual inspection of the underwater components should be conducted to ensure compliance with the contract documents and to make any corrections to the record drawings for future reference and inspections. In addition, a visual inspection may be performed at critical stages during the construction sequence to ensure that components and/or systems are being installed correctly.

Issues may arise during construction that warrant more detailed inspection or testing to verify installed components and conditions. As an example, placement of a concrete caisson may require two phases of inspection: (1) to ensure proper preparation and leveling of the site prior to the installation of the structure, and (2) to inspect the caisson after placement to ensure that no damage has occurred during placement. Alternatively, a new construction inspection of a timber pile–supported structure may require only one inspection to confirm no damage has occurred to the members, such as breaching of the protective treatment, allowing for possible accelerated deterioration due to marine borer infestation. Inspecting piles in a pile-supported structure is generally advantageous to ensure that no broken or damaged piles are present prior to construction of the deck.

3.4.3 Methods of Inspection for Steel Components

An inspection of the above-water elements should include visual observation on a daily basis to ensure that the construction is being performed in accordance with the contract plans and specifications.

A Level I underwater effort is all that should be required for new construction. A Level I inspection effort is conducted using visual/tactile means. Marine growth should be nonexistent to minimal for a new construction inspection, so a Level II effort should not be required; likewise, Level III efforts are not typically required. If steel thicknesses must be verified after installation, measurements may be taken by using a micrometer where feasible or by an ultrasonic thickness measuring device. This can be accomplished on above-water and underwater elements. Material submittals should be reviewed to ensure that they comply with the contract documents.

Additional inspections may be warranted for specific structures or elements, such as inspecting welds using magnetic particle testing, coupon samples, and voltage potential measurements.

An electrical potential survey should be conducted for steel members with cathodic protection. Potential measurements should be taken at points throughout the structure to determine the effectiveness of the cathodic protection system. In general, an Ag–AgCl half-cell, dry electrode immersed in seawater is used in the marine environment, and readings between −0.800 and −1.200 V indicate adequate protection. These ranges may vary depending on factors such as water salinity and type of half-cell. Readings less negative than −0.800 V may indicate inadequate protection and readings more negative than −1.200 V may indicate overprotection, which could lead to accelerated anode depletion, coating debonding, and possible accelerated corrosion for amphoteric metals such as aluminum.

3.4.4 Methods of Inspection for Concrete, Timber, Masonry, and Polymer Composite Components

An inspection of the above-water elements should include visual observation and material testing to ensure that the construction is being performed in accordance with the contract plans and specifications.

A Level I underwater effort is all that should be required for new construction. A Level I inspection effort is conducted using visual/tactile means. Marine growth should be nonexistent to minimal for a new construction inspection, so a Level II effort should not be required; likewise, Level III efforts are not typically required.

Material submittals should be reviewed to ensure that they comply with the contract documents.

3.4.5 Methods of Inspection for Slope Protection

A Level I underwater effort is all that should be required for new construction inspection of slopes. Inspection efforts on slope protection, such as armor stone, riprap, gabions, concrete liners, and scour protection mattresses are often conducted using visual/tactile and/or survey scanning means.

Performing Level I efforts, visually without technology, may not be feasible or practical due to the size of the installation, waves, currents, or restricted visibility. This is particularly true for the offshore face of breakwaters. In such cases, performing the inspection using techniques, such as sonar, may be more cost-effective or technically advantageous. The results of surveys may indicate areas of potential problems that may be further investigated by diving.

3.4.6 Methods of Inspections for Mudline

An inspection of the mudline near and around structural elements is conducted using lead-line soundings and potentially visual/tactile means to evaluate scour or changes in the bottom conditions.

In some cases, performing an underwater dive inspection may not be feasible or practical due to waves, currents, restricted visibility, or the scale of the task. In such cases, performing or supplementing the inspection using techniques, such as multibeam or side-scan sonar, fathometers, lead lines, or similar depth measurement equipment, may be more cost-effective or technically advantageous. The method(s) of inspection used must be able to detect undermining if it is identified as a potential concern. The results of the survey may indicate areas of potential problems that an underwater inspection may further investigate.

Establishing a baseline for the mudline is essential for assessing future changes. As-built bathymetry surveys are generally required for new construction. If the development of this survey is not in the scope of the construction, it should be completed by others with a method that will provide comprehensive and reliable data.

3.4.7 Evaluation and Rating

A condition assessment rating should be assigned to each inspected facility in accordance with Section 2.6, "Overall System Ratings." Significant damage or improper installation observed should be included in the list of deficiencies given to the owner's representative. These deficiencies could affect the capacity of the element, whether it is a utility, load-bearing structure, mooring, and so on.

The purpose of a new construction inspection is not to conduct a detailed analysis of the facility; however, if the deficiency is not corrected, an engineering evaluation of the effect of the deficiency on the capacity should be conducted prior to assigning a condition assessment rating. Such an evaluation should typically be limited to the assessment of individual or typical components and should also be limited to an evaluation of the effect of the damage on the individual component capacity, without considering the actual or anticipated loading (structural demand). Should the need for a more rigorous analysis be apparent as a result of the new construction inspection, an engineering evaluation should be recommended as a follow-up action.

3.4.8 Recommendations

Upon completion of the new construction inspection, a list of deficiencies should be prepared and presented to the owner's representative for contractor resolution. This should be done continuously or periodically throughout the construction project. Documentation is important to track resolution, assess outstanding issues, or provide records for warranty requirements or legal support. A baseline inspection, if not already scheduled, should be recommended prior to acceptance of the structure by the owner. If a baseline inspection is not warranted, the recommended interval to the next regularly scheduled routine inspection should be assigned using the guidelines of Chapter 2, "Standards of Practice."

3.5 BASELINE INSPECTIONS

3.5.1 Objectives

A baseline inspection should be conducted on a newly constructed structure prior to acceptance by the owner, or it may be conducted on an existing structure that has not been previously inspected. The objectives of a baseline inspection include

- Verifying general compliance with design drawings, if available;
- Ensuring no significant defects prior to owner acceptance;
- Establishing a structure/component datum for future reference and comparison;
- Developing recommendation(s) for follow-up action; and
- Determining the recommended interval to the first routine inspection.

In addition, for existing structures without available drawings or design details, the baseline inspection may be used to gather information sufficient to support a structural evaluation. Alternatively, the baseline inspection may be used to develop record drawings or as-builts of the facility. These can be of value for future modifications or repairs to the structure.

3.5.2 Method of Inspection and Documentation

3.5.2.1 General Considerations. A baseline inspection is conducted to confirm dimensions, component locations, existing conditions, and other physical features. For new construction, the scope and level of effort expended depends on the specific type of construction and material type and whether the facility has undergone a new construction inspection. The information documented during this inspection can be utilized as a basis for the baseline inspection, thus reducing field time requirements. Sufficient information should be collected during the baseline inspection to allow future comparisons to be made. Levels of inspection would not apply to a baseline inspection for a new structure.

When conducting a baseline inspection of an existing facility for which no previous inspection information is available, the scope of the inspection will depend on the availability of design drawings.

For new construction and existing construction where design drawings are available, the baseline inspection should be conducted to verify the general accuracy of the drawings. For existing structures where design drawings are not available, the baseline inspection should be conducted to gather sufficient information to generate drawings or develop structural details of the facility. The level of detail required for this effort will depend on what the information will be used. In some cases, documenting the dimensions of individual members and of the overall facility may be sufficient.

Where a structural analysis must be performed using the results of the baseline inspection, the inspection must be much more detailed. Detailed measurements and testing should be used in such cases. Establishing the location, size, and cover of reinforcing steel may be necessary for concrete structures. In addition, establishing the compressive strength of the concrete by testing of core samples by NDT, such as Schmidt Hammer or Windsor Probe testing or by a combination of core sample testing and NDT, may be necessary. Collecting coupon samples of structural or reinforcing steel may be required to determine tensile strength and other properties relevant to the structural analysis. Similarly, it may be necessary to collect and test timber samples to establish the strength and ductility characteristics of timber portions of the structure. Dissecting typical connections to document details may be needed where connection details are unknown or are not readily apparent. At a minimum, a visual inspection of all underwater elements in the scope should be performed by verifying size, location, and material. The inspection should also include water depths at defined locations to aid in determining scour and/or sediment deposition rates and possibly debris surveys.

Similar to the underwater inspection, the above-water inspection for a baseline inspection should include a visual inspection of all elements, verifying size, location, and material.

3.5.3 Recommendations

Upon completion of the baseline inspection, recommended actions should be assigned to the structure in accordance with Section 2.7, "Recommended Action Guidelines." In addition, the recommended interval to the first regularly scheduled routine inspection should be assigned using the guidelines of Section 2.1.1, "General."

3.6 DUE DILIGENCE INSPECTIONS

3.6.1 Objectives

A due diligence inspection is an assessment of a facility to determine its value and fitness for purpose and confirm the status of its assets and liabilities. A due diligence inspection is performed when a transfer of ownership or other financial transaction related to the facility is being considered. Due diligence inspections are typically requested by the principals in a transaction but can also be requested by third parties, such as investment banks, pension funds, private equity firms, and others, which may have a stake in the transaction. The objectives of a due diligence inspection include

- Forming an engineering opinion of the general condition of a structure;
- Estimating order-of-magnitude replacement costs and/or repair costs;
- Supporting the performance of a financial, investment, or insurance valuation process; and
- Developing recommendations for follow-up actions.

A due diligence inspection can include a significant degree of engineering effort beyond just the inspection, such as reviewing past records, analyzing known or discovered deficiencies, and considering the future life-cycle costs of the facility, including the calculation of the present value. The due diligence inspection is only one piece of a broader due diligence exercise, the result of which is ultimately reduced to a financial determination as to whether the investment or transaction should proceed and, if so, whether the cost is reasonable. The engineering inspection element of a due diligence inspection is discussed in this section.

3.6.2 Methods of Inspection and Documentation

A due diligence inspection can be performed at various levels of detail, depending on the type of facility being evaluated, the needs of the requester, and schedule. For instance, the needs of a major terminal investor/operator will be different from a financial investor looking to acquire a stake in an asset. However, each needs to determine the value of the facility and what liabilities are associated with it.

The due diligence inspection includes a condition assessment that results in the documentation and assessment of each facility's physical condition. The condition assessment should be performed as generally described in this manual, with the scope defined by discussions with the requester. Because these types of inspections may be requested later in the negotiation process, and generally with the expectation that the facility is as advertised, schedule (i.e., available time to complete the inspection) may become the driving factor in developing the scope and completing the inspection. This may require multiple teams working on short notice, or a reduced level of inspection detail. Reports should document the scope of work performed during the inspection.

3.6.3 Recommendations

Upon completion of the due diligence inspection, condition ratings should be assigned to the structures in accordance with Section 2.5, "Element-Level Damage Ratings," and overall ratings should be assigned per Section 2.6, "Overall System Ratings." Service life modeling may be developed in accordance with Section 2.3, "Service Life Modeling: Purpose and Value." If follow-up repair design is required at a future date, recommendations of specific repair methods based on the results of the inspection should be provided to establish future costs. If repair design is not warranted, the recommended interval to the next regularly scheduled routine inspection should be assigned using the guidelines of Section 2.1.1, "General."

3.7 SPECIAL-PURPOSE INSPECTIONS

3.7.1 Objectives

Special-purpose inspections are conducted with a specific purpose. Such purposes may include

- Further quantification of the extent of observed deterioration, where not readily apparent;
- Determination of the root cause of observed deterioration to develop the appropriate repair method or to prevent future deterioration;
- Determination of structural significance that has resulted or will result from observed deterioration;
- Determination of the significance of observed deterioration on future durability and maintenance requirements; and
- Development of recommendation(s) for follow-up action.

3.7.2 Methods of Inspection and Documentation

Special-purpose inspections typically include field sampling and testing, laboratory testing and analysis, or a combination thereof. The scope should be dictated by the type of testing or analysis to be performed. The scope may need to be expanded depending on initial results. Standard test methods exist for many special-purpose inspection types. Such methods are typically defined by organizations such as ASTM or AASHTO.

Documentation will also vary greatly depending on the specific investigation technique and objective. Recorded information may include field-testing results, laboratory-testing results, or materials analysis results.

3.7.3 Recommendations

Upon completion of the special-purpose inspection, recommended actions should be assigned to the structure in accordance with Section 2.7, "Recommended Action Guidelines." In addition, if follow-up repair design is required, recommendations for specific repair methods, based on the results of the special-purpose inspection, should be provided. If a repair design is not warranted, the recommended interval to the next regularly scheduled routine inspection should be assigned using the guidelines of Section 2.1.1, "General."

3.8 REPAIR CONSTRUCTION INSPECTIONS

3.8.1 Objectives

A repair construction inspection should be conducted during the construction of repairs and should be ongoing throughout the construction, providing inspections at strategic points during the work. In general, these inspections are conducted by the owner's construction management (CM) team that manages the repair construction contract on a day-to-day basis, with or without the assistance of a professional CM firm and/or consulting engineering firms with specialized marine inspection and repair experience. In some cases, the design engineer of record, or a specially contracted firm, conducts the inspections. Timely execution of repair construction inspections can often prevent costly remobilization costs, back charges, unintentional acceptance of nonconforming work, and so on. The primary objectives of the repair construction inspection are as follows:

- Evaluating quality to verify contractor compliance with repair design documents,
- Verifying repair quantities for contractor payment,
- Responding to field questions and resolving field problems, and
- Developing a list of deficiencies for the contractor to correct.

3.8.2 Methods of Inspection and Documentation

The scope of repair construction inspection will be dictated by the repair methods employed for specific projects. Repair construction inspection should typically be conducted throughout the repair process rather than waiting until the end of the project to perform an acceptance inspection. This is because the success and longevity of repairs often depend on proper surface preparation and repair application, and such preparation is typically visible only in the repair process.

As an example, placement of a timber pile post with a concrete collar around the mudline connection may require three phases of inspection: (1) to ensure the soundness and elevation of the existing cut-off pile stub, (2) to inspect the placement of formwork, internal reinforcing, and attachment of the timber sections, and (3) to inspect the final concrete placement and condition. Alternatively, a repair construction inspection for patching a steel sheet pile wall may require only two phases of inspection: the first to confirm the proper surface preparation and the second to visually confirm the quality of the weld.

Documentation obtained during a repair construction inspection should provide details on the activities of the contractor, any modifications to the design documents, and the quantity of repairs completed. In addition, all field problems and associated resolutions should be documented for future reference.

3.8.3 Recommendations

Upon completion of the repair construction inspection, a punch list of deficiencies should be prepared and presented to the contractor for resolution. This should ideally be done continuously or periodically throughout the repair project.

3.9 POSTEVENT INSPECTIONS

3.9.1 Objectives

Postevent inspections should be conducted following significant events such as a major storm, earthquake, flood, vessel impact (allision), fire, observation of a significant structure change, or other similar incidents. The objectives are as follows:

- Conduct a rapid investigation to determine whether damage resulting from the event may have significantly affected the integrity of the structure either locally or globally.
- Determine whether the observed damage is a life safety concern and/or requires immediate action.
- Assign a postevent condition rating to the structure.
- Develop recommendation(s) for follow-up action.

3.9.2 Methods of Inspection and Documentation

Postevent inspections should be somewhat rapid visual/tactile inspections. The focus of the postevent inspection should be damage, which was likely to have been caused by the event. Such damage may include breakage, overstressing cracks, buckling, settlement, slope failure, and so on. General observations of long-term or preexisting deterioration, such as significant corrosion-related damage or other deterioration, may be reported to the owner as appropriate, but should not be the focus of the inspection nor documented in the report. Postevent damage can often be observed and differentiated by noting fresh breaks or changes in the material.

The decision to execute a postevent inspection should depend on the type and severity of the event and the structure type and its vulnerability to damage. Above-water observations often provide the best clues as to whether an underwater inspection is prudent. Such observations may include shifting or differential settlement, misalignments, significant cracking, bulging, breakage, and so on.

Major storms or tsunamis may trigger an underwater postevent inspection of breakwaters or other structures in the path of the event. Many US waterways have USGS websites with real-time and historical flow data. Similarly, USGS marine buoys record historical wave height, wave period, wind speed, and direction.

Damage caused by earthquakes or allisions (vessel impact) on structures such as pile-supported waterfront facilities often occurs both above and below water. The underwater postevent inspection may, therefore, be reserved only for structures that exhibit above-water damage.

Ultimately, the decision to execute an underwater postevent inspection is a matter of judgment and should be weighed against the importance of the structure in terms of public safety and/or economic impact.

The methodology of conducting a postevent inspection should depend on the structure type and the type of damage anticipated. Although slope failures or scour may be readily apparent in waters of adequate visibility, overstressing cracks on piles covered with marine growth will not be readily apparent. Where such hidden damage is suspected, marine growth removal should typically be performed on a representative sampling of components in accordance with the Level II inspection requirements described in Section 3.1.3.

Because of the rapid nature of the inspection, documentation should be minimal, consisting primarily of brief notes describing observed conditions and summarized into a postevent condition rating.

3.9.3 Ratings and Recommendations

A postevent condition rating should be assigned to each inspected structure in accordance with Section 2.6.3, "Postevent Damage Ratings." This rating system is different from that used for other inspections because the focus of the postevent inspection is to identify significant damage that has resulted from the event. However, this information should be documented if significant damage or deterioration is observed, which does not appear to have resulted from the event. The purpose of a postevent inspection is not to conduct a detailed analysis of the structure. Should the need for additional inspection or for a more rigorous structural analysis be apparent as a result of a postevent inspection, recommendations for follow-up action such as a special-purpose inspection, structural repair or upgrade design inspection, or engineering evaluation should be assigned as appropriate in accordance with Section 2.7, "Recommended Action Guidelines."

REFERENCE

APWA (American Public Works Association). 1999. *Uniform color code.* Washington, DC: APWA.

CHAPTER 4
SERVICE LIFE ESTIMATION

4.1 GENERAL

This chapter provides guidance on estimating the remaining service life of a marine facility. In general, two terms are used in the engineering industry to define the life of a structure: design life and service life. The chief difference between these two terms is that *design life* is a theoretical life span estimated during the planning and design stages, and *service life* is the actual life span of the structure. The service life of a structure, and what remains at any given point in time, is of particular interest when inspecting and evaluating existing structures, as it is a key piece of information for facility stakeholders to use in programming maintenance, capital improvements, and budgetary funding.

Service life can be defined as the length of time a structure or facility may be used economically before

- Emergent damage or deterioration causes a significant reduction in capacity,
- Condition of the structure results in a threat to public safety, or
- Structure is functionally obsolete for intended use.

When applying this definition, the economical use of a structure is generally considered the point where the cost of continued maintenance and repairs exceeds the replacement cost, measured over the projected design life of the replacement structure. It is also important to convey to stakeholders that the remaining service life assumes that structures are used and maintained in a reasonably prudent manner. This includes repairing damaged elements promptly, completing cyclical inspections, and performing routine maintenance in accordance with industry standards.

When reduced capacity controls the remaining service life, it is also important to note that this is in reference to the original design capacity. The service life of a facility may be extended through a reduction in allowable loading or operations if it meets the facility's current and future use requirements. It is recommended that an updated analysis be completed to ensure that the proposed loading is within the structure's reduced capacity and that the allowable loading is appropriately documented and posted.

This chapter covers the three basic construction materials most commonly used for waterfront facilities: concrete, steel, and timber. Polymer composite materials have also been recently added to this chapter as they are being used more frequently in waterfront construction. As the remaining service life is typically controlled by primary structural elements or those that make up large portions of the structure, this chapter intends to address material components only when used in this capacity. For example, polymer composite materials are often used for discrete areas of a facility or secondary elements such as fendering, with a minor impact on overall service life, and would not be covered under this chapter.

It is important to understand that predicting the remaining service life is not an exact science but rather a tool to aid the owner in facility planning. The accuracy is largely driven by the level of data collected in the field during an inspection. Once the service life is estimated, the remaining service life may be extended through increased maintenance and repairs

Figure 4-1. Concrete pier with pile jackets incorporating cathodic protection to increase service life.
Source: Courtesy of Appledore Marine Engineering, LLC, reproduced with permission.

focused on the controlling members. Preventative measures such as coatings used to provide a barrier from the environment or cathodic protection that arrests the corrosion process through an electrochemical process (Figure 4-1) are examples of service life extension.

4.2 IMPORTANCE OF ACCURATE ESTIMATIONS

Waterfront facility managers need to understand the remaining service life of a facility with and without repairs. A structure's service life may be determined by evaluating original design plans, maintenance records, and the most recent inspection and condition assessment. The service life estimate accuracy depends on the quality of available information, and the estimate may be updated after the completion of each assessment.

Basic service life estimates can be prepared using the information obtained during a routine-level inspection. As this information is primarily based on Level I inspection techniques with Level II and Level III performed at limited locations, engineering judgment is often used in lieu of more in-depth analyses. If a more detailed and accurate estimate is desired, additional investigation efforts, testing, and analyses may be required, especially when structures are nearing the end of their service life.

The better the estimate of the remaining service life, the more feasible the decisions made concerning short- and long-term planning, maintenance, repair, and possible replacement of the facility. However, predicting the remaining service life is an estimate and requires advanced knowledge and experience. Factors affecting the service life of a structure include climate, construction materials and details, exposure, and original material treatments and/or composition.

4.3 STATE OF THE ART

In the last 20 years, the science of predicting the remaining service life of marine structures has advanced because of technological changes and a better understanding of deterioration mechanisms. The body of knowledge for concrete structures has advanced with a better understanding of deterioration mechanisms, transport properties, and modeling.

The service life estimation approach to steel, timber, and polymer composite structures is more elemental; basic principles are important in making defensible predictions. For steel structures, the estimated rate of corrosion is typically the key metric for steel deterioration. For timber structures, the key metrics are related to fungi, insects, bacteria, and marine borers, and the locations and environments in which they thrive.

For example, a timber pile may be susceptible to marine borers, bacteria in the submerged and tidal zones, fungi, insects, and bacteria in the splash and atmospheric zones. Timber structures often deteriorate from the inside, making it difficult to assess the extent of deterioration. Experience with similar structures is important, and specialized tools, such as resistance drills, aid in assessing the level of deterioration.

The use of polymer composite materials as primary structural elements is still in its infancy with respect to the structure life spans. Although initial data generally indicate good durability with limited modes of deterioration in the marine environment, there is a limited sample size to validate the actual service life compared to predicted design life, and the actual results may depend on the specific manufacturer and material composition. It is important to understand these limitations and unknowns when estimating the service life of polymer composite structures, and the accuracy level should be presented accordingly.

Although the technology of service life prediction is advancing, experience and knowledge of similar structures are an important foundation in making accurate predictions. Also important is understanding the variety of factors that affect the service life; therefore, developing an inspection history of a structure provides a firsthand knowledge of deterioration rates.

4.4 DETERIORATION ZONES

For concrete and steel, and to some degree timber, deterioration rates are, in part, related to where they exist vertically in the water column. The definitions of these deterioration zones are provided in Table 4-1, from highest to lowest.

The accelerated rates of corrosion of steel structures in the splash and low water zones highlight the importance of zones. The splash zone is one of the more severe zones, where chloride, moisture, and oxygen combine in greatest quantity. Chloride availability is the most abundant in the splash zone, where salt accumulates and crystallizes just above the tidal zone and is not dispersed by tides. Timber structures are typically affected the greatest by fungal decay and marine borer deterioration. Fungal species are active only above the reach of salt spray, and marine borers can be active from the tidal zone to the mudline.

The primary mode of deterioration for concrete structures is corrosion of the embedded reinforcing steel. This type of deterioration is most prominent in the splash zone and is generally minimal below low water because of the reduced oxygen levels. Historically, some concrete structures that lacked proper attention during construction have experienced dramatically shortened service lives because of chemical degradation. Chemical degradation requires advanced testing to fully define the root cause.

Table 4-1. Deterioration Zone Definitions.

Zone	Description
Atmospheric zone	The area above the upper limits of the splash zone, which remains consistently dry. However, the area may be subject to salt-laden air.
Splash zone	The area above the high-water mark (MHHW MHW, MHWS, etc.) that is subject to constant wetting and drying due to splashing of water. Except for rare extreme events, the area is never continually immersed for any period. The area is characterized by the abundance of oxygen and the buildup of chlorides in high concentrations. The upper limit is defined by local conditions of splashing by wake, wave, and chop
Tidal zone (intertidal)	The area between the low water mark (MLW, MLLW, MLWS, etc.) and the lower limit of the splash zone. The zone is subject to daily tidal wetting and drying, but chloride buildup is not as severe as in the splash zone, and average oxygen availability is lower
Low water zone	The area is approximately 1.5 ft (0.5 m) below the lower limit of the tidal zone (MLW) where certain very aggressive steel corrosion activities are known to occur in some locations. Does not occur in all locations. Very aggressive corrosion in this zone is known as ALWC. This zone can vary by location and profiling is important to locate it
Submerged zone (immersed)	The area between the mudline and the lower limit of the tidal zone. Except for rare extreme events, the area is never exposed to the atmosphere and is constantly wet. The average oxygen availability is lower than in the tidal zone
Disturbed soil zone	Defined as those first few feet of loose soil that is disturbed or put in motion by waves, currents, propeller/thruster wash, etc. The zone is characterized by soil densities far less than the firmer soils below. The upper limit is usually referred to as the mudline
Undisturbed soil zone	The zone underlying the disturbed soil zone extends to the lowest elevation of the structure (pile tip elevation). Characterized by low oxygen availability

When estimating a structure's service life, mainly when governed by reductions in structural capacity, it is important to understand the locations of maximum deterioration and how they relate to the member's area of maximum stress. For example, peak corrosion losses are often observed in the splash zone of steel structures; however, bulkheads and quay walls are often designed on the basis of maximum moments that occur well below the mudline for cantilever wall structures, or in the submerged zone of anchored walls. Therefore, there may be significant reserve capacity in the areas of peak deterioration.

4.5 CONCRETE FACILITIES

The service life of concrete structures is a function of various exposure variables and deterioration modes. One of the more common modes affecting remaining service life is chloride ingress, leading to reinforcing steel corrosion. Other modes of deterioration include delayed ettringite formation, sulfate attack, or reactive aggregate. To better understand and predict the deterioration process of concrete, representative samples need to be collected and analyzed to identify the inherent material properties of the concrete matrix. This is important, as two concrete mixes made around the same time may be visually similar but could have significantly different properties. These differences could be attributed to the mix design, but also the construction means and methods. Once the properties are known, the specific concrete elements may be modeled with software. Some examples of such software include STADIUM, Life 365, and COMSOL.

A primary mode of deterioration in the marine environment is corrosion of the embedded reinforcing steel caused by chloride exposure. The exposure and transport properties of the concrete greatly affect this process. For instance, concrete generally deteriorates more rapidly in tropical environments with higher concentrations of chlorides in seawater. The transport properties of concrete are a measure of how quickly deterioration mechanisms, such as chlorides, are transported from the surface of the concrete to the reinforcing steel.

A concrete structure goes through several stages, including full chloride penetration, time to initiation of corrosion, and the critical corrosion threshold that results in damage to the structure. Given that the matrix of the concrete, location, and exposure zone may significantly affect the time of each phase, modeling is the best tool to determine time from corrosion to visible damage.

4.5.1 Sampling

Once it has been determined that an estimation of service life is desired, sampling, testing, and service life modeling based on concrete cores extracted from a structure should be performed. One of the first steps is to collect representative samples from the structure to define the material properties. Once collected, the samples are used for material testing to define the parameters for modeling.

The samples should be extracted from areas within the structure that best represent its structural load-carrying capabilities, integrity, and relevant modes of deterioration. Obtaining concrete core samples that are statistically representative is important. A small number of cores may be designated to evaluate specific areas of more advanced deterioration that may not be representative of the entire structure.

The testing laboratory should be involved in determining the diameter and number of cores required to complete the testing program, as some tests require a specific volume or size of material. Careful thought should go into the development of a coring plan, to minimize damage to reinforcing, collection of wastewater, and targeting specific locations for modeling. Advanced planning, with consideration of access and tides, maximizes the opportunity and reduces costs for the owner. For example, coring of piles often requires a larger and more specialized crew and a vessel or other similar type of floating working platform (Figure 4-2).

Figure 4-2. Coring operations over water to extract samples from prestressed piles.
Source: Courtesy of Appledore Marine Engineering, LLC, reproduced with permission.

The coring plan should include the following specific characteristics:

- Plan depicting overall locations; the selection of sampling locations should account for the various concrete types and exposure conditions,
- Core diameter (typically 2 to 4 in., depending on the element, testing protocol, and reinforcing location and spacing),
- Depth of core,
- Element to be cored (i.e., deck, pile, pile cap, etc.),
- Boundary conditions (i.e., submerged, tidal, splash, or atmospheric),
- Handling of core after extraction, and
- Core labeling.

Checking local environmental regulations is an important part of the planning process to assess any requirements for containment and collection of coring wastewater (Figure 4-3). Record documents (plans and specifications) should be reviewed prior to developing the coring plan for a basic understanding of reinforcing steel density; the final field core layout

Figure 4-3. Coring operation of a dry dock wall; note wastewater containment. Source: Courtesy of Appledore Marine Engineering, LLC, reproduced with permission.

should be completed with the assistance of a reinforcing steel locating instrument.

Once extracted, the cores should be assembled as extracted and labeled for identification by the laboratory (see Figure 4-4 shown following Section 4.5.2). Within 1 h of extraction, cores should be wrapped in clear plastic and surrounded in bubble wrap. The wrapped cores should be further wrapped in wet toweling and placed in the storage container.

Wooden core boxes (assembled with screws) are the preferred containers for storage and shipping. Containers should have an interior plastic liner to provide a short-term hermetic barrier. Cores should be packaged and shipped to the laboratory within seven days of extraction.

All storage, handling, and shipping of cores should be in compliance with ASTM C42/C42M (ASTM 2020d) and best practices followed so as to maintain specimen integrity and to prevent dryout.

4.5.2 Laboratory Testing and Analysis

Laboratory analysis is required to ascertain the specific characteristics of the concrete, chemistry profiles (typically chlorides), and the boundary conditions.

The transport properties of the concrete identify the pathways and rate of migration for deterioration species. This testing allows the engineer to

Figure 4-4. Extracted core sample.
Source: Courtesy of Appledore Marine Engineering, LLC, reproduced with permission.

better understand the concrete's deterioration mechanisms and the contamination migration. The testing regime provides the necessary input constituents to complete the concrete's modeling and service life prediction.

The following tests are generally completed to characterize the concrete. However, coordinating with the modeling software developers is essential, as many of these tests have been modified to reflect actual conditions more accurately.

1. Compressive strength in accordance with ASTM C39/C39M, *Standard Test Method for Compressive Strength of Cylindrical Concrete Specimens* (ASTM 2024a). The preferred length/diameter ratio for cores is 2; however, the test can be completed down to a ratio of 1. This may be required for slender elements such as prestressed piles.
2. Air void network determination in accordance with ASTM C457/C457M, *Standard Test Method for Microscopical Determination of Parameters of the Air-Void System in Hardened Concrete* (ASTM 2024b). This test is used to aid in modeling the rate of migration of deterioration species.
3. Petrographic examination in accordance with ASTM C856, *Standard Practice for Petrographic Examination of Hardened Concrete* (ASTM 2020b). This test identifies the water-cement ratio and possible degradation mechanisms.
4. Chloride profiling. Chloride ion profiling determines the extent and limit of chloride contamination. Chloride determinations are made in accordance with ASTM C1152, *Standard Test Method for Acid-Soluble Chloride in Mortar and Concrete* (ASTM 2020c).

The latest modeling software version defines contaminants' migration on the ionic level. To describe the migration, specific testing of these parameters is completed. The following tests are required:

1. Porosity in accordance with ASTM C642, *Standard Test Method for Density, Absorption, and Voids in Hardened Concret*e (ASTM 2021a).
2. Rapid chloride penetration test in accordance with ASTM C1202, *Standard Test Method for Electrical Indication of Concrete's Ability to Resist Chloride Ion Penetration* (ASTM 2022a).
3. Drying laboratory test in accordance with ASTM C1585, *Standard Test Method for Measurement of Rate of Absorption of Water by Hydraulic-Cement Concretes* (ASTM 2022b).

Note that the methods to provide a longer concrete service life constantly evolve. This can include any combination of concrete mix proportion variations, surface sealers, reinforcing coating or treatment, cathodic protection methods, and chemical admixtures, which may have an impact on the modeling input and tests required. The addition of corrosion inhibiting admixtures has become a popular method of delaying reinforcing steel corrosion. The addition of these admixtures can also increase the conductivity of the concrete even though the permeability is unchanged, alering the results of tests that rely on resistivity such as the Rapid Chloride Penetration Test (ASTM C1202) (ASTM 2022a). The applicability of each test used, interpretation of the results, and how they may impact modeling results need to be thoroughly understood.

4.5.3 Service Life Modeling

One of the primary modes of concrete structure deterioration in the waterfront environment is corrosion of the internal reinforcing steel. Corroding reinforcing steel may expand up to seven times its original volume, exerting significant pressure, which may result in mechanical damage, typically as cracks or spalls. This mechanical damage results in an acceleration of the deterioration because of exposure of fresh substrate and additional avenues of ingress for the deleterious chemicals. This is one example of a deterioration mode, and, due to the heterogeneous nature of concrete, many modes need to be modeled.

Many technological advances have occurred in concrete modeling software to determine the remaining service life of a structure, with one of the latest evolutions being the development of the software STADIUM. This software is a finite-element model that accounts for the composition of the concrete, the transport properties of the concrete, the exposure of the site, and the structure's dimensions to estimate the remaining service life. For instance, the software models the exposure conditions in the tidal zone differently from below low water because of the difference in available chlorides and oxygen. The software also accounts for elements that may have different environments on each side, such as a deck element. STADIUM is an example of commonly used modeling software; others are used in the industry, including Life 365 and COMSOL.

The key output parameter is expressed in duration to corrosion initiation (Figure 4-5). Whereas some debate is ongoing, 500 ppm of chlorides in the concrete at the location of reinforcing is usually accepted to represent a reasonable value for the critical corrosion initiation threshold value for black steel with some incremental increases in threshold value for treatments such as epoxy or galvanizing, or the addition of corrosion inhibiting admixtures in the mix design. Other chloride content criteria, such as the ratio of chloride to hydroxide, can also be examined to further refine the predicted corrosion initiation threshold. Once chlorides reach the level for corrosion initiation, corrosion propagates over time. The generally accepted propagation period is 10 years for black steel to cause significant cracking, eventually resulting in deterioration and ultimately component failure.

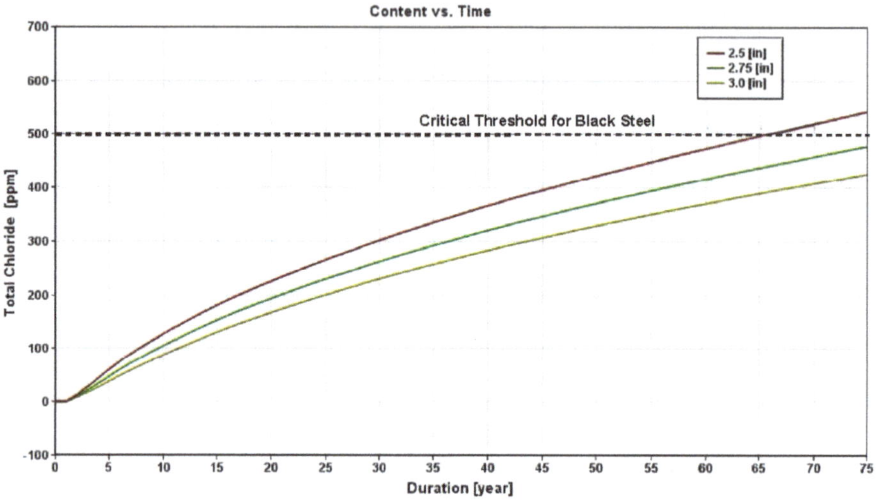

Figure 4-5. STADIUM output data depicting time to corrosion initiation as a result of chloride contamination.
Source: Courtesy of Appledore Marine Engineering, LLC, and Simco, reproduced with permission.

Although engineering judgment and experience continue to be important factors in evaluating the remaining service life of concrete structures, modeling software development is a valuable tool in the decision-making process. In addition to predicting the end of the service life of a structure, modeling tools can also simulate the effects of various types of repairs, including supplemental concrete and sealers. This can be used to help not only estimate the extended service life that the repairs provide but also determine the approximate year that the repairs should be performed.

4.6 STEEL FACILITIES

4.6.1 Background

Understanding that no single accepted value exists for the corrosion rate of uncoated steel in seawater is important. The rate of corrosion depends on site-specific environmental factors that can vary greatly. These factors include

- Chloride content,
- Dissolved oxygen,
- Moisture content,

- Chemical or pollution content,
- Oxygen,
- Nutrient and bacteria content,
- Hardness,
- pH,
- Soil resistivity,
- Sulfate content,
- Range of tide,
- Temperature,
- Current velocity,
- Wave action, and
- Cathodic protection.

The corrosion rate varies significantly along the vertical length of the element and where it lies within the corrosion zones, as defined in Table 4-1. For example, the International Navigation Association's (2005) publication "Accelerated Low Water Corrosion" (ALWC), defines a special zone, the low water zone [0.5 m below Mean Low Water (MLW) to the Lowest Astronomical Tide], where microbial induced corrosion (MIC) in the ALWC zone is prone to occur. MIC and ALWC are localized but aggressive forms of corrosion that occur worldwide and are becoming more frequent. However, the exact reasons for its occurrence and susceptible areas are not well understood, and ALWC is not present in every location and must be investigated locally at each site.

In addition to contributions from MIC, other chemicals, pollution, and high nutrient content have been shown to increase the corrosion rate by "an order of magnitude," as quoted in a University of South Florida study for FDOT (Sagüés, et al. 2009). For example, Beech and Campbell (2008) found that the corrosion rates were 10 times higher in a particular part of a southern England harbor where sulfate-reducing and sulfate-oxidizing bacteria and higher concentrations of organic carbon were found.

The Port of Rotterdam performed a recent study (Jongbloed 2019) that noted premature corrosion holes in steel sheet pile and steel combination-wall quay walls that had been in service for less than 20 years, much faster than the predicted corrosion rates using Eurocode guidelines. MIC was suspected to be one of the primary causes, due to the recognizable bright orange spots and severe pitting in the low water zone. A research program resulted in a list of 23 probable causes; however, sulfate-reducing bacteria was consistently present in the water samples collected, resulting in a conclusion that MIC was the predominant cause of premature damage. Additional discussion on the results of the study and applicable corrosion rates are included in Section 4.6.2, and equations are used for calculating the remaining service life of steel structures in Section 4.6.6.

4.6.2 Variation in Published Unprotected Steel Corrosion Rates

The steel corrosion rate must be estimated to determine an economical service life, and appropriate measures must be designed to extend that life. As previously mentioned, many factors affect the rate of steel corrosion in a marine environment. Multiple studies have been undertaken to empirically document rates in different locales. A compilation of reference documents, studies, and consultations addressing steel corrosion in the seawater splash zone was examined and is presented in Table 4-2. Rates were expressed in various units, and the three most common units—mil per year (mpy), mm per year (mm/y), and in. per year (ipy) are used along with the conversions.

The basis of these studies varies considerably, including location, sample size, material composition and profile, and the method of rate determination. The compiled data in the table illustrate the variability in the reported corrosion values; the data should not be used as a direct corrosion rate to apply to design life and remaining service life estimation without understanding the full parameters of the study and adjusting accordingly.

The California Department of Transportation Division of Engineering Services Materials Engineering and Testing Services published the following corrosion rates for steel piling exposed to corrosive soil and/or water or marine exposures in their Version 3.2 Corrosion Guidelines (CALTRANS 2021):

- Soil Embedded Zone: 0.001 in. (0.025 mm)/year,
- Atmospheric Zone (marine): 0.002 in. (0.051 mm)/year,
- Immersed Zone (marine): 0.004 in. (0.102 mm)/year, and
- Splash Zone (marine): 0.006 in. (0.152 mm)/year.

These corrosion rates apply only to the exterior surface of the steel pipe pile. Typically, the interior surface of the pile (soil plug side) will not be exposed to sufficient oxygen to support significant corrosion. This may also be the case when oxygen is depleted, such as behind bulkheads. However, depending on the nature of the backfill, corrosion may still progress although at a diminished rate.

Eurocode 3 (BSI 2007) provides corrosion rates for structures in temperate waters, as given in Table 4-3.

Table 4-4 provides the recommended value for the loss of thickness due to corrosion for piles and sheet piles in soil, and Table 4-5 provides the recommended value for the loss of thickness due to corrosion for piles and sheet piles in water per Eurocode 3.

Other entities have developed their own corrosion rates to implement on the basis of site-specific data. The recent study performed by the Port of Rotterdam, as discussed in Section 4.6.1, included more than 500,000 measurements that were collected and statistically processed.

Table 4-2. Summary of Estimated Corrosion Rates for Unprotected Steel in the Seawater Splash Zone.

#	Source	Rate (mpy)	Rate (mm/year)	Rate (ipy)
1	FHWA Driven Pile Manual (Hannigan, et al. 2016)	7	0.18	0.007
2	Peruvian Port (Farro et al. 2009)	21.86	0.56	0.022
3	Swedish Commission Report (Camitz 1998)	11.8	0.3	0.012
4	US Environmental Protection Agency Report (CH2M HILL 2004)	12–20	0.30–0.51	0.012–0.020
5	Fontana (1986)	10–16	0.25–0.40	0.010–0.016
6	Port Engineering textbook references (Bruun 1989):			
	Chellis (1961)	25–40	0.64–1.02	0.025–0.040
	Edwards (1963)	4–14	0.10–0.36	0.004–0.014
	Creamer (1970)	55	1.4	0.055
7	FDOT Seawall Study (FDOT 2014)[a]	16	0.41	0.016
8	Tomlinson (1987)	3.54	0.09	0.004
9	Campeche, Mexico (Balasubramanian et al. 1999)	3.07	0.078	0.003
10	Eurocode (BSI 2007)	3	0.075	0.003
11	Chef Menteur Pass (Romanoff 1962)[b]	5	0.127	0.005
12	Seymour Coburn (Coburn 1988)	27.9	0.709	0.028
13	BSI 6349 (BSI 2000)	6.69	0.17	0.007
14	US Navy Study, Key West (Brouillette and Hanna 1960, p. 17)	10	0.254	0.01
15	Cook Inlet, AK (Daley and Ingraham 1996)[c]	30–60	0.76–1.50	0.03–0.06

[a]Uncoated steel sheet piling on Florida Atlantic Ocean beach exposed directly to seawater at high tides only.

[b]1961 observations of four lengths of coated sheet pilings pulled after 32 years in service at a site nearby IHNC, Chef Menteur Pass.

[c]Cathodic Protection in Cook Inlet Waters, Materials Performance, February 1991.

Table 4-3. Corrosion Rates from the European Building Code.

Location	Total loss in 25 years	Equivalent annual loss
Splash zone	1.9 mm (0.075 in.)	0.076 mm/year (0.003 ipy)
Immersed zone	0.90 mm (0.035 in.)	0.036 mm/year (0.0014 ipy)
Atmospheric zone	0.25 mm (0.010 in.)	0.010 mm/year (0.0004 ipy)

Source: BSI (2007), reproduced with permission.

Table 4-4. Recommended Value for the Loss of Thickness (mm) due to Corrosion for Piles and Sheet Piles in Soils, with or without Groundwater.

	Required working life (years)				
Soil type	5	25	50	75	100
Undisturbed natural soils (sand, silt, clay, schist, etc.)	0.00	0.30	0.60	0.90	1.20
Polluted natural soils and industrial sites	0.15	0.75	1.5	2.25	3.00
Aggressive natural soils (swamp, marsh, peat, etc.)	0.20	1.00	1.75	2.50	3.25
Noncompacted and nonaggressive fills (clay, schist, sand, silt, etc.)	0.18	0.70	1.20	1.70	2.20
Noncompacted and aggressive fills (ash, slag, etc.)	0.50	2.00	3.25	4.50	5.75

Source: BSI (2007), Table 4-1, reproduced with permission from British Standards Institution.

Notes: Corrosion rates in compacted fills are lower than in noncompacted ones. In compacted fills, the figures in the table should be divided by 2.

The values given for 5 and 25 years are based on measurements, whereas the other values are extrapolated.

Table 4-5. Recommended Value for the Loss of Thickness (mm) due to Corrosion for Piles and Sheet Piles in Fresh Water or in Sea Water.

	Required working life (years)				
Water type	5	25	50	75	100
Common fresh water (river, ship canal, etc.)	0.15	0.55	0.90	1.15	1.40
Very polluted fresh water (sewage, industrial effluent, etc.) in the zone of high attack (water line)	0.30	1.30	2.30	3.30	4.30
Seawater in temperate climate in the zone of high attack (low water and splash zones)	0.55	1.90	3.75	5.60	7.50
Seawater in temperate climate in the zone of permanent immersion or in the intertidal zone	0.25	0.90	1.75	2.60	3.50

Source: BSI (2007), Table 4-2, reproduced with permission from British Standards Institution.

The processed results established a more refined corrosion rate for use in designing and evaluating structures at the Port.

Their recommended corrosion rates include nine scales (Jongbloed's curves) (Figure 4-6b), depending on the presence of the contributing conditions identified in the area. These included electrochemical effects, biological activities, galvanic influence, fabrication type, aeration, local waterflow, vertical salinity gradients, vessel cathodic protection presence, and relative elevation on the structure (Figure 4-6b).

The Jongbloed's curves represent the equivalent mean value of corrosion, which equals the sum of the mean uniform and pitting corrosion with an accuracy of approximately 0.4 in. (1 mm) during an exposure period of 50 years. The worst-case corrosion scale at the Port has a mean corrosion rate of 0.35 in. (9 mm) over 50 years, but maximum peak corrosion rates of up to 1.06 in. (27 mm) over 50 years can be present at areas of maximum pitting.

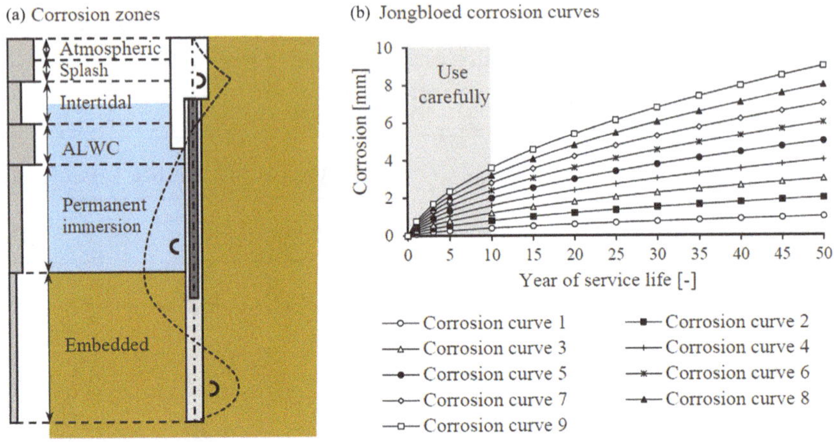

Figure 4-6. Uniform corrosion scales at Port of Rotterdam (Jongbloed curves). Source: Roubos et al. (2020). Courtesy of the Port of Rotterdam, reproduced with permission.

4.6.3 Data Summary

A summary of the various recommendations for corrosion rates in the splash zone from the sources noted previously is provided in Table 4-6.

Although this table summarizes the data from the various published recommended corrosion rates, application of these rates should be used carefully and conservatively, as without site-specific data, actual corrosion rates may fall significantly outside the published ranges, especially if subjected to localized contributing conditions such as electrochemical effects, significant vessel cathodic protection presence, and ALWC.

Table 4-6. Summary of Corrosion Rates in the Splash Zone.

Zone	in./year (ipy)	mils/year (mpy)	mm/year (mm/year)
High	0.06	60	1.50
Low	0.003	3	0.08
Average	0.016	16	0.41

4.6.4 Corrosion Protection Systems

Corrosion protection is typically used for steel marine structures. This method of protection is effective and economical in particular zones. The various approaches often used include

- Cathodic protection,
- Jacketing or encasements,
- Coatings and wrappings, and
- Combinations of the aforementioned.

Cathodic protection (active or passive) can be used in the soil and submerged zones, but it is partially effective only in the tidal zone and not applicable in the splash or atmospheric zones. The design of a cathodic protection system is complicated and can have unwanted and unintended negative consequences if not performed correctly, especially active systems. Cathodic protection systems (CPSs) require periodic measurement, adjustment, and maintenance. (Note: annual maintenance and/or inspection after significant events are typically required for effectiveness.)

A passive CPS consists of bulk zinc or aluminum alloy blocks electrically connected to the structure typically by welding of a steel core. An active CPS uses a rectifier to convert the power source from alternating current to a direct current that powers an underwater anode sled, deep well anodes, or both. In this scenario, the negative side of the cathodic protection battery is the steel structure, and the positive side is the impressed current through the anode.

It is difficult to predict how effective the system has been over the years (especially true with active impressed current systems versus passive systems) without actual remaining steel thickness measurements, because of the absence of any visual changes to the anodes.

Plastic coating is limited to use on pipe piles and H piles. Concrete encasements and jackets are an effective but expensive means of providing protection.

Protective coatings offer a means of providing additional life to piles in marine environments at moderate additional first cost and can be reapplied as needed to those portions of the piles exposed to severe corrosion. For example, several of the agencies providing corrosion rates in Section 4.6.2 indicate that steel intended for service in seawater tidal or splash zone locations should be coated. These include FDOT and CALTRANS. The FDOT Structures Design Guidelines state that uncoated steel Pipe Piles without internal redundancy and uncoated H-piles shall not be used in extremely aggressive environments, which includes marine exposure (FDOT 2025).

The CALTRANS Memorandum to Designers states, "Steel piles should not be used in the splash zone unless alternative mitigation measures such as protective barrier coatings and/or cathodic protections are considered" (CALTRANS 2014). Coatings generally have a service life of 20 years or less in the marine environment.

Table 4-7 provides *Unified Facilities Criteria 4-151-10* approximate expected periods of protection (negligible loss of metal) afforded by various coating systems for marine exposure assuming that the coatings

Table 4-7. Period of Protection for Steel to Expect from Various Common Coating Systems.

Coat description	Period of protection (years)
Coal tar epoxy (15–20 mils thickness)	10–20
Galvanizing (7–9 mils thickness)	10–15
Metalized aluminum	15–20
Concrete encasement	25

Source: DoD (2012).

Note: Coating thickness shown is dry film thickness.

are applied properly (DoD 2012). Note that fusion-bonded epoxy coating is not specifically included in this table but is also a popular alternative steel coating with an expected period of protection of 15 to 20 years at 14 to 16 mils dry film thickness.

4.6.5 Original Service Life Calculation

The first step in calculating remaining service life is to calculate the expected life at the time of original construction using the data and recommendations provided in the aforementioned sections. Consider the following three possibilities for a steel pipe pile that was designed for service in the seawater splash zone where the bare steel corrosion rate is 5 mpy:

- Bare steel,
- Epoxy coating and sacrificial steel, and
- Concrete encasement and sacrificial steel.

4.6.5.1 Bare Steel Only. In this case, corrosion of sacrificial steel is assumed to be average, at 5 mpy (0.005 ipy, 0.127 mm/year). The calculation is complete if the record documents indicate how much steel was added for sacrificial purposes (sacrificial steel thickness/5 mpy = design service life). If the sacrificial steel thickness is not provided or known, a structural analysis would need to be performed to determine the required thickness to support the various imposed loads. In either case, if the use of the structure has changed (e.g., the live load has increased or decreased), more or less sacrificial thickness may be currently available, which should be calculated. For this example, assume 1/8 in. (3 mm) of sacrificial steel thickness was provided in the original steel pipe pile design; a 5 mpy corrosion rate would yield a design service life of 25 years.

4.6.5.2 Epoxy Coating and Sacrificial Steel. A review of the record documents may indicate the original coating thickness. (Note: The original coating may have been augmented over the years.) However, if this information is not available, a coating thickness must be assumed. From Table 4-7, for this example, a coating thickness of 20 mils of coal tar epoxy will be assumed, which has a 15-year period of protection. The expected design service life would be the period of protection provided by the sacrificial steel plus the period provided by the coating 25 + 25 = 40 years.

4.6.5.3 Concrete Encasement and Sacrificial Steel. Similar to the previous coating example, a concrete encasement offers 25 years of service life. The encasement has to extend far enough below the intertidal zone to ensure that the expected life of the bare steel at the termination of the encasement is greater than the encasement in the splash zone. If this is not the case, then the corrosion rate of the steel for the zone in which the encasement ends is the lone factor in determining the service life.

For the previous example, the expected service life is just 25 years, because the sacrificial steel within the encasement is never utilized. The reinforcement of the encasement itself is the governing factor. Caution should be exercised, as in some instances, a partial jacket without an accompanying cathodic protection system may significantly accelerate corrosion because of the potential corrosion cell from the change in environment just below the jacket.

4.6.6 Remaining Service Life Calculation

Roughly speaking, the remaining service life is the expected service life at design less the current age. However, many assumptions are made when calculating the original service life. For a more accurate estimate of life expectancy, the simple model presented herein can be calibrated and modified with field measurements and results of site-specific studies. Additional data that could be collected and used include

- Measurements of actual site-specific bare steel corrosion rates;
- Measurements of remaining steel thicknesses in the splash zone, above and below;
- Measurements of remaining coating thicknesses; and
- Following data to a lesser degree as the effects of these are not well understood:
  - o Currents, waves, and prop wash;
  - o Measurements of salinity, water temperature, pollution, and other environmental data (including any changes over the time period in question); and
  - o Stray current measurements.

If the data collected are representative of what was the expected age, then the service life model can be expected to be fairly accurate. If the data collected differ from expectations, two possibilities exist: (1) the assumptions were not correct and should be adjusted, or (2) something else is contributing to the anomalous data that must be reconciled before accepting.

Some agencies have developed nonlinear empirical equations to provide a more refined method of estimating the remaining service life of steel structures. The Port of Rotterdam, using the aforementioned Jongbloed's corrosion curves and measured remaining steel thickness at a known exposure period, uses the following equation to determine the predicted corrosion loss Y (in mm) after T years:

$$Y(T) = \frac{-2 \cdot a \cdot Y_p + \sqrt{(2 \cdot a \cdot Y_p)^2 + 4 \cdot a \cdot T}}{2 \cdot a} \qquad (4\text{-}1)$$

wherein

$$Y_p = \frac{\frac{50}{a} - y_{50}^2}{2 \cdot y_{50}} \qquad (4\text{-}2)$$

$$a = \frac{\left[50 - \left(\frac{y_{50}}{y_{10}} \cdot 10\right)\right]}{y_{50}^2 - y_{10}^2 \cdot \frac{y_{50}}{y_{10}}} \qquad (4\text{-}3)$$

where

$Y(T)$ = Predicted corrosion loss (mm),
T = Time in years,
Y_p = Site-adjusted corrosion loss parameter (mm),
a = Calculated slope constant ($1/\text{mm}^2$),
y_{10} = Measured or anticipated corrosion losses at year 10 (mm), and
y_{50} = Measured or anticipated corrosion losses at year 50 (mm).

Although the Port concluded that the equations for producing corrosion curves and remaining service life can be applied to structures worldwide, they must be adjusted for the anticipated corrosion losses at year 50 (y_{50}), which will vary on the basis of local conditions and needs to be determined for the site.

4.7 TIMBER FACILITIES

4.7.1 Background

Research and development of methods and modeling to predict the service life of timber have not received the level of attention concrete and steel have received. Specific methods and criteria for determining timber's estimated or remaining service life is highly dependent on engineering judgment, coupled with known environmental parameters and historical data for the specific structure and location being evaluated.

As with structures constructed of concrete and steel, several items affect the service life of timber structures, including design details, exposure variables, and modes of deterioration. Timber can be affected by one or more of the following modes of deterioration: mechanical, chemical, or biological. Localized mechanical and biological degradation are the most common forms of timber member deterioration and often work in concert with one another, resulting in significant damage.

Mechanical degradation results from induced stresses caused by structural loading, which can be instantaneous, short term, medium term, long term, or permanent. Except for cyclical abrasion and repetitive overstressing, mechanical damage is typically caused by an instantaneous and short-term load such as berthing or debris impact forces. Chemical deterioration generally consists of corrosion or loosening of connection hardware, not the actual timber elements, and occurs perpetually. Biological degradation results from damage by various destructive agents on timber, including fungi, insects, bacteria, and marine borers. These agents damage timber, resulting in changes to the member cross section and/or material strength, which progresses continuously once initiated.

4.7.2 Sampling

For service life estimation of timber structures, the required sampling data are generally needed to determine applicable load restrictions or apply engineering judgment on when load capacity or operational may be required. Loss of capacity of deteriorated timber structures is most often a result of section loss in primary structural elements, bearing loss, and/or corroded or loose connections that limit intended load distribution. Some deterioration modes, including mechanical damage and external degradation, can be quantified through visual observation and physical measurements obtained using tactile inspection techniques.

However, concealed internal section loss of timber members such as deterioration from *Teredo* marine borers, requires more advanced Level III inspection techniques. In some instances where a more precise load rating or service life estimate is needed, additional sampling and testing may also be performed to validate material properties, determine preservative penetration and retention, or identify the presence of deleterious biological organisms.

The required sampling size for Level III inspections is typically 5%, but should be adjusted on the basis of facility size, observed condition of the structure, and the desired confidence level for the service life estimate. A larger sample size results in improved confidence in the estimate. An important consideration is that timber is an anisotropic material, meaning that its properties vary depending on the direction of the load. There is naturally an increased degree of variation in the actual material makeup and properties, which carries through to each member's individual susceptibility and rate of deterioration. Therefore, engineering experience and judgment are necessary when determining an appropriate sample size, evaluating deterioration levels and potential impacts on member capacity, and estimating a facility's remaining service life.

The different types of Level III inspection techniques commonly used for timber elements are discussed in more detail in Chapter 3, "Scope of Inspection Work." The two most effective methods currently used are coring and resistance drilling to determine the remaining timber section, related deterioration rates, and estimate remaining service. Coring uses a small 1/4 in. diameter by 6 in.-long incremental coring tool to bore into the timber member and extract a core sample. This core sample is then evaluated, in the field by the inspector, for section loss and may also be sent for laboratory analysis. This technique is considered a partially destructive technique as it leaves a hole in the timber member that requires plugging with treated wood dowels or steel bolts to prevent a remnant entry point for additional marine borers. This technique is falling out of favor because of the difficulty of extracting cores, especially below water, and subjectiveness with the evaluation of void and decay presence.

Resistance drilling has been a method for Level III inspections for quite some time; however, the tools, detail collected, and processed data have made significant advances in the last 10 to 15 years. As a result, resistance drilling is becoming increasingly popular. Contemporary methods of resistance drilling utilize a special resistograph tool in place of an ordinary drill. This tool has a small-diameter needle bit that bores into the timber and measures and records a quantifiable resistance value, which is plotted either digitally or on an analog display showing how much resistance the wood is providing versus the distance traveled. An example of a diver performing resistance drilling with a resistograph unit and an output plot is shown in Figures 4-7 and 4-8.

Figure 4-7. Diver performing resistance drilling with resistograph.
Source: Courtesy of Collins Engineers Inc., reproduced with permission.

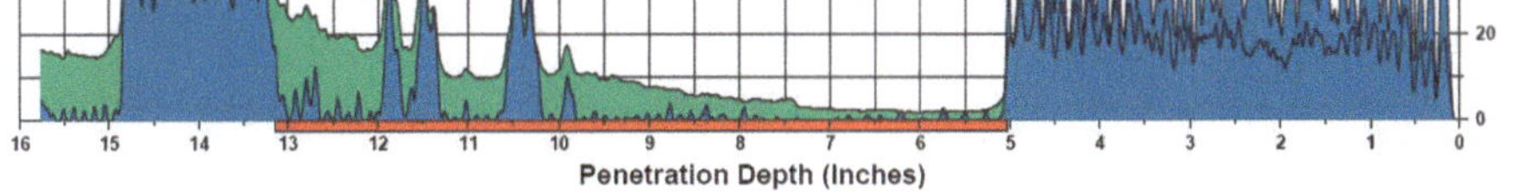

Figure 4-8. Example of a resistograph graph of resistance versus penetration depth.
Source: Courtesy of Collins Engineers Inc., reproduced with permission.

Level III testing and sampling is an essential technique for collecting data to complete a service life estimate, when a higher level of granularity is warranted. Determining whether sampling and testing is necessary, the quantity required, and how the data are applied needs to consider timber's anisotropic and nonhomogeneous material properties. Additional engineering judgment and adjustments may be required to account for the variability of timber properties present throughout the structure, compared to the small sample size of the material tested. Some examples of Level III sampling techniques can include the following, and subsequent testing and analysis are discussed in Section 4.7.3:

- Coring for determining timber species, material properties, and moisture content, typically consisting of minimum 2 in. diameter by 3/4 in. Cores;
- Visual grading for determining strength properties; and
- Cores for determining creosote or other treatment penetration and retention, typically consisting of similar-sized cores performed for Level III inspections.

4.7.3 Laboratory Testing and Analysis

Laboratory testing and analysis can be performed to obtain additional in situ characteristics and/or verify material properties. Core samples that have been collected can be tested in a laboratory to determine timber species, material strength properties, preservative treatment type, and retention, or other evaluations used for determining the remaining service life.

Some of the laboratory tests performed may include

- ASTM D143, *Standard Test Methods for Small Clear Specimens of Timber* (ASTM 2023b);
- ASTM D2395, *Standard Test Methods for Density and Specific Gravity (Relative Density) of Wood and Wood-Based Materials* (ASTM 2022c);
- ASTM D2555, *Standard Practice for Establishing Clear Wood Strength Values* (ASTM 2024c);
- ASTM D2899, *Standard Practice for Establishing Allowable Stresses for Round Timber Piles* (ASTM 2017);
- ASTM D4442, *Standard Test Methods for Direct Moisture Content Measurement of Wood and Wood-Based Materials* (ASTM 2020a); and
- Analysis for the Presence, Penetration, and Retention of Various Wood Preservatives per American Wood Protection Association Standards (AWPA 2024).

The two more common tests performed are timber species identification and timber protective treatment retention. Timber species identification testing can help categorize the species of wood, which, in conjunction with visual grading, may provide valuable information for the service life estimate. Timber protective treatment retention rate testing identifies the rate of loss of the protective treatment and when the protective treatment is no longer sufficient to prevent marine borer or fungi attacks. Timber cores can also be submitted for laboratory analysis to determine the presence of marine borers or other detrimental biological organisms when the information cannot be visually identified in the field.

4.7.4 Service Life Estimation

Determining the service life of timber structures requires engineering experience and judgment, along with collecting data in the field. When investigating the existing condition of a timber structure, one must remember that, unlike concrete and steel, timber is a nonhomogeneous and anisotropic material. Therefore, timber degradation typically occurs randomly. Although patterns of deterioration sometimes emerge, the damage may concentrate in random and/or localized areas of the structure. This can result from minute differences in environmental factors such as areas of hard driving due to rocky subsurface conditions, which can damage the outer portions of piling. This produces breaches in the protective treatment and possible accelerated deterioration from marine borers.

In addition, a portion of a structure may consist of timbers from a different treatment plant or batch, resulting in slightly different levels of protection. Consequently, sampling to determine the existing condition of a timber structure must be conducted carefully, understanding the limitations of representative sampling. Valuable information can be gleaned from historical records, design information, and past inspection records.

Once information about a structure has been obtained, it can be used to estimate a rate of deterioration. Because timber is susceptible to several deteriorative processes, this estimate should account for all the deteriorative processes identified. In some instances, the remaining service life may be based on more than one rate of deterioration. As an example, structures in a marine environment often have a much different rate of deterioration affecting the substructure piles than that affecting the superstructure members (i.e., pile caps, stringers, and decking).

Often, the expected life of these two structural components is significantly different. However, once determined, these rates of deterioration can be used, along with engineering judgment and experience, to obtain an estimated remaining service life for the structure. Something that should also be considered with timber deterioration in the marine environment, more specifically the biological degradation, is that the rate tends to increase over time instead of occurring linearly. In general, once the degradation advances and repair programs have been initiated, the deterioration and rate of repairs are accelerating and the end of the structure's useful service life approaches more rapidly.

When determining the remaining service life of a timber structure, consideration should be given to the amount and type of anticipated maintenance repair work that will be conducted. The remaining service life of a structure that will not be maintained will obviously be less, in many cases, much less than that of a structure that is routinely inspected and repaired (i.e., ongoing maintenance repair will reduce the effective rate of deterioration, thus extending the service life).

Therefore, three types of analysis might be considered: (1) the current state of the structure, including damage and deterioration both above and below water, is assessed using an average rate of decay to date; (2) the structure can be evaluated assuming maintenance of only the worst case, critical damage; and (3) the structure can be evaluated on the basis of current conditions and assuming regular maintenance is conducted.

Several models exist to address and estimate the strength of timber under load. Because these models deal only with the strength of the timber over time, modifications must be made to incorporate the influence of deterioration and maintenance. Some success has been made by combining durability and strength models to develop estimates of the remaining service life; research has been conducted in Canada and Australia.

The Forestry and Wood Products Research and Development Corporation formulated a major national project in Australia in 1998 to address issues of fungal decay, termites, corrosion of fasteners, and marine borers. This work was developed utilizing various hardwood species and pine treated with chromated copper arsenate that were subjected to coastal waters in Australia.

The results were then compared to data on in-service marine pilings that, in a marine environment, are frequently the members that limit the structure's service life. The studies indicate that despite several inherent uncertainties in determining the service life of timber structures, the findings generated by the study model provided good agreement with field data. This line of research may hold promise for the future and other locations but will require significant effort to determine estimated service life data for specific areas, regions, and environmental conditions.

It is important to consider the relationship between maximum deterioration and maximum internal forces in the members. Fungal decay occurring at the top of piles may result in significant section loss at that location. However, timber pile capacity is often geotechnically or buckling controlled, and no structure capacity reduction may be warranted. Similarly, the total area loss and impacts on section properties must be considered. Because of buckling effects, the total section loss on the exterior of a pile will have a greater impact on member capacity than an equal amount of section loss on the interior of the pile.

In summary, by understanding a timber structure's historical information, an estimate can be made of its residual service life with the assistance of engineering experience and reasonable judgment. Historical information can include design documents, exposure, usage and rehabilitation, current inspected condition, and anticipated future use of the structure,

4.8 POLYMER COMPOSITE MATERIALS

4.8.1 Background

Polymer composite materials, for structural use in waterfront facilities, encompass various types of polymeric materials, including vinyl, fiber-reinforced polymers (FRP), and new and recycled plastics. These materials are often used for secondary members and fender components in plastic lumber, fender piles, camels, sacrificial wear, and rub blocks. However, advances in manufacturing and production technologies have produced higher strength and more durable polymer composites. As a result, their use has expanded to being used as primary structural elements such as bearing piles, sheet pile walls, superstructure members, and decks.

Service life is a function of the structure's ability to support the required loads and operations, governed mainly by primary structural elements and components that make up large portions of a facility. Consequently, this section should be used only for estimating the service life of structures that are primarily constructed of polymer composite materials. It is not intended to address structures primarily constructed of timber, steel, or concrete, where polymer composites are used only as secondary, supplementary, or accessory elements.

Uses of polymer composite materials as primary structural elements for marine facilities typically include

- Piles [defined by ASTM D7258 (ASTM 2023a)]: Solid polymeric piles (Type I), polymeric with fiber or mineral reinforcement (Type II), polymeric with metallic reinforcing bars (Type III), polymeric with nonmetallic reinforcement (Type IV), FRP shell with a concrete core (Type V), and any other polymeric pile where the pile strength or pile stiffness requires the inclusion of the polymer or a minimum of 50% of the weight or volume is derived from the polymer (Type VI);
- Beams, pile caps, and structural members: Solid polymeric shapes, solid polymeric with nonmetallic reinforcing, solid polymeric with metallic reinforcing, FRP shell with concrete or polystyrene core and reinforcing;
- Decking: FRP shell with concrete or polystyrene core and reinforcing, FRP grating;
- Bulkhead walls: FRP sheet pile, FRP wales, vinyl sheet pile, FRP tie-rods; and
- Floating docks: FRP floating dock modules, polymeric shell floating dock modules, FRP access gangways.

4.8.2 Sampling and Testing

Compared to traditional materials, polymer composites have limited historical use as primary structural components in the marine environment. As a result, there are no known standardized sampling methods or techniques that would be performed as part of a routine inspection, load rating, or service life estimate. As the chemical and material makeup of the polymer composite material can vary significantly on the basis of type and manufacturer, if no as-built information is available, localized destructive sampling and testing may be necessary to obtain the required properties.

In addition, interior degradation and capacity loss may be a function of the material strength properties and not just a measurement of changes in section properties. In these cases, material testing may also be warranted.

Sampling and testing of polymer composite materials would include removing coupons from elements on-site and performing laboratory analysis to determine material strength properties, abrasion resistance, and ultraviolet (UV) resistance. Possible tests performed may include

- ASTM D7290, *Standard Practice for Evaluating Material Property Characteristic Values for Polymeric Composites for Civil Engineering Structural Applications* (ASTM 2022d);
- ASTM D6109, *Standard Test Methods for Flexural Properties of Unreinforced and Reinforced Plastic Lumber and Related Products* (ASTM 2024d);
- ASTM D6108, *Standard Test Method for Compressive Properties of Plastic Lumber and Shapes* (ASTM 2024f);
- ASTM D5379/D5379M, *Standard Test Method for Shear Properties of Composite Materials by the V-Notched Beam Method* (ASTM 2021b);
- ASTM D2344/D2344M, *Standard Test Method for Short-Beam Strength of Polymer Matrix Composite Materials and Their Laminates* (ASTM 2022e);
- ASTM D638, *Standard Test Method for Tensile Properties of Plastics* (ASTM 2022f);
- ASTM D6641, *Standard Test Method for Compressive Properties of Polymer Matrix Composite Materials Using a Combined Loading Compression (CLC) Test Fixture* (ASTM 2024e); and
- ASTM D4329, *Standard Practice for Fluorescent Ultraviolet (UV) Lamp Apparatus Exposure of Plastics* (ASTM 2021c).

4.8.3 Service Life Estimation

Historical information for estimating the service life of polymer composites, such as case studies, sampling methods, and testing, are not well documented. This is likely attributed to the recent developments in materials and limited use as primary structural members. As a result, the specific methods and criteria for determining polymer composites' estimated or remaining service life is highly dependent on engineering judgment, supported by an understanding of the site characteristics, environmental exposure, and loading conditions.

For polymer composite structures, it is also especially important to understand the material type, composition, and/or the manufacturer due to the broad range of polymer composite types, properties, susceptibility to degradation, and deterioration rates. Similar to concrete, timber, and steel, the modes of deterioration for polymer composite materials can be mechanical, physical, chemical, or biological. Because of a predominantly synthetic makeup and the use of little quantity of natural materials in today's practices, biological deterioration is extremely rare for polymer composites with limited known causes that impact the service life.

Physical and mechanical deterioration primarily consists of manufacturing defects or damage that has occurred and is often confined to localized and random areas. Chemical degradation is likely to be a more widespread issue with impacts on the elements' polymer or resin matrix.

As polymer composites cover a broad range of polymeric materials, susceptibility to these deterioration mechanisms is highly dependent on their polymer composite makeup such as the type of resin, fiber type and orientation, and manufacturing process. The following provides a summary of deterioration modes that have been observed in polymer composite structures located in the marine environment, and how they may impact the service life.

- Moisture Ingress: This type of physical deterioration can be localized because of a manufacturing defect or crack, or widespread if the material composition has small voids or delamination. The moisture can penetrate the polymer composite matrix, leading to its degradation, and in colder climates, it results in freeze–thaw damage. The result is a loss of material strength, stiffness, and dimensional stability of the polymer composite, reducing the load-carrying capacity and service life.
- Ultraviolet Degradation: Exposure to UV radiation from the sun can cause degradation of the matrix and the reinforcement, leading to discoloration, cracking, and loss of mechanical properties. This type of deterioration is more prevalent in older polymer composite systems, as newer materials typically contain additives for UV stabilization and protection, meeting ASTM standardized testing requirements. Similar to corrosion in steel, this type of degradation is usually widespread with more advanced locations based on exposure locations, and it gradually changes material behavior over time.
- Chemical Degradation: Polymer composite materials are often susceptible to reactions with petroleum products, solvents, or other chemicals that can dissolve the resins or polymers. Typically, this deterioration is localized and has minimal impact on the integrity of the structure. However, if exposure is widespread, it can lead to a complete loss of member capacity. Unless there is repetitive exposure, this degradation generally occurs over a short period of time.
- Mechanical Damage: Like all materials used in marine construction, polymer composite materials can be susceptible to impact damage in

the marine environment, such as allisions from floating debris or vessels, or cyclical abrasion. Polymer composites are usually anisotropic, meaning that their properties vary depending on the direction of the load. They may be more prone to damage from unexpected loading conditions or directions, resulting in overstressing, delamination, cracking, fracture, reducing the structural integrity of the polymer composite. Some polymer composite materials may be more brittle, especially when coupled with UV degradation, exacerbating susceptibility to mechanical damage. Most mechanical damage is instantaneous, and capacity reductions are applied as necessary on the basis of the condition observed.

The process for estimating the service life of polymer composites is similar to that of other materials, starting with an expected design life and adjusting the remaining service life on the basis of the time in use and the conditions observed. As mentioned previously, the initial predicted design life should be used cautiously without proven data to substantiate these predictions. A key component to the remaining service life estimation is based on the nature of deterioration/damage, which may be a one-time event, and extrapolation may not be applicable.

When determining impacts on service life, the modes of deterioration are separated into three main categories: widespread material degradation; random localized damage or overstressing; and prevalent manufacturing defects affecting the performance of the polymer or resins.

Determining the remaining service life needs a consideration of whether the impacts are in localized areas or affect widespread portions of the structure. For example, the service life may not be affected if localized deterioration has been arrested and there is sufficient structural redundancy. In some instances, the remaining service life may also be based on more than one rate of deterioration, for example, elements exposed to direct UV, versus underdeck components, or elements exposed to frequent abrasion.

In conclusion, estimating the service life of polymer composite structures in the marine environment relies heavily on understanding the actual material composition and applying sound engineering judgment. If no as-built or existing information is available, contacting possible manufacturers of the materials or obtaining samples for laboratory testing may be warranted to determine material properties.

Like wood, polymer composites are often anisotropic, and damage and overstressing in random areas may be present that impact the structural integrity of only isolated elements. Alternatively, material degradation located within the resin or polymer matrix may have more widespread impacts, but this type of deterioration is less common and would be difficult to detect without performing material testing.

For more extended design and service lives, which is likely the case for polymer composite structures, consideration must be given to how long the structure will continue to be functional for intended use. In some cases, the actual remaining service life may be governed by the structure becoming functionally obsolete, instead of deterioration and loss of capacity. In addition, the presentation of the estimated service life should reflect the level of confidence attributed to the limited data available on polymer composite materials with prolonged marine exposure.

REFERENCES

ASTM International. 2017. *Standard practice for establishing allowable stresses for round timber piles*, ASTM D2899. West Conshohocken, PA: ASTM.

ASTM. 2020a. *Standard test methods for direct moisture content measurement of wood and wood-based materials*, ASTM D4442. West Conshohocken, PA: ASTM.

ASTM. 2020b. *Standard practice for petrographic examination of hardened concrete*, ASTM C856/C856M. West Conshohocken, PA: ASTM.

ASTM. 2020c. *Standard test method for acid-soluble chloride in mortar and concrete*, ASTM C1152/C1152M. West Conshohocken, PA: ASTM.

ASTM. 2020d. *Standard test method for obtaining and testing drilled cores and sawed beams of concrete*, ASTM C42/C42M. West Conshohocken, PA: ASTM.

ASTM. 2021a. *Standard test method for density, absorption, and voids in hardened concrete*, ASTM C642. West Conshohocken, PA: ASTM.

ASTM. 2021b. *Standard test method for shear properties of composite materials by the V-notched beam method*, ASTM D5379/D5379M. West Conshohocken, PA: ASTM.

ASTM. 2021c. *Standard practice for fluorescent ultraviolet (UV) lamp apparatus exposure of plastics*, ASTM D4329. West Conshohocken, PA: ASTM.

ASTM. 2022a. *Standard test method for electrical indication of concrete's ability to resist chloride ion penetration*, ASTM C1202. West Conshohocken, PA: ASTM.

ASTM. 2022b. *Standard test method for measurement of rate of absorption of water by hydraulic-cement concretes*, ASTM C1585. West Conshohocken, PA: ASTM.

ASTM. 2022c. *Standard test methods for density and specific gravity (relative density) of wood and wood-based materials*, ASTM D2395. West Conshohocken, PA: ASTM.

ASTM. 2022d. *Standard practice for evaluating material property characteristic values for polymeric composites for civil engineering structural applications*, ASTM D7290. West Conshohocken, PA: ASTM.

ASTM. 2022e. *Standard test method for short-beam strength of polymer matrix composite materials and their laminates*, ASTM D2344/D2344M. West Conshohocken, PA: ASTM.

ASTM. 2022f. *Standard test method for tensile properties of plastics*, ASTM D638. West Conshohocken, PA: ASTM.

ASTM. 2023a. *Standard specification for polymeric piles*, ASTM D7258. West Conshohocken, PA: ASTM.

ASTM. 2023b. *Standard test methods for small clear specimens of timber*, ASTM D143. West Conshohocken, PA: ASTM.

ASTM. 2024a. *Standard test method for compressive strength of cylindrical concrete specimens*, ASTM C39/C39M. West Conshohocken, PA: ASTM.

ASTM. 2024b. *Standard test method for microscopical determination of parameters of the air-void system in hardened concrete*, ASTM C457/C457M. West Conshohocken, PA: ASTM.

ASTM. 2024c. *Standard practice for establishing clear wood strength values*, ASTM D2555. West Conshohocken, PA: ASTM.

ASTM. 2024d. *Standard test methods for flexural properties of unreinforced and reinforced plastic lumber and related products*, ASTM D6109. West Conshohocken, PA: ASTM.

ASTM. 2024e. *Standard test method for compressive properties of polymer matrix composite materials using a combined loading compression (CLC) test fixture*, ASTM D6641/6641M. West Conshohocken, PA: ASTM.

ASTM. 2024f. *Standard test method for compressive properties of plastic lumber and shapes*, ASTM D6108. West Conshohocken, PA: ASTM.

AWPA (American Wood Protection Association). 2024. *Book of standards*. Clermont, FL: AWPA.

Balasubramanian, R., D. C. Cook, T. Perez, and J. Reyes. 1999. "Study of the initial corrosion products formed on carbon steel exposed along the Gulf of Mexico." In *Hyperfine Interactions: Proc., 6th Latin American Conf. on Applications of the Mossbauer Effect*, edited by A. G. Bohorquez, L. E. Zamora, G. A. Perez Alcazar, and J. A. Tabares, 37–41. Colombia: Cartagena of Indias.

Beech, I., and S. Campbell. 2008. "Accelerated low water corrosion of carbon steel in presence of a biofilm harboring sulphate-reducing and sulphur-oxidising bacteria recovered from a marine sediment." *Electrochimica Acta*. 54 (1): 14–21.

Brouillette, C. V., and A. E. Hanna. 1960. *Corrosion survey of steel sheet piling*, TR-097. Port Hueneme, CA: US Naval Civil Engineering Laboratory. Accessed March 14, 2025. chrome-extension://efaidnbmnnnibpcajpcglclefindmkaj/https://apps.dtic.mil/sti/tr/pdf/AD0249658.pdf.

Bruun, P. 1989. *Port engineering*. Houston: Gulf Publishing.

BSI (British Standards Institution). 2000. *Maritime structures, Part 1: Code of practice for general criteria*, BS 6349-1. 2nd ed. London: BSI.

BSI. 2007. *Design of steel structures: Pilings*, BS EN 1993-5:2007. Eurocode 3. London: BSI.

CALTRANS (California Department of Transportation). 2014. *Deep foundations*. Memo to designers 3-1. Sacramento, CA: CALTRANS.

CALTRANS. 2021. *Corrosion guidelines*. Version 3.2. Sacramento, CA: California Department of Transportation Division of Engineering Services Materials Engineering and Testing Services Corrosion Branch.

Camitz, G. 1998. *Corrosion and protection of steel piles and sheet piles in soil and water. Report 93*. Solna, Sweden: Swedish Commission on Pile Research. Geo Risk & Vibration Scandinavia AB. Archived.

CH2M HILL. 2004. *Wyckoff sheet piling corrosion issues*. Bellevue, WA: Technical Memorandum Draft Prepared for US Environmental Protection Agency Region 10. Archived. http://yosemite.epa.gov/R10/CLEANUP.NSF/1f1b668c32575908882570ad006d6644/46f42e6cc3c7c488882570ae0080c655/$FILE/Sheet%20Pile%20Wall%20Corrosion%20Memo%20-%20Final%20Draft-.pdf.

Coburn, S. K. 1988. *Corrosion factors to be considered in the use of steel piling in marine structures*. Vero Beach, FL: Pile Buck.

Daley, J. C., and D. Ingraham. 1996. "Underwater maintenance and inspection of cathodic protection systems (Cook Inlet, Alaska)." *Mater. Perform.* 35 (1): 23–32.

DoD (US Department of Defense). 2012. *General criteria for waterfront construction*, UFC 4-151-10. Washington, DC: Government Printing Office.

Farro, N. W., L. P. Veleva, and P. Aguilar. 2009. "Mild steel marine corrosion: Corrosion rates in atmospheric and seawater environments of a Peruvian port." In *ECS meeting abstracts*. Pennington, NJ: Electrochemical Society. http://ma.ecsdl.org/content/MA2009-01/14/645.abstract.

FDOT (Florida Department of Transportation). 2014. "Florida Atlantic ocean seawall study." Results reported by FDOT State Corrosion Engineer Mario Peredes. Archived. http://www.fdot.gov/research/completed_proj/summary_smo/fdot-bdk79-977-02-rpt1.pdf. http://www.fdot.gov/research/completed_proj/summary_smo/fdot-bdk79-977-02-rpt.pdf.

FDOT. 2025. *Structures manual*, Topic No. 625-020-018. Tallahassee, FL: FDOT Structures Design Office.

Fontana, M. 1986. *Corrosion engineering*. 3rd ed. New York: McGraw-Hill.

Hannigan, P. J., F. Rausche, G. E. Likins, B. Robinson, et al. 2016. Vol. 1 of *Design and construction of driven pile foundations*, FHWA-NHI-16-009, FHWA GED 012. Washington, DC: FHWA.

International Navigation Association. 2005. *Accelerated low water corrosion*, PIANC/AIPCN MarCom Working Group 44. Brussels, Belgium: International Navigation Association.

Jongbloed, B. P. 2019. *Inspection to prediction of corrosion-induced degradation of quay walls in port of Rotterdam.* Rotterdam, Netherlands: Port of Rotterdam Authority.

Romanoff, M. 1962. "Corrosion of steel pilings in soils." *J. Res. Natl. Bur. Stand. C. Eng. Instrum.* 66C (3): 1–22.

Roubos, A. A., D. L. Allaix, T. Schweckendiek, R. D. Steenbergen, et al. 2020. "Time-dependent reliability analysis of service-proven quay walls subject to corrosion-induced degradation." *Reliab. Eng. Syst. Saf.* 203: 107085. Accessed March 14, 2025. https://www.sciencedirect.com/science/article/abs/pii/S095183202030586X?via%3Dihub

Sagüés, A., N. Poor, L. Cáseres, and M. Akhoondan. 2009. "Development of a rational method for predicting corrosion rates of metals in soils and waters." *Tech. Rep. BD497.* Tallahassee, FL: Florida Department of Transportation.

Tomlinson, M. J. 1987. *Pile design and construction practice.* 3rd ed. London: Viewpoint.

CHAPTER 5
DOCUMENTATION AND REPORTING

5.1 GENERAL

An engineering report, including documentation, generally provides a systematically organized record of the inspection; the conditions encountered; the analyses, assessments, and judgments made by the responsible engineer; and recommendations for future use and remedial measures. Depending on the needs of the client, the report may range from a simple summary letter to a highly detailed narrative supplemented by extensive documentation of existing conditions, testing, and analyses. The overarching goal of reporting and documentation is to communicate physical deficiencies/findings to the client and provide a record for future inspections and projects.

The extent of the report should be determined by the complexity of the structure, the type of inspection, and the needs of the owner. Factors that may influence the choice of report format and detail include the owner's standards; the size, complexity, and importance of the structure; the existing condition; the anticipated use and distribution of the final report; and the potential for public involvement or litigation related to the facility. Further, securing, handling, storage, and distribution of reports, supporting documentation, and raw data (hard copies and electronic files) should be accomplished in accordance with client security and contractual requirements.

The type of report may influence the extent of data collection and documentation work necessary. Even simple letter reports, which summarize the results of inspection and analysis activities, may require broad inspection and analysis activities as a basis for the report. The documentation of these activities may not be included in the final report but should be available to support the contents. Report authors should be aware of their readers' expectations and emphasize conciseness and explanation/elaboration as appropriate to their audience.

In the case of a structural repair or upgrade design inspection, a formal report may be unnecessary, because the data collected are typically used to prepare a set of plans and specifications for construction. Regardless of inspection or deliverable type, engineering reports should be objective and fact-based, with repeatable assessments presented clearly and on the basis of sound engineering judgment. Further, inspection reports are typically a *snapshot in time* of observations during the field inspection phase of a project; limitations encountered during the inspection, such as access to elements, should be expected and documented.

5.2 ROUTINE INSPECTION REPORT

This section describes the contents of a comprehensive report for a routine inspection. It is beyond the scope of this manual to provide a narrative for reports that discuss the findings of other types of inspections (design/upgrade, construction, etc.). Reports for these inspections may contain many of the same items discussed herein. Some items may not apply to all reports, and additional items may be essential to certain reports. Although the length, detail, organization, and arrangement of specific items within a report may vary, reports should include the following sections:

- Executive summary,
- Introduction,
- Existing conditions,
- Evaluation and assessment,
- Recommendations, and
- Supporting appendixes.

Deliverable format, for example, hard copies and electronic copies, is typically defined in the scope of work. However, the author should prepare the report to accommodate the ways in which the medium can be read. Electronic documents intended to be reviewed on a screen should incorporate hyperlinks for document navigation. Reliance on color variations alone to depict conditions in figures will be lost if printed in black and white, so symbols and hatches should be used to convey findings regardless of printer capability.

5.2.1 Executive Summary

If the report is a comprehensive document, then an executive summary is commonly provided. This summary provides an up-front snapshot of the contents of the report. By definition, the executive summary is brief, stating only the most basic information about the inspection, findings, and recommendations. Although concise, the executive summary is often comprehensive to provide the reader with a thorough snapshot of the report's contents. Depending on client preferences, the executive summary can also be tailored to focus on critical findings and observations that require immediate action or operational restrictions. Further, the executive summary can instead be a management summary in which findings and recommendations are understandable to nontechnical readers. Ultimately, reporting should be tailored to maximize the utility of the final product for the client.

5.2.2 Introduction

The following items are typically provided in the introductory section of a routine inspection report:

- Scope of work with enumeration of any items specifically excluded, limitations on inspection, or analysis dictated by the owner or site conditions.
- Makeup of the inspection team and the period during which the inspection was performed.
- Description of the facility generally includes drawings depicting the structure's structural configuration. These drawings typically consist of a location plan, general plan(s), and cross sections through the overall structure sufficient to describe the basic features of the facility. A description of the functions and uses of the facility and design loads should also be provided. Where this description is extensive or covers multiple facilities, generating a separate facility description report section is sometimes necessary.
- Listing of pertinent documents that include original drawings, drawings of subsequent repairs or alterations, and previous reports pertaining to the facility. A summary of the facility's background is beneficial in understanding the full history of a facility.
- Description of inspection, testing, and analysis methods may be included in one or more appendixes if lengthy.
- Administrative details including coordination and point-of-contact information (this information is sometimes provided in an appendix).

5.2.3 Existing Conditions or Observations

This section should include the results of the inspection, special testing accomplished in the field, and results of any laboratory testing. The inspection information should be reported factually without comment or analysis. This section serves as the basis for further assessment and evaluation and contains only repeatable data that would be accepted by another competent engineer conducting a similar inspection. These results should be documented with notes, sketches, photographs, and video as appropriate.

Environmental conditions such as water clarity, depth, current, biofouling, or other relevant site characteristics should be documented. Particular emphasis should be placed on documenting existing conditions that could affect future repair or inspection activities at the site. Buried or otherwise inaccessible elements should be documented with appropriate caveats.

5.2.4 Evaluation and Assessment

This section should include an evaluation of the structure, based on the information described and gathered in the introduction and sections of existing conditions. Each condition described in the previous section should be evaluated and a statement made as to the effect of the condition on the capacity and serviceability of the structure. If the results of prior inspections are available, comments on changes since the previous inspections and estimates of the rate of deterioration should be included. It should also be stated whether a reported condition does not adversely affect the structure's load capacity. Any assumptions made in quantifying the assessment should be clearly indicated and substantiated with appended calculations or other backup information, as appropriate.

This section may also contain a narrative describing causes contributing to the observed conditions. The engineer's assessment of the structure and the rationale for that assessment should also be included. The rating should be assigned if a numerical condition assessment rating system is used, similar to that described in Chapter 2, "Standards of Practice." The assessment should be based on engineering judgment and accepted engineering practice standards. Some waterfront facility owners require predicted condition assessment ratings at a predefined planning horizon (e.g., 5 and 10 years) for use in planning. These predictions are most commonly prepared when adequate historical information on facility condition is available and rates of deterioration can be ascertained. Caution is warranted when inadequate information or extended planning horizons can yield inaccurate guidance.

5.2.5 Recommendations

This section should contain recommendations for future use (or restrictions on the use) of the structure; recommendations for repairs or replacement, including estimates of costs; recommendations for structural upgrades (when appropriate); and recommendations for the type of inspection and timing thereof. Recommended actions to be considered are elaborated upon in Section 2.7, "Recommended Action Guidelines," while a risk-based approach to repair prioritization is included in Section 2.7.1, "Repair Priority." Note that the application of the repair priority matrix requires close coordination with the facility owner to appropriately gauge risk tolerance to provide accurate risk categorizations. A routine inspection report does not typically provide an exhaustive array of repair or replacement options, and where this is warranted, a separate engineering study should be undertaken.

5.2.6 Appendixes

One or more appendixes containing data, analyses, and supporting information may be included. They may include items such as environmental data, record drawings, detailed inspection procedures, a description of the rating system utilized, lists of defects, field and laboratory testing results and procedures, calculations, life-cycle cost analyses, detailed cost estimates, subconsultant reports, and references. Avoid burying important information in appendixes where the reader may miss it. Accentuate relevant information in the body of the report to ensure that key observations and conclusions are not missed.

5.3 DOCUMENTATION

The type of report and extent of documentation should be defined in the scope of work. The type of documentation required can significantly affect the inspection cost. Chapter 3, "Scope of Inspection Work," provides specific documentation guidance for a range of inspection types.

5.3.1 Field Notes

Detailed field notes and sketches should always be prepared and maintained as a record of the inspection. Written field notes should be prepared with the care that would be used for survey notes as they could become evidence in legal proceedings. The required level of detail will depend on the type of inspection being conducted. Notes prepared for an inspection made in anticipation of developing repair plans, specifications, and cost estimates (such as a repair or upgrade design inspection) would typically be prepared in greater detail than notes taken in support of a routine inspection.

5.3.2 Photographs and Video

Digital photography can be used to depict typical conditions and defects. Underwater photography in clear, calm water is simple when compared to taking photographs in turbid water—where special equipment such as a clear water box must be used. The importance of good-quality above- and below-water photographic documentation cannot be overstated. This is especially true when laypersons will digest the description of findings. Photographs are also extremely useful for illustrating existing conditions when developing repair documents. Secondary graphics must be overlaid to emphasize defects.

Typical conditions should be photographed, but documenting every defect may not be necessary. Where appropriate, photographs of defects should include a scale to capture dimensional properties. A minimum number of photographs may be required at each structure, including general views of the facility. These are usually most effective when the photograph is taken from a heightened perspective (*bird's eye view*). Photograph quality and clarity should be assessed before leaving the inspection's geographic area. Visual aids must be included to help the reader understand where photographs were captured and at what orientation; this can be accomplished through site plans depicting photograph locations or the inclusion of wide-frame photographs to orient the reader.

Video recording may also be used for documentation, either alone or in combination with still photography. Additional time is required to view and edit the digital files. The video system should record both the inspector's comments and descriptions. Follow-on editing may be used to supplement the original comments. Scrutiny of video recordings should not be considered a substitute for hands-on inspection by qualified personnel, particularly for underwater inspections.

Recording footage not pertinent to the inspection report findings such as starting, stopping, moving, adjusting lighting, malfunctions and so on, should be deleted. An engineer or owner may quickly become disinterested in video recordings if they contain recorded material that is not pertinent. Originals of the video files should be maintained, but editing can be used to produce a video executive summary and to remove unessential footage. The use of an on-screen clock or counter is useful when reviewing the files and finding specific items.

Sonar and LiDAR imaging can similarly provide the reader with an enhanced perspective of structure configuration and conditions, particularly when stitched together to provide a composite view of the structure above and below water.

Recent advances in field photography and videography include the application of small unmanned aircraft systems, commonly known as drones. The drone can be positioned anywhere in the sky, giving the engineer a better perspective of the inspected assets. Drone cameras can be adjusted remotely by the drone operator at a safe location, reducing the safety risk for ground personnel inspecting difficult and tough-to-reach assets. The drone platform increases the quality of work with more precise, comprehensive data and high-resolution aerial photographs.

5.3.3 Calculations

Supporting calculations should be prefaced with a summary of assumptions, code references, and results. This summary is especially important to orient the reader where output from analysis software is included. Further, as required, calculations should be substantiated with sketches or figures depicting relevant geometry. All calculations should be dated and indicate both the preparer and the checker.

5.3.4 Cost Estimates

Similar to calculations, supporting cost estimates should be prefaced by a summary of assumptions. Further, as required, cost estimates should be substantiated with key data for context, such as the estimated date of construction, escalation, if any, contingencies, design and permitting fees, owner administration costs, and inspection costs.

5.3.5 Other Documentation

Inspection and assessment reports are sometimes appended with additional documentation as dictated by the scope of work or the client. Examples include the following:

- References,
- As-built documentation,
- Subconsultant reports of findings,
- Survey data, and
- Inspector credentials.

5.4 ELECTRONIC RECORD KEEPING

Hard copies of inspection records have been traditionally retained for the record. Although simple and to the point, the information is subject to loss, destruction, or misinterpretation due to the individuality of the inspector and the method of recording and maintaining the information. Inspection records are now often collected and maintained electronically.

Furthermore, with various tools embedded in the software applications, the possibility of misinterpretation and data loss (or duplication) is greatly reduced if data are entered directly into the application in the field.

5.4.1 Spreadsheets

Spreadsheets (such as Microsoft Excel) are the easiest way to set up and record information and are commonly used. Spreadsheets can be rapidly configured to meet the requirements of the inspection project. Spreadsheets have certain limitations: (1) the ease of setup and modification allows for potential corruption of the data, (2) they do not provide the capability to analyze the input information to ensure validity, (3) they have limited search capability, (4) they are considered flat files that have no relationship between various tables, meaning each record must contain all common information of the structure and offer no hierarchical presentation of data, and (5) they do not efficiently handle photographs and other images.

5.4.2 Databases

Databases store and manage data according to the configurations of the particular application. The most prevalent small database application suitable for personal computers is Microsoft Access. Although the current version has many more features and operates with fewer problems than the original, it is still limited to a database of approximately 4 GB in size. When high-quality photographs are stored, the storage capacity can be easily exceeded.

Other types of mainframe server implementations, such as SQL server, are virtually limitless in the number of records that can be handled. These databases are accessed from personal computers over a network or the Internet. They require IT expertise to set up and maintain. These servers easily store photographs of any size, although transmission rates across the Internet may set a practical size limit.

5.4.3 Basic Architecture of the Inspection Database

The fundamental building block of the inspection database is the table. Like a spreadsheet, it is simply an aggregate of columns and rows, typically referred to as fields and records. Unlike the spreadsheet, the fields (columns) are uniquely defined and categorized according to the type of data. An inspection database would most probably include multiple tables. By relating the tables to one another, powerful search and sorting functions can be utilized to select specific data from the mass of information. Inspection data ranging from a simple structure to the entire inventory of a port can be stored and intelligently retrieved.

5.4.4 Reporting Data

Spreadsheet applications boast a simplistic setup conducive to printing, reflecting an inherent "what you see is what you get" paradigm. However, their versatility in presenting data in varied formats is notably limited. Conversely, numerous database client applications excel in furnishing robust report-generation functionalities primed for immediate printing. These reports are rendered in read-only formats, diverging from data input forms, meticulously formatted, and fine-tuned for optimal print output. They often integrate advanced capabilities such as dynamic sorting and grouping of data, attributes typically absent in input forms. Moreover, reports exhibit proficiency in showcasing a broad spectrum of data types stored within the database, encompassing even multimedia elements like images and photographs.

5.4.5 Trends

With the increased use of in-the-cloud data storage and unlimited secure access over the Internet, the opportunities for setting up remote databases to warehouse an owner's inspection data are nearly limitless. Standardization of inspection protocols, such as that provided by this manual, allows for the rapid development of standard architectures, routines, and applications to be used in the database systems. Having a central server allows instant updates of the forms the user sees and the reports that can be generated. Multiple users can be easily accommodated at one time.

Although most applications present data in two dimensions using currently available technology, there is no reason why data cannot be presented in three dimensions, with fly-through or time-stepping through the data showing changes between various inspections. The development of inexpensive tablet computers with very good, embedded cameras, Internet access via WiFi or cell, and inexpensive custom applications, will allow inspectors to upload field data directly to the database. This approach will also allow the owner's maintenance or repair staff to access the information and expected inspection activities while on site via GPS-enabled tablets.

Geographic information systems (GISs) integrate hardware and software to help visualize facility conditions and much more. Geographic information system tools using map-based database navigation yield enormous potential for facility owners, inspectors, and end users. Preferably, these systems should be located at and operated by the asset owner. A state-of-the-art GIS is connected to the Enterprise Resource Planning system and the asset owner's document management system, so any information or documentation can be accessed with just a simple click on the map.

Over the years, 3D has become the standard for GIS. Also, the development of the Building Information Model (BIM) has opened up the possibility for the asset owner to have all the information at hand. All records and reports can be easily stored in either the BIM or the Document Management System (DMS). When needed, the documents can be easily retrieved from the system.

Using mobile applications, having a GIS also introduces the possibility to store the inspection results directly in the system. These applications are able to show the inspector the former inspection results and the spot where they were taken. This enhances the repeatability of the inspection. GIS information of assets can also be used by advanced systems that show, for instance, the degradation of assets.

Artificial intelligence tools like ChatGPT or Microsoft Pilot are fast becoming additional tools in the inspector's toolkit. Facilitating expedited word processing or assisting with calculations, these tools will almost certainly find space in the inspection documentation workflow. Caution is warranted when overreliance on generated information can distort the engineer's reporting of findings.

5.4.6 Dashboards

Data analysis and dashboarding tools (i.e. PowerBI and SAS) give meaningful insights into the data collection results and provide the first steps to data-driven decision-making and asset management. These applications are already used in a variety of industries and are in their infancy for waterfront structures.

Dashboards are tools that provide a visual representation of key performance indicators or metrics in real time, allowing users to identify trends, patterns, or anomalies quickly. These can be customized to display information relevant to specific roles or departments such as asset management or project development.

5.4.7 Digital Twins

Technological advancements continue to add to the engineer's toolkit, and one such innovation is the ability to rapidly develop digital models, or twins, of existing facilities. Through rapid data capture in the field employing photogrammetry and LiDAR equipment, facility documentation can be supplemented with scaled digital replicas for use in planning, communication, and further assessment. Ultimately, the application of such technologies and processes can support the goal of providing usable information to enable decision-making, visualization, and change tracking.

5.5 DATA COMPARISON OVER TIME

Various inspections may have taken place for a specific facility over years or decades. This creates an opportunity to track the damage and develop information that can help form opinions as to future degradation and recommendations for repairs. If data are compiled in such a manner that allows for modification during future inspections, these data can be configured to provide snapshots of information at any particular time while allowing for subsequent alteration of the records to incorporate future degradation. Data that are developed to support this type of analysis must be prepared in a well-thought-out and organized manner using software designed for this purpose.

In addition, as-built conditions and repair history documentation should be referenced, retained, and shared with the inspection team. These types of documents provide an important basis for comparison and enable more thorough assessments and recommendations.

CHAPTER 6
ADMINISTRATIVE CONSIDERATIONS AND PLANNING

The facility owner's employees or independent engineering consultants may conduct inspections. When consultants perform the inspection, a detailed request for proposals (RFP) is essential to obtaining precise proposals with comparable and accurate scope of work statements. This is an important step that limits cost increases because of unforeseen circumstances. Although this is true for projects in any environment, it is particularly important in the marine environment, as many conditions affecting the work are at or below the waterline with limited access. In some circumstances, the owner must consider the merit of a detailed RFP in cases where unforeseen conditions will not affect the overall cost of the project. Details for the scope of work are provided in Chapter 3, "Scope of Inspection Work."

Prior to issuing notice to proceed, the development of a well-defined contractual agreement including a detailed scope of services is essential to a successful project. In many cases, the detailed scope of services from the proposal is included as an appendix to the agreement. In addition, working on and adjacent to the water creates specialized needs for insurance coverage and permitting. This chapter discusses the variations in the contracting process that are unique to the marine work environment.

6.1 REQUEST FOR PROPOSALS AND AGREEMENTS

An RFP generally contains the following items:

6.1.1 Defining Owner Requirements

- **Intent:** A statement of the purpose and type(s) of inspection(s) (i.e., initial, routine, construction, impact/damage, etc.). The owner may consider including limits for the levels of underwater inspection (Section 3.1.3) as well as limits for any required testing, imaging, or surveys. Alternatively, the owner may opt to provide only a list of deliverables and leave the means and methods of investigation up to the consultant.
- **Deliverables:** The deliverables drive the scope of work, and as such, the deliverables must be actionable and fit for purpose. The person/group in need of the deliverables should review the RFP prior to advertising. The RFP should clearly state what constitutes completion for the contract, and this is typically directly related to the deliverables for inspection contracts.

  The contractual agreement will also typically include termination conditions. A list of deliverables should be provided, which may include engineering reports, repair drawings, cost estimates, images, survey data, construction bid documents, permit applications, and specifications. Format types should be included for all digital deliverables (i.e., required image file type).
- **Schedule:** Timeline for completion of the fieldwork, including a stipulation regarding the potential variability of weather conditions and ship schedule and the effects of same on the ability to execute the work. Timeline for submission of all deliverables, including critical findings, draft deliverables, and addressing of client comments, if applicable. It is common to include statements recognizing that the owner must provide timely access to the facility to be inspected, and conversely, that the engineer should schedule the work to minimize disruption to other operations at the facility.
- **Pricing Structure:** Above- and below-water inspection of waterfront facilities is specialized work conducted in a challenging environment, leading to uncertainty in inspection costs. The efficiency of the work can be impeded by weather-related conditions or limitations to access. This is particularly true for diving work, which can be greatly affected by cold water, currents, poor visibility, waves, and in-water boat traffic. Consideration should be given to the use of a time and materials contract to compensate the consultant for variations in working conditions that are beyond control.

  Alternatively, a lump sum contract could increase risk to the architect/engineer, which could, in turn, increase costs. The pricing structure for the contract should be made clear in the RFP (i.e., lump sum, not to exceed, cost plus, time and materials, contingency, etc.). It should also be noted whether retainage will be withheld, or whether there are cost savings to the client for early payment. Wage criteria and requirements such as prevailing wage and union should also be included. If necessary, the rate sheet providing the rate structure (hourly rates) for each staffing category or individual proposed on the project, as well as all markups on materials and equipment to be utilized, should be included.

The best practice for the RFP is to include a sample rate sheet to define the level of detail required. In some cases, the funding source can alter the agreement and pricing structure. For instance, portions of an investigation may be funded by insurance or federal set-aside funds, the details of which may be pertinent to include in the RFP. Emergency work may also change the bidding scenario (i.e., noncompetitive bid), schedule, and scope of work.

6.1.2 Owner Considerations for Project Control and Risk Management

- **Staffing Requirements:** A statement that all work shall be conducted under the direction and direct control of the engineer of record who should be a civil/structural engineer licensed or registered in the jurisdiction of the facility. An owner may also consider defining the minimum qualifications for staff including key personnel resumes, similar project experience, years of experience, and organizational structure (i.e., org chart). The owner may consider awarding bonus points where the owner would prefer staff to have certain specializations, such as when the project manager or lead engineer-diver is a board-certified port engineer (Civil Engineering Certification | ASCE). For additional information concerning the qualifications of the inspection team, see Section 2.4 of this manual.
- **Selection Criteria:** In general, the owner will benefit most from the selection of a consultant based on criteria that focus on qualifications and experience. Other factors, such as safety record, availability, technology, and disadvantaged business, might also be important to the owner. Although cost is a consideration, it is generally not advisable to make cost the primary criteria (provided that is allowed for by the owner's procurement policy).
- **Security Requirements:** A description of any mandatory security clearance or specialized training required for personnel, vehicles, and/or boats to access the facility. This could include Maritime Security (MARSEC) rules that apply to the facility. Maritime Security is a three-tiered system used by the United States Coast Guard and Canada to communicate and respond to maritime security threats. Level 1 is the normal level of operations where minimum security measures are maintained. Level 2 is a heightened level of risk with additional protective security measures for a period of time. Level 3

is an exceptional level of risk that applies when there is a probable or imminent risk of a security incident, which requires further protective security measures.

- **Safety Requirements:** A description of any mandatory safety qualifications or on-site briefing required to work on site. This could include details of site safety plans that must be submitted and any other related documents/forms.
- **Insurance Requirements:** Statements listing the types and minimum limits of insurance required by the owner. See details in Section 6.2.
- **Meetings:** The number and types of meetings that will be required during the project and whether the meetings will be conducted in person/on site or virtually. Consideration should be paid to the need for coordinating meetings for on-site operations such as vessels and railroads.
- **Changes to the Scope of Work:** The need for additional testing and/or inspection may arise as the work progresses. In addition, the application of good engineering practice may not require an inspection of all elements in some situations. An initial field assessment may lead to redirection of the work to areas determined to have more significance on the condition of the structure. An additional services provision is commonly used to allow for adjustment to scope and fee should these situations arise. It is common to include the change-order mechanism in a contractual agreement.

6.1.3 Other Supporting Information

- **Description of Facility and Limits of Inspection:** A description of the facilities to be inspected, including extents and an estimate of the number and type of structural components (piles, fenders, etc.). Basic plans and photographs are of great value during the proposal process. When possible, this description should include basic information regarding marine environmental conditions (current, water visibility, major hazards such as contaminants, etc.), access restrictions, and any other information relevant to the execution of the inspection work. There may be some cases where portions of a structure may be omitted from the inspection, such as where remediation, demolition, or repairs are already planned.
- **Pertinent Documents Providing Facility Information:** A description of previous inspection reports and/or repair projects should be included. Pertinent documents should be included in their entirety to aid in developing a detailed scope of work such as record drawings, previous load rating analysis computations, location maps, geographic location surveys, hydrographic surveys, underwater

imaging, previous repair details, pertinent environmental studies/reports, geotechnical borings, or other studies/investigation reports in the area. Digital record keeping of all such documents is imperative to provide sufficient information for future work.

6.2 INSURANCE

The liability issues related to fieldwork typically involve the potential for accident and/or injury, whereas the liability associated with office-related work may involve issues concerning professional liability. Careful planning and conduct of the work can reduce liability, but insurance is highly recommended and often required by law. A thorough understanding of the appropriate insurance requirements is essential to protect employees, the public, consultants, and owners/clients involved in the project.

The types of insurance coverage that will be required for a particular project vary but often include the following:

- Commercial general liability and property damage,
- Automobile liability and physical damage,
- Workers' compensation,
- Longshoremen and harbor workers' insurance,
- Jones Act Maritime Insurance (for work performed from vessels),
- Commercial Hull and Protection and Indemnity (for consultants with commercial vessels),
- Professional liability (errors and omissions),
- Contractor's pollution liability,
- Vessel pollution insurance, and
- Railroad protective liability insurance (RPL).

Certain clients, including governmental agencies, may also require special types of insurance, such as owner's and consultant's protective liability insurance. The insurance coverage, with appropriate monetary limits, is provided by individual policies for each of these categories or is sometimes provided by basic policies in conjunction with an umbrella policy to raise the limits of the underlying policies.

6.2.1 Commercial General Liability and Property Damage Insurance

Commercial general liability insurance can provide a wide range of coverage to protect consultants and owners from losses and claims alleging liability for bodily injury or damage to the property of others. This policy would normally cover damage caused by the consultant's operations, but special endorsements may be necessary to cover watercraft operations. Typical coverage limits range between \$1 million and \$5 million.

6.2.2 Automobile Liability and Physical Damage

Automobile liability insurance provides coverage to protect consultants and owners from losses and claims associated with vehicular accidents. This is important coverage, as divers and dive equipment are often transported to the project site by vehicle and, in many cases, dive equipment is directly operated out of a vehicle to support dive operations. Automobile liability insurance typically comes with two components to address bodily injury and property damage.

Bodily injury liability coverage addresses claims alleging liability for injuries and lost wages associated with a vehicular accident. This coverage is typically only for injuries to other people, as coverage does not apply to the consultant's own injuries.

Property damage coverage addresses claims alleging liability for property damage resulting from a vehicular accident but does not apply to the consultant's own property. Typical coverage minimums are $1 million for bodily injury and property damage liability.

6.2.3 Workers' Compensation Insurance

Workers' Compensation (WC) Insurance is a no-fault type of insurance that protects workers who suffer occupational injury, disability, or disease. If a worker is injured on the job, they do not have to prove that the employer was negligent to be compensated for medical expenses and lost time, or for partial or complete disability. This is a state-mandated insurance requirement, and its benefits are administered by individual state boards. Workers' compensation provides coverage for work on land, and on, over, under, or adjacent to waters that are not considered to be navigable. Workers' compensation is not required for sole proprietorships with no employees or subconsultants with employees.

A waiver of subrogation endorsement, which prohibits the insurer from attempting to seek restitution from a third party that causes any kind of loss to the insured, may be required by the contract. The cost of the endorsement is typically in addition to the workers' compensation premium. Workers' compensation rates for diving and vessel operations are typically more expensive than other industries. These rates apply even to somewhat low-hazard types of underwater investigations, because engineers performing shallow underwater inspections are rated in the same classification as salvage divers dismantling a sunken ship under hazardous conditions.

6.2.4 Longshoremen and Harbor Workers' Insurance

Longshoremen and harbor workers' insurance (USL&H) is similar to workers' compensation insurance but provides coverage for employees working on, over, and adjacent to navigable waters of the United States.

It is federally mandated coverage and generally provides greater benefits to the employee than workers' compensation insurance. Its benefits are monitored by a federal board, but the coverage must be purchased from a private insurance company.

Workers' compensation insurance coverage does not include USL&H coverage, unless it is specifically added as an endorsement to the workers' compensation policy. The determination of applicability for state workers' compensation laws versus the USL&H Act is not clear-cut, especially in cases where the location of the injury or disability occurrence is not clear or where the limits of the navigable waterway are not well defined.

6.2.5 Jones Act Maritime Insurance

Jones Act Maritime Insurance is a no-fault coverage for employees, similar to WC and USL&H, except that it covers employees who are members of vessel crews. Because divers and other inspection-related support personnel may work from a boat or a floating work platform, Jones Act Maritime Insurance may apply. Jones Act Maritime Insurance has usually not been considered applicable to inspection-related work, unless the diver or inspector was regularly assigned to a vessel. Court decisions on this coverage have not been consistent, and prudence dictates that the coverage be provided for inspection work to prevent a possible gap in coverage.

6.2.6 Marine Protection and Indemnity Insurance

Marine Protection and Indemnity (P&I) and Charterer's Liability Insurance are required should any of the services require marine operations. The consultant shall provide for owned vessels, or obtain from the vessel owner for chartered vessels, Protection and Indemnity Insurance with respect to third-party bodily injury and/or property damage arising from marine operations, including damage to piers, wharves, other fixed or movable structures, and loss or damage to any other vessel, craft, or property on such vessel or craft. This insurance typically varies with the size of the vessel. When the vessel is rented or chartered in connection with the contractual services, the consultant should also maintain Marine Liability insurance with limits applicable to the estimated replacement value of the chartered vessel.

6.2.7 Professional Liability Insurance (Errors and Omissions)

Professional liability insurance, sometimes called *errors and omissions insurance*, provides owners with a means of protecting themselves from negligent or erroneous acts of engineering consultants on a project. It also serves to protect a consulting engineer from excessive losses resulting from claims by the owner or third parties. Professional liability insurance is not typically available to commercial divers or consultants attempting to provide engineering services.

If an engineer was negligent in the inspection and evaluation of a structure that was later associated with injuries, property damage, consultant's claims for extras, or owner's claims, the professional liability insurance would provide coverage against claims for the consultant and protection for the client or owner. In these cases, a nominal deductible typically applies. Professional liability insurance limits of at least $1 million are typically considered to be the minimum coverage.

6.2.8 Contractor's Pollution Liability

The contractor's pollution liability insurance (CPL) covers events occurring in the field during inspection or similar work. This insurance covers pollution caused by the inspectors themselves or the equipment they are using. Examples include breakage of a hydraulic line from a hydraulic pack used for inspection equipment, or the accidental dislodging of contaminated materials into the water while cleaning or probing the structure. Pollution caused during or after construction because of a faulty design by the engineer would be covered under the professional liability insurance policy, providing pollution coverage has not been excluded. Contractor's pollution liability insurance does not cover pollution caused by the engineer while they are in the field.

6.2.9 Vessel Pollution Insurance

Vessel pollution insurance covers pollution that is caused by the vessel. An example includes oil discharge from the motor on the inspection boat.

6.2.10 Railroad Protective Insurance

When work takes place close to the right of way of a regulated railroad, railroad protective insurance (RPL) will most likely be required. The railroad will need to be listed as an additional insured under the policy. The insurance provides coverage for damage to the railroad's property and rolling stock and liability under the Federal Employer's Liability Act (should a railroad employee get injured). Alternatively, and at the discretion of the railroad, a *hold harmless* agreement indemnifying the railroad may be used instead of RPL. The railroad exclusion found in most general liability policies can also be deleted.

6.3 INDEMNITY

Indemnity, duty to defend, and hold harmless clauses are important considerations for contractual agreements. Limitations to indemnity/defense are important considerations, as they can affect the applicability of insurance policies. The consultant's obligation to indemnify is triggered by a determination of fault and a judgment being entered in a court of law. The consultant's obligation to defend is initiated as soon as a claim is filed against the indemnitee.

The obligation to defend is sometimes excluded from professional liability policies. In that event, contracts should be revised to remove the obligation to defend claims relating exclusively to the negligent performance of their professional services. A hold harmless clause is a further level of protection/obligation because it means a party is blameless even if they are indeed at fault. It is important to review these clauses in a contract with risk management and legal counsel.

6.4 CERTIFICATES OF INSURANCE

Certificates of insurance are issued by insurance companies or their authorized representatives, generally insurance brokers, as evidence of the insurance coverage provided by a company. The certificates show the types of insurance, the insurance company's name and policy number, the limits for each type of coverage, and the expiration date for each policy. The certificate should be issued in the client's name as the certificate holder and indicate the project for which it is issued. The certificate will also indicate the notification process that will be used to provide the certificate holder notice prior to the cancelation of the insurance. Certificates may also indicate special endorsements such as the inclusion of other parties as additional insureds. This can include grant funding agencies or other parties when applicable.

Providing certificates of insurance is a routine procedure that is without cost to the insured. Facility owners should not accept certificates of insurance that are not accompanied by copies of the applicable endorsements and are not addressed specifically to them; otherwise, the insurance could be canceled without their knowledge.

6.5 PERMITTING

Permits for work in and around the water can be expensive and time-consuming to obtain. Permits are not typically required for inspection work; however, they are a crucial part of the planning process for construction work. Therefore, the consultant should be aware of various permits that might be required in order to remediate and/or replace a structure in order to better inform the inspection recommendations.

6.6 NOTICE TO PROCEED

Notice to proceed is required prior to any action on the contract. Notice to proceed is a letter from the owner or director of a company or business to the consultant. The notice informs the consultant of the date when work can begin. An email should never be considered sufficient legal notice to proceed. The person signing the notice to proceed should have legal authority to award, which is typically based on thresholds in revenue value for procurement.

APPENDIX A

SPECIAL CONSIDERATIONS FOR SPECIFIC STRUCTURE TYPES AND SYSTEMS

A.1 INTRODUCTION

This appendix serves as a reference document for users and providers of waterfront facility inspection services. Its purpose is to describe common problem areas associated with structural components for specific structure types. Given the variety of structures in use and the conditions under which these structures are expected to perform, presenting an all-inclusive listing of what to look for during an inspection at a specific site is impractical. Instead, the information in this appendix may serve as a starting point for developing the scope and methodology of a particular investigation and location. In addition, the common problems described herein refer primarily to considerations during a routine inspection and generally do not apply to other types of inspections, which may be focused on specific areas.

A.2 OPEN-PILED STRUCTURES

A.2.1 General

Open-piled structures generally consist of a solid deck supported on piles driven or drilled into site soils. They are commonly used in various configurations, such as piers, wharves, jetties, and dolphins (Figure A-1). Open-piled structures can be constructed of concrete, steel, timber, polymer composites, or any combination of these materials. The superstructure (decking, beams, joists, etc.) and substructure (piles, bracing, pile caps, etc.) may be of different materials depending on the design and use of the

Figure A-1. Typical, open-piled concrete wharf with a timber pile fender system. Source: Courtesy of Simpson Gumpertz & Heger Inc., reproduced with permission.

structure. To a large extent, deterioration patterns for the various structures are directly related to the materials used.

Depending on construction type and configuration, horizontal loads may be distributed by either rigid or flexible deck diaphragms. These loads are resisted by axial loading of batter (sloped or raked) piles or by a combination of bending and frame action of the plumb (vertical) piles. In some cases, bracing is provided to improve pile performance.

An underwater inspection team is typically responsible for inspecting the portions of the piles below the high-water mark, although the exact demarcation can vary. An above-water inspection team will usually inspect deck framing and decking of an open-piled structure. The underwater inspection team may inspect the entire pile for structures that do not extend far above the waterline. For tall ocean piers or similar structures with significant freeboard, inspection of the upper portions is not feasible from the water or by boat. In these cases, an above-water team may employ special equipment to provide access from the deck above.

A.2.2 Timber Structures

Timber structures are typically constructed with timber piles supporting pile caps, deck beams (stringers), and timber decking. On a typical timber dock, timber stringers are laid above and perpendicular to the pile caps, and the decking is fastened directly to them. The structure is primarily held in place by gravity, with lateral restraints between each member. Some steel connectors may provide some load continuity. Examples of steel connectors include drift pins and brackets between piles and pile caps, brackets between deck stringers and pile caps, and spikes between decking and stringers (Figures A-2 and A-3). Timber piles can be supported by bracing to reduce the effective length of the piles for vertical loads.

Timber structures are subjected to rot, fungus, deterioration from marine borers, and other environmental factors. Timber elements are typically treated with preservatives or wrapped to extend service life.

Table A-1 summarizes what to look for when inspecting timber open-piled structures.

Figure A-2. (Left) Typical connection between plumb and battered piles to a pile cap on a timber wharf, connected with steel elements; (right) steel strapping securing a mooring cleat to the pile cap.
Source: Courtesy of Simpson Gumpertz & Heger Inc., reproduced with permission.

Figure A-3. Underside of a typical timber wharf.
Source: Courtesy of Simpson Gumpertz & Heger Inc., reproduced with permission.
Note: Stringers spanning between pile bents. Bridging is provided at the midspan of the stringers for stability.

Table A-1. Open-Piled Timber Structures: Checklist for Inspections.

Section or part	What to look for	Comments
Piles	Damaged or missing piles; alignment (straightness) of piles from top to bottom; scour pits at mudline; loss of bearing at pile head; fungal rot; wrap conditions	Accelerated rates of deterioration in the splash zone and wet areas
Pile caps, stringers, and braces	Damaged, loose, or missing members; alignment of members along length (rotation); signs of distress at bearing areas; fungal rot on top surfaces or wet areas; deterioration at connections; wrap conditions	The underside of low decks may need to be inspected by a diver The undersides of high decks may need to be inspected by aerial lifts, snoopers, or other inspection access equipment
Deck	Damaged, loose, or missing members; alignment of members along length (rotation or sagging); rot; wear; excessive deflection	Localized removal of deck coatings, surfaces, or overlays may be necessary to assess the condition of supporting members The underside of low decks may need to be inspected by a diver The undersides of high decks may need to be inspected by aerial lifts, snoopers, or other inspection access equipment
Mudline and dredged berth faces	Excessive dredging at the face of the structure; mechanical damage to piles caused by dredging equipment	Measure mudline depths at the structure's berthing face and compare with design dredge depths for the structure

A.2.2.1 Timber Piles and Bracing. Timber piles are found on older structures or lightly loaded modern docks. Timber piles are commonly used for breasting and fender systems as well. Timber piles are naturally limited in length and diameter by the available size of trees. The typical upper bound length is about 75 ft, with nominal diameters ranging between 12 and 16 in. Timber piles naturally taper with length, with the smaller diameter driven into the soil. Because of the limitation in element size, marine structures supported by timber piles often have a comparatively large number of closely spaced piles. This is often on the order of 10 ft on-center. Timber structures in deep waters are also typically braced.

To determine the physical condition, inspect piles for rot; checking or splitting; abrasion; shell peeling; damage from marine borers; and vertical, lateral, or rotational displacement. Check for scour and undermining at the mudline, especially for piles along the berthing area subject to propeller wash and piles in strong currents. Check exterior and fender piles, particularly corner piles, for damage or abrasion from vessel contact. Examine fender piles closely for breaks or splintering associated with hard impacts. Examine fasteners to ensure that they are not loose, damaged, missing, or protruding at locations with the potential of damaging a berthing vessel. Check piles for bearing against the pile cap. Large gaps between the members might indicate damage to the pile below water or the loss of shims.

Inspection for rot and marine borer damage may require drilling into and/or sounding the pile as some damage is not evident from the exterior. For piles constructed of marine borer-resistant timber or not exposed to borer damage, the key inspection areas are at the top of the piles and connections to bracing members. For piles not resistant to marine borer damage, the zone of highest deterioration is typically near the mudline and within the splash zone [in general, about 5 ft above and below mean low water (MLW)]. Pay close attention to locations where the piles are cut or altered for structural connections (e.g., holes for bolts, notches, tapers), as these locations directly expose the timber to the environment.

Inspect the overall condition of pile wraps and assess their effectiveness at keeping marine borers away from the timber. Inspect the pile wrap for damage or deterioration, as these areas may allow marine borers to penetrate behind the wrap. Critical areas include the seams and lap splices of the wrap materials (Figure A-4). Check steel bands or nails that are used to fasten the wrap for damage or deterioration. Loose or missing fasteners and bands may prevent a tight wrap. Depending on the wrap thickness, assessing the condition of the pile can be difficult. In some cases, hammer sounding or other inspection techniques such as resistance drilling can indicate the condition of the pile immediately under the wrap. Thin wraps may also deteriorate from UV exposure; wrapped exterior piles with southern exposure are especially vulnerable to this type of damage.

Inspect bracing for damage or deterioration, misalignment or rotation, and evidence of overloading. Check connections between the bracing members and piles/pile caps for looseness or missing hardware. Braces are susceptible to damage at the location of the connection hardware and at the ends. Like piles, check the condition of the wrapping for damage or deterioration.

Figure A-4. Typical timber pile with wrap damage.
Source: Courtesy of Simpson Gumpertz & Heger Inc., reproduced with permission.
Note: Wrap tear that originates from the nailed seam.

A.2.2.2 Timber Pile Caps and Stringers. Perform a general observation of the pile caps and stringers for damage or deterioration, misalignment or rotation, and evidence of overloading. Evidence of cap or stringer overloading may appear at points of maximum bending stress and maximum compression stress, appearing as cracking or sagging at midspan and buckling or crushing directly over supports. The upper surfaces of these members are also most susceptible to environmental deterioration. Timber pile caps and stringers are susceptible to rot at the bearing points because water will dry slower at the contact surface between members. Inspect connections, looking for loose, damaged, or missing hardware. Locations where these elements have been cut or altered for structural connections (e.g., holes for bolts, notches, tapers) are also vulnerable to damage, as these locations directly expose the timber to the environment.

A.2.2.3 Timber Decking. Timber decking is typically fastened to timber stringers with spikes. Nominal sizes of 3 × 12 or 4 × 12 timbers are often used as decking on timber docks. Inspect the decking for damage, deterioration, misalignment or rotation, and evidence of overloading. Note loose planks, areas of rot, damage caused by insects, checking, splitting, and missing connection hardware. Decks supporting vehicles, especially forklifts, may sustain wear damage in the most heavily trafficked areas.

Water ponding on the deck may be a sign of damaged stringers below. If the timber is covered with a layer of asphalt, inspect the asphalt for cracks, holes, or other obvious damage that may indicate distress of the underlying decking. Pay close attention to locations with expansion joints, railings, mooring equipment, drainage elements, and other features or attachments to the deck. These locations are vulnerable to damage and deterioration. The inspection team should look for missing/broken/loose connections, obstructions, and other hazardous conditions of curbs, handrails, and catwalks. As shown in Figure A-5, steel plate patches placed over damaged deck areas are only cosmetic and do not provide any protection for the deck members. Patches should be removed as necessary to assess the extent of damage hidden underneath.

Figure A-5. The deck of a typical timber wharf.
Source: Courtesy of Simpson Gumpertz & Heger Inc., reproduced with permission.
Note: Numerous steel plate patches placed over worn, damaged, or deteriorated deck locations.

A.2.3 Concrete Structures

Concrete open-pile structures are typically constructed with driven piles supporting a monolithic deck or framed deck system (Figure A-6). Driven concrete piles are typically prestressed to increase strength and performance. The deck of a typical concrete marine structure may be constructed of precast elements, cast-in-place elements, or some combination of the two.

Figure A-6. Open-pile mooring dolphin with monolithic deck and prestressed piles. Source: Courtesy of Simpson Gumpertz & Heger Inc., reproduced with permission.

It is often challenging to distinguish mild reinforced concrete elements from prestressed concrete elements when the extent of damage does not readily expose the reinforcement for observation. Reference structural drawings when available and note the more stringent damage ratings for prestressed elements.

Table A-2 summarizes what to look for when inspecting concrete open-piled structures.

Table A-2. Open-Piled Concrete Structures: Checklist for Inspections.

Section or part	What to look for	Comments
Piles	Damaged or missing piles; spalling; alignment (straightness) of piles from top to bottom; scour pits at mudline; impact damage; cracking	Accelerated rates of deterioration in the splash zone and wet areas. Structural cracks and spalling are common at the tops of piles near deck connection

(Continued)

Table A-2. (*Continued*).

Section or part	What to look for	Comments
Pile caps, beams	Damaged or missing members; alignment of members along length (rotation); signs of distress at bearing areas; corrosion spalling, concrete delamination, impact damage; cracking at midspans and ends	The underside of low decks may need to be inspected by a diver The undersides of high decks may need to be inspected by aerial lifts (Figure A-2), snoopers, or other inspection access equipment
Deck	Damaged or missing members, alignment of members along the length (rotation or sagging), corrosion spalling; concrete delamination; impact damage; cracking; expansion joint condition	The underside of low decks may need to be inspected by a dive The undersides of high decks may need to be inspected by aerial lifts (Figure A-2), snoopers, or other inspection access equipment
Mudline and dredged berth faces	Excessive dredging at the face of the structure; mechanical damage to piles caused by dredging equipment	Measure mudline depths at the structure's berthing face and compare with design dredge depths for the structure

A.2.3.1 Concrete Piles. Inspect concrete piles for structural cracks and spalls, corrosion of steel reinforcement, honeycombing, chemical deterioration, and abrasion damage. Check for vertical, lateral, or rotational displacement that may indicate underwater structural damage or overstressing. Structural cracks or spalls can typically be found in areas of high moment for the piles, at the connection of the piles to the pile cap or deck, or a location close to the middle of the exposed length of the pile. Horizontal or diagonal cracking may indicate overstress of the pile in either bending or shear, respectively. Corrosion spalling and chemical deterioration are typically found in the splash zone and other areas where water may intermittently contact the piles. Longitudinal corner cracking is a common deficiency, where the corner reinforcing has corroded and expanded, popping off the unrestrained concrete corner. Check for scouring and undermining at the mudline, especially for piles along the berthing face subject to propeller wash and strong currents. Check exterior piles, particularly corner piles, for damage from vessel contact (Figure A-7).

Figure A-7. Severely damaged prestressed concrete pile from vessel impact. Source: Courtesy of Simpson Gumpertz & Heger Inc., reproduced with permission.

A.2.3.2 Concrete Pile Caps, Beams, and Deck Elements. Concrete pile caps, beams, and deck elements form the superstructure of an open-pile concrete structure (Figure A-8). Concrete pile caps are typically cast in place to accommodate construction and pile-driving tolerances. Concrete beams and decks can be either precast or cast in place. Some precast elements may also be prestressed. Precast beams supporting flat deck panels often span between cast-in-place pile caps. Hollow core panels may also bridge between pile caps, especially across long spans. Cast-in-place concrete decks are typically used to form continuous monolithic deck slabs for structures supporting large deck loads such as container wharves.

Inspect all concrete components of the superstructure for structural cracks and spalls, corrosion, honeycombing, chemical deterioration, and abrasion damage. For concrete deck and beam elements, note the sizes of cracks, spalls (open and closed), staining, exposed reinforcing bars, delamination, or other signs of corrosion of the reinforcement. Inspect the underside of the deck, as this area is especially vulnerable to corrosion damage (Figure A-9). The underside of the deck should be sounded at any locations suspected to be hollow. The concrete cover from representative closed spalls should be removed to observe the level of corrosion of the underlying reinforcing bars.

Connections of precast elements require special attention. Connections are typically vulnerable to corrosion deterioration and require special attention during inspection. Loss of cover, spalling, and heavy staining from connection locations may warrant further investigation to confirm the adequacy of a connection.

Figure A-8. Soffit of a concrete deck showing piles with capitols.
Source: Courtesy of Simpson Gumpertz & Heger Inc., reproduced with permission.

Figure A-9. Corrosion spalling along soffit of reinforced concrete deck.
Source: Courtesy of Simpson Gumpertz & Heger Inc., reproduced with permission.

A.2.4 Steel Structures

Steel structures are typically constructed with piles supporting pile caps, deck beams, and steel grating for decking. On a typical steel-framed dock, steel beams may be bolted perpendicular to and between the steel pile caps and used to support the deck. The structural members are typically connected by welding or bolting. Steel piles may be braced to reduce the effective length of the piles. Steel structures are subject to corrosion, abrasion, and other environmental factors. Steel expands as corrosion takes place and can often be observed in the form of sheeting or delamination. These thin outer layers of steel no longer provide structural support and should be removed to sound metal before taking thickness measurements. Localized buckling may also be observed on steel elements with thin wall thicknesses.

Table A-3 summarizes what to look for when inspecting the condition of steel open-piled structures.

Table A-3. Open-Piled Steel Structures: Checklist for Inspections.

Section or part	What to look for	Comments
Piles	Damaged or missing piles; alignment (straightness) of piles from top to bottom; scour pits at mudline; corrosion; pitting; impact damage; condition of coatings and wraps	Accelerated rates of deterioration in the splash zone and wet areas
Pile caps, deck framing, and bracing	Damaged or missing members; alignment of members along length (rotation); signs of distress at bearing areas; corrosion; pitting; impact damage; condition of coatings	The undersides of low decks may need to be inspected by a diver The undersides of high decks may need to be inspected by aerial lifts, snoopers, or other inspection access equipment
Mudline and dredged berth faces	Excessive dredging at the face of the structure; mechanical damage to piles caused by dredging equipment	Measure mudline depths at the structure's berthing face and compare with design dredge depths for the structure

A.2.4.1 Steel Piles. Steel piles can be found in various sizes, shapes, and steel grades. Pipe piles and H piles are the most commonly used sections. Inspect all steel piles for vertical, lateral, or rotational displacement. Check for scouring and undermining at the mudline, especially for piles along the berthing face subject to propeller wash and strong currents. Check exterior piles, particularly corner piles, for damage from vessel contact.

Check steel piles for corrosion (Figure A-10), local and global buckling, and cracks or damage at the connection of the steel piles to other members. Use calipers and scales to determine the remaining thickness of flanges, webs, and stiffeners of H-piles. Use ultrasonic testing to determine the remaining thickness of pipe piles or to record a more accurate thickness of H-piles.

Figure A-10. Pitting and corrosion of a steel pile within the splash zone. Source: Courtesy of Simpson Gumpertz & Heger Inc., reproduced with permission.

If the structure has a cathodic protection system, test and document the system's condition with an underwater voltmeter after removing marine growth. Key inspection elevations for corrosion deterioration are at MLW and MHW levels. These areas typically experience the highest rate of corrosion loss, especially in seawater. Other locations vulnerable to corrosion include periodically wet locations, welded connections with poorly matched materials, or connections where dissimilar metals can cause galvanic corrosion processes. The latter condition can often be found on metallic appurtenances, for example, a stainless-steel ladder connected to a carbon steel pile. Also, inspect connections for evidence of overstress or damage (e.g., fracture of welds).

A.2.4.2 Steel Framing, Bracing, and Decking. Perform a general observation of the steel framing and bracing for damage or deterioration, misalignment or rotation, and evidence of overloading. At the caps and beams, evidence of overloading may appear at points of maximum bending stress and maximum compression stress. Examples of overstressing include buckling or sagging at midspan or directly over support points such as piles. Inspect welded or bolted connections between pile caps and deck beams. Typically, these members are located above the waterline and can be inspected from a workboat as part of the above-water inspection.

However, if the deck is located significantly above the water, other means of accessing these areas for inspection may be required.

Check all steel members for corrosion (Figure A-11). Use calipers and scales to determine the remaining thickness of flanges, webs, and stiffeners. Use ultrasonic testing to determine the remaining thickness of hollow steel section (HSS) elements or to record a more accurate thickness of other steel elements.

If the structure has a cathodic protection system, test and document the system's condition. Locations vulnerable to corrosion include wet locations, welded connections with poorly matched materials, and connections where dissimilar metals can cause galvanic corrosion processes. The latter condition can often be found at metallic appurtenances (e.g., a stainless-steel ladder connected to carbon steel framing). Also, check connections for evidence of overstress or damage (e.g., fracture of welds).

Steel decks are used on some structures and often include steel grating. For steel grating, areas of corrosion, overloading, or loss of paint or other protective coatings should be documented, especially at grating connections, which are easily damaged.

Figure A-11. Corrosion of steel bracing.
Source: Courtesy of Simpson Gumpertz & Heger Inc., reproduced with permission.

A.2.5 Polymer Composite Structures and Elements

Polymer composites have commonly been used as an alternative to timber for the load-bearing parts of light structures and in other areas where timber is commonly used. Polymer composite materials typically have higher strengths than timber and are resistant to corrosion and marine borers. Polymer composite members usually consist of a recycled plastic matrix with glass fiber reinforcement and a low-friction skin.

Table A-4 summarizes what to look for when inspecting the condition of polymer composite structures.

Polymer composite piles can be used as fender piles, load-bearing piles for light structures, floating camels, or as elements in pile cluster dolphins. Polymer composite beams can be used as load-carrying members in light structures, rubbing surfaces to protect structures, or as elements in large built-up camels.

Inspect polymer composite members for signs of overloading, chemical degradation, abrasion, and general wear. Vulnerable areas include connections and locations where the structural elements have been cut, drilled, or otherwise modified. Elements with significant exposure to sunlight may show signs of UV deterioration (Figure A-12). Polymer composite decking elements and panels often show signs of deterioration at support points, frequently requiring removal of restraining clips and select elements for closer examination.

Table A-4. Open-Piled Polymer Composite Structures: Checklist for Inspections.

Section or part	What to look for	Comments
Piles	Damaged or missing piles; alignment (straightness) of piles from top to bottom; scour pits at mudline; impact damage; abrasion; wear; UV damage, chemical degradation	
Other structural elements	Damaged or missing members; alignment of members along length (rotation); signs of distress at bearing areas; impact damage; abrasion; wear, UV damage	Connections are vulnerable to deterioration The polymer composite decking elements are vulnerable to deterioration at support points

Figure A-12. Polymer composite pile with UV damage.
Source: Courtesy of Creative Pultrusions, reproduced with permission.

A.3 RELIEVING PLATFORMS

A.3.1 General

Relieving platforms are a variation on the open-pile structure (Figure A-13). While a typical open-pile structure has an elevated deck, which also serves as the working surface, a relieving platform has a low-level deck, usually located in the tidal zone. Soil or sand is placed over the deck and filled to the working elevation. A topping may be placed over the soil to form the working surface.

The seaward edge of the relieving platform has a seawall to retain and protect the fill against erosion, typically constructed with sheet piles. This type of structure can be used for wharf or pier construction. Vertical piles support the dead and live loads in the fill area, thereby partially relieving the seawall from the lateral pressures associated with the heavy vertical surcharge. Materials used to construct relieving platforms include steel, concrete, and timber. When used to build a wharf, a sheet pile bulkhead is usually either at the face of the wharf or at the inshore side of the wharf. Lateral earth pressure from the retained soil of a sheet pile is resisted by batter piles or an anchorage system (e.g., tiebacks, deadman system, or soil anchors).

Relieving platforms inherently have high dead loads due to the soil fill. For timber-relieving platforms built in fresh water, the entire structure can be built underwater or just above the low tide mark, thus helping to preserve the timbers. At the beginning of the twentieth century, this type of structure was quite popular for pier and wharf construction. Many timber-relieving platform structures are more than 100 years old and still in use.

Figure A-13. Deteriorated relieving platform.
Source: Courtesy of Childs Engineering Corp., reproduced with permission.

A.3.2 Typical Components and Problem Areas

Relieving platforms should be inspected in a manner similar to open-piled structures. However, unique considerations for relieving platforms are presented in the following sections.

Table A-5 summarizes what to look for when inspecting the condition of these structures.

Table A-5. Relieving Platforms: Checklist for Inspections.

Section or part	What to look for	Comments
Piles	See checklists for open-pile structures	Look for overstressing at bearing areas and indicators of lateral movement and determine the cause of deficiencies
Deck	Spalling, cracking, fungal rot, separation, fill loss	Excavation will be required for the examination of the topside of the deck. The underside of the deck typically requires inspection by a diver because of low overhead clearance

(Continued)

Table A-5. (*Continued*).

Section or part	What to look for	Comments
Seawall	Alignment, settlement, spalling, cracking	Check the face for plumbness and check joints for alignment and settlement
Sheet pile	Corrosion, spalling, interlock separation, impact damage, cracking, alignment, cathodic protection, coating, fungal rot, talus cones at toe of sheet pile	Look for gaps and spaces between concrete or timber sheets. Check for interlock separation on steel sheets. Check for corrosion holes and associated sink holes. Talus cones may form at the toe of sheet piles, indicating a loss of soil fill
Pavement	Sinkholes, cracking, settlement	Cracking parallel to the structure face indicates lateral movement. Sinkholes may indicate washout of soil fill
Mudline and dredged berth faces	Excessive dredging at the face of the structure; mechanical damage to piles or seawall caused by dredging equipment	Measure mudline depths at the structure face and compare with design dredge depths for the structure

A.3.2.1 Access. Because of the low deck nature of this type of structure, an inspection of the exposed components can be difficult and often requires divers to use penetration-diving techniques (Figure A-14). On older timber structures, pile spacing is typically very tight, limiting diver access. Debris tends to collect and get entrapped under these structures, further restricting access.

Figure A-14. Underside Soffit of a relieving platform at extreme low tide. Source: Courtesy of Childs Engineering Corp., reproduced with permission. Note: Access and inspection of deck components can be difficult and often requires penetration diving techniques.

A.3.2.2 Lateral Stability. This structure type resists lateral loads from earth pressure and can be susceptible to collapse if early problems remain unmitigated. On timber structures, signs of lateral instability include the presence of nonbearing piles, overstressed batter piles, crushing of pile caps, and separation of pile cap splices. On all relieving platforms, monitor the alignment and settlement of the seawall or face of the structure for movement that would indicate impending failure. Document cracking in paved deck surfaces. Cracks running parallel to the structure's face indicate lateral movement.

A.3.2.3 Overdredging. Determine the design dredge depth at the face of the structure and check for compliance. Take soundings along the wall to determine that the mudline at the toe is at the proper elevation. A deeper dredge elevation compared to the design increases the effective column length of the piles and could overstress the piles. Overdredging also increases lateral loading on structures with sheet piles retaining fill. Timber-relieving platforms are particularly susceptible to failure as a result of overdredging.

A.3.2.4 Settlement and Sinkholes. Localized settlement and/or formation of sinkholes on the deck of a relieving platform occur when fill migrates through the deck. On timber structures, this is often caused by the failure of a deck plank or separation of two adjoining planks, allowing fill to be lost. On concrete decks, the fill could migrate through an expansion joint or possibly a weep hole. Sinkholes observed along the face of the walls may indicate washout from the face of the seawall or sheet piles. Excavation may be required to determine the cause of sinkholes.

A.4 WALLS AND RETAINING STRUCTURES

A.4.1 General

Bulkheads, sheet pile walls, and seawalls are all examples of wall and retaining structures found along shorelines. These structures create a barrier between land and water by providing an elevated surface of retained fill where a natural slope would otherwise exist. These structures are typically exposed to wind, wave, and ice forces.

Seawalls sometimes extend above grade to form elevated barriers against floods, waves, or tsunamis. The outside shape of seawalls may vary and can be designed to reflect or redirect the energy of the waves away from the shoreline.

Wall and retaining structures rely on varying components for strength and stability. In general, three main types of structural systems are used to retain fill and resist loads, but a combination of multiple systems can be used to provide strength for the structure. These include the following:

- Cantilevered systems,
- Anchored systems, and
- Gravity/mass systems.

Cantilevered walls rely on the flexural rigidity of the walls themselves to retain soil. Formed concrete walls or driven sheet pile walls are embedded sufficiently deep into the soil to achieve the required fixity to resist the retained loads.

Anchored walls rely on an anchorage system combined with the strength of the wall to maintain stability. The anchors carry lateral loads through tension by tying the sheets back to more competent soils, usually outside of the anticipated soil failure wedge. Anchor systems can consist of steel tie rods or strands connected to a deadman or drilled/grouted anchors set in earth or rock. Typically, only the wall face and heads of the anchors are exposed for visual inspections.

Gravity walls rely on self-weight to resist the overturning forces from the retained fill. Gravity walls can be constructed from stacked blocks, stone, masonry, or be constructed monolithically from concrete. Gravity walls require a component underlying substrate to resist the large bearing forces associated with both gravity loads and overturning loads. The substrate may consist of firm, competent native soils or of engineered substrates such as compacted fill or cribbing.

In all cases, proper drainage of water behind wall structures along waterfronts is important to prevent washout and buildup of hydrostatic pressures on the structure. Drainage is often achieved using weep holes at the face of the wall.

A.4.2 Common Types of Walls and Retaining Structures

A.4.2.1 Concrete Walls. The most common material used to build seawalls is concrete. Mass concrete gravity seawalls were traditionally the most common form of wall structures along waterfronts due to their performance against harsh environmental conditions (Figure A-15). Advancements in the industry have allowed cantilevered and anchored concrete wall systems to become more common today.

Figure A-15. Mass concrete seawall.
Source: Courtesy of Childs Engineering Corp., reproduced with permission.

A.4.2.2 Sheet Pile Walls. Sheet pile walls are constructed of interconnected vertical sheets driven into site soils. The sheets are backfilled on one side and can be either anchored or cantilevered. Sheets can be constructed from steel, concrete, fiber-reinforced polymer (FRP), vinyl, and timber (Figure A-16).

Figure A-16. Steel sheet pile bulkhead structure with a concrete cap beam. Source: Courtesy of Childs Engineering Corp., reproduced with permission.

A.4.2.3 Soldier Pile Walls. Soldier pile walls consist of intermittent soldier piles with horizontal panels spanning between each pile to retain fill. Soldier piles most commonly consist of steel H-piles driven at regular intervals along the bulkhead line with panels spanning between the flanges of the H-piles. This type of structure is often used as shoring for deep excavations.

Soldier pile walls can be constructed as a cantilevered system or as a combined system with tiebacks and walers.

A.4.2.4 Gravity Block Walls. Gravity block walls consist of stacked stone or masonry set in mortar to form seawalls, bulkheads, or quay walls. Cut granite, limestone, or precast concrete are common materials used for the blocks. Gravity block walls were historically built with cut stones founded on sunken timber mats or cribs. Prior to techniques allowing walls to accommodate greater freestanding heights, this type of structure was widely considered more appropriate in areas subject to severe wind, wave, ice, and other environmental conditions that were formerly considered too harsh for timber or steel bulkheads.

Utilizing its own weight to resist overturning, gravity block systems typically are founded directly upon the bearing substrate or on a mat or crib-type foundation, distributing the somewhat large dead load of the structure to the underlying bearing stratum. Gravity block walls are traditionally used as retaining structures to form seawalls or quay walls but can be used to form piers, jetties, marginal wharves, breakwaters, and numerous other applications in the marine environment. A typical gravity wall section is shown in Figure A-17.

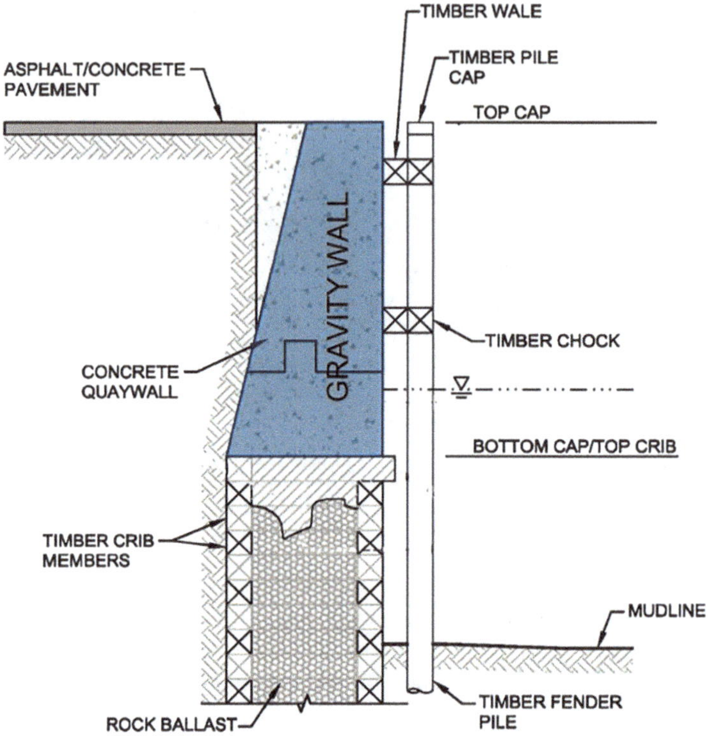

Figure A-17. Cross section of a typical gravity wall on timber cribbing. Source: Courtesy of Marine Solutions Inc., reproduced with permission.

A.4.3 Typical Components and Problem Areas

The inspection of retaining walls and bulkheads should be performed using a method similar to that of open-piled structures by inspecting as much of the wall as possible during the above-water inspection at low tide and performing an underwater inspection of the remainder of the wall. Observe the wall for misalignment of the overall structure and plumbness of individual elements making up the bulkhead or wall system. Document differential settlement between elements and displacement or damage by vessel impact or other means. The general observation of the wall should include an observation of the fill behind the wall, noting any signs of loss of fill, such as depressions or sinkholes.

Tables A-6 through A-8 summarize what to look for when inspecting the condition of these structures.

Table A-6. Sheet Pile Walls: Checklist for Inspections.

Section or part	What to look for	Comments
Sheet piling	Corrosion, interlock separation, impact damage, cracking, local overstressing, alignment, cathodic protection, coating, clear weep holes, rot	Identify component sizes, quantify deficiencies noted, and determine the cause of deficiencies
Anchorage system	Corrosion, cracking, local overstressing, displacement	Excavation may be required for the examination of anchors and walers on the fill side
Backfill	Sinkholes, settlement	Look for associated holes or damage on adjacent sheet piles
Geometry	Plumbness of face, bulges, overdredging, scour	Changes in structure geometry are indications of a failure of one or more structural systems
Concrete cap	Spalling, cracking, alignment	Cracking and misalignment indicate wall displacements associated with a failure of other structural components

Table A-7. Seawall Structures: Checklist for Inspections.

Section or part	What to look for	Comments
Seawall face	Erosion, spalling, cracking, missing blocks, cracked blocks	Assess the material condition for structural integrity; additional testing, such as concrete coring, may be warranted
Seawall top	Plumbness of face, bulges, misalignment, settlement	Identify causes of deficiencies Additional investigation, such as surveys, soil borings, or other testing, may be required. Monitoring over time may be required to determine whether the anomaly is active or stable
Seawall toe	Scour, undermining, armor stone displacement	The mudline in front of the seawall should be evaluated to ensure that design parameters are maintained; survey and document loss of material in front of the seawall
Backland or paved areas	Sinkholes, settlement, drainage	The deck surface behind a seawall is susceptible to loss of fill through openings in the wall or erosion of soil by overtopping water; drains and scuppers should be inspected to make sure they are able to vent floodwater
Weep holes	Clogging	Weep holes are placed to relieve hydrostatic pressure on the wall and should be observed to make sure they are free draining

Table A-8. Gravity Block Walls: Checklist for Inspections.

Section or part	What to look for	Comments
Wall system	Sweeping, bowing, leaning, misalignment, settlement, localized collapse	If identified, the cause should be determined, and the inspection should specifically look for secondary damage, such as cracking and loss of blocks. Consider the landward area affected by the loss of lateral support; a survey and periodic monitoring may be required to determine whether wall movement is active
Backfill	Depressions, sink holes, surface tensile cracking parallel to wall, joint separation, drainage conditions	Causes of backfill problems may not be evident by visual inspections of the wall; excavations, geotechnical borings, or ground-penetrating radar may be utilized
Blocks	Weathering, cracks, common precast concrete defects, erosion, displacement, reinforcement corrosion, interlock shear	Cracking and displacement of blocks are often the secondary result of other problems; conditions that may affect block interlock or friction should be considered
Mortar	Loss, degradation, shrinkage, crushing, leakage	The results of settlement and differential stresses placed on blocks from mortar loss should be considere

(*Continued*)

Table A-8. (*Continued*).

Section or part	What to look for	Comments
Drains	Obstructions, evidence of activity, direction, adequacy	Drain conditions may be a contributing factor to other problems
Joints	Separation, displacement, leakage, filler condition, vegetation	Water staining at joints may indicate the extent of hydrostatic levels behind the wall; loss of fill through joints may be the cause or contributor to backfill problems
Mudline	Backfill deposits, scour, undermining, heaving	The depths and general contours of the mudline should be considered to be at least 1× the wall height outboard from the wall. The actual or likely original design elevation of the mudline at the toe of the wall should be compared to current elevations
Foundation	Integrity of members, loose or missing members, voids, loss of ballast, settlement, bearing contact	Both vertical and lateral stability and load-bearing capacity should be considered; degradation of foundation members should also be considered

A.4.3.1 Access. Many seawalls are located in highly exposed locations, subject to significant wind, current, and wave action. Underwater inspection of these structures can be extremely hazardous, requiring specialized diving techniques.

A.4.3.2 Weathering and Damage at Face. The exposed face of a seawall is typically a flat or curved surface. Concrete seawalls are susceptible to erosion and corrosion-related spalling.

Steel components of bulkheads should be investigated for section loss caused by corrosion. Measure the steel thickness using nondestructive methods, such as ultrasonic testing, and implement a procedure that will result in a representative view of the steel losses caused by corrosion. Pay special attention to high-stress areas such as tie rod and wale connections. Key inspection elevations are within the splash and tidal zones. These areas typically experience the highest rate of corrosion metal loss, especially in seawater. Also, check the condition of coatings and cathodic protection systems (Figure A-18).

The general appearance of the blocks in gravity block walls should be considered to indicate how well they have resisted weathering. Quality stone and concrete will retain their sharp edges and corners. In some cases, stone tool marks may be present.

Erosion of the blocks near the water surface, especially when subjected to heavy surface water currents and freeze–thaw cycles, is common (Figure A-19). This condition can cause significant section loss at the waterline.

Cracks through blocks should be observed for weathering to evaluate whether the cracking is caused by a recent condition or is ongoing. Vertical or diagonal cracks forming through blocks are often caused by differential settlement, changes in loading or load distribution, or failure of lower masonry courses (Figure A-20). For interlocking precast concrete blocks, visible cracks may be evidence of shear of the block's key or interlock, which are typically not visible.

Figure A-18. Severely deteriorated steel sheet pile bulkhead.
Source: Courtesy of Jacobs, reproduced with permission.

Figure A-19. Erosion of masonry blocks at the water surface.
Source: Courtesy of Marine Solutions Inc., reproduced with permission.

Figure A-20. Cracking through masonry blocks.
Source: Courtesy of Marine Solutions Inc., reproduced with permission.

A.4.3.3 Seawall Toe. The toe of the seawall is susceptible to wave action and moving water and should be observed for the effects of scour and undermining. Take soundings along the wall to compare the mudline elevation at the toe to the design mudline elevation.

Armor stone, if present, should be observed for displacement. For the armor stone to be effective, it needs to be maintained in position. If settlement is present because of scour or if wave forces have moved the stone, document the locations. The general size and type of stone should also be determined to verify that unplanned deposits have not replaced the planned protection.

A.4.3.4 Foundations. Pile foundations for seawalls should not be exposed. If scouring and undermining expose the piles, take measurements to monitor further erosion.

Foundations for block walls typically consist of a timber mat or timber crib type that may be filled with stone ballast. However, some gravity block walls may not have a separate foundation, only having a base masonry course laid directly on a leveled bearing stratum. Exposure of the foundation of a gravity block wall beneath the waterline is not uncommon. Timber mats or cribbing should be observed for common types of timber deterioration but should also be specifically observed for loose or missing members, direct bearing contact between the mat or crib and the lowest block course, and loss or settlement of ballast material. Figure A-21 shows a collapsed masonry block wall where the foundation no longer provides sufficient support.

Figure A-21. Collapse in masonry block wall due to insufficient foundation support. Source: Courtesy of Marine Solutions Inc., reproduced with permission.

A.4.3.5 Washout and Drainage. Settlement and/or the development of sinkholes directly behind a bulkhead indicate loss of fill through the structure. This could be caused by corrosion holes in the sheets, separation of the sheets, or failure of the drainage system behind the wall.

Look for signs of settlement and sinkholes behind the seawall. This is evidence of loss of expansion/construction joint fillers or broken/displaced drainage piping, which allows the fill to wash away (Figure A-22).

Figure A-22. Hole in pavement due to erosion behind sheet pile bulkhead. Source: Courtesy of Moffatt & Nichol Inc., reproduced with permission.

A.4.3.6 Anchorage System. The anchorage system for a bulkhead is critical to its function. If exposed, the tie rods and wales should be examined to determine loss of steel, distress such as localized buckling, and deformation. Hidden components, such as tie rods, interior wales, and deadman anchors, require excavation to examine. Bolts attaching interior wales should be closely examined.

A.4.3.7 Geometry and Alignment. Observations should be made of the structure's plumbness and alignment. Variations in the geometry of the structure can indicate failure of one or more vital components. Out-of-plumb sheet piles may be caused by failure at the toe of the wall or failure of the anchorage system, depending on the direction of the movement. Isolated bulges in the bulkhead can indicate local failure of one or more tie rods.

For gravity block walls, loss of foundation support, loss of mortar, changes in load distribution, and other factors can lead to localized displacement or collapse of blocks. In cold regions, damage to blocks by ice pile-ups or movement of masonry units by surface ice may also be a concern. Furthermore, vegetation can grow from ledges and voids created from mortar or block loss when soil collects in these areas. Vegetation with significant root structures can be highly detrimental, inducing stresses and even causing block displacement.

Because of the nature of gravity structures, problems with global stability, differential settlement, and foundation support are some of the most common causes of failure. These problems are often first manifested as vertical (bowing) or horizontal (sweeping) bulging, misalignment, and/or a change in the plumbness (leaning) of the wall, and they can occur gradually over time or suddenly at any point in the structure's life. Figure A-23 shows a masonry block wall that has settled and is bowing.

Figure A-23. Settlement and bowing of masonry block seawall.
Source: Courtesy of Marine Solutions Inc., reproduced with permission.

The wall's elevation, alignment, and plumbness should be specifically monitored during periodic inspections such that small movements can be monitored. The landward area supported by the wall should be monitored for settlement, depressions, and sinkholes. Lateral movement can often be observed first by increases in the width of the joint between the upland surface and cap or top course of the wall. Often, tensile cracking parallel to the wall will also develop in the pavement or ground surface.

A.4.3.8 Concrete Cap Defects. Cast-in-place concrete caps are often used to maintain alignment at the top of a bulkhead. The concrete should be checked for corrosion-related spalling. Also, check the concrete cap for cracking and misalignment at cracks that may be an early indication of settlement, lateral failure, anchor system failure, or impact.

A.4.3.9 Overdredging. Bulkheads are highly sensitive to overdredging at the toe of the wall. Loss of foundation support at the toe of the structure from overdredging or excessive scour can initiate instability and excessive stresses in the wall, resulting in bulging, misalignment in sheet piling, or failure of the overall bulkhead. Soundings should be taken along the wall to compare the mudline elevation at the toe to the design mudline elevation.

A.4.3.10 Impact Damage. Impact damage is common in bulkheads because they are used on the waterfront as berthing structures for ships. This may result in cracking of the concrete cap or dented, deformed, or broken sheets.

A.4.3.11 Hydrostatic Relief Ports. Some bulkheads are designed with hydrostatic relief ports or weep holes. These ports relieve water pressure from behind the sheets and reduce the overall pressure on the wall. The ports should be observed to ensure that they are not clogged and allow water to freely flow through them.

Filter fabric is typically used behind these weep holes to prevent washout of the retained fill. Look for sinkholes in the working surface of the bulkhead above weep holes.

A.4.3.12 Loss of Mortar. Mortar loss between units can produce minor settlements and displacements, affecting the weight distribution of the block matrix. Mortar is usually less durable than the stone or concrete block material and, as such, requires maintenance. Figure A-24 shows the loss of mortar between blocks in a masonry block wall.

Figure A-24. Loss of mortar between masonry block units.
Source: Courtesy of Marine Solutions Inc., reproduced with permission.

A.5 REVETMENTS

A.5.1 General

Revetments are slope protection systems typically constructed of stone, concrete, or other materials that will provide sufficient armoring for the slopes to prevent erosion from current or wave action (Figure A-25). A properly constructed revetment will consist of an outer armor layer, filter layer(s), and toe protection. The visible armor layer can be random stones, concrete rubble, or a more well-ordered array of interlocking structural elements. A properly constructed armor layer will have at least two full layers of rock. The filter layer, mostly not visible, will consist of smaller rocks, filter fabric, or both. The filter layer provides drainage and retention of the underlying soil. Toe protection provides stability against undermining at the bottom of the structure.

The armor layer of revetments can show damage as an uneven level of the revetment. This can be caused by missing or displaced rocks, where some filter layers can be visible underneath. An uneven revetment layer can also indicate subsidence of the underlying soil, often caused by improper or missing filter layers under the armor layer. Also, watch for other signs of settlement or instability, such as slip failure or erosion at the toe of the slope.

Figure A-25. Revetment armored with a stone riprap.
Source: Courtesy of Simpson Gumpertz & Heger Inc., reproduced with permission.

Damage to revetment can happen from scour or wave action from the waterside or from drainage of surface water from the top. It is important that an inspection aims to identify the source of the damage.

A.5.2 Typical Components and Problem Areas

Table A-9 summarizes what to look for when inspecting the condition of revetments.

Table A-9. Revetments: Checklist for Inspections.

Section or part	What to look for	Comments
Armor layer	Uneven layers, missing or displaced rocks, signs of subsidence	
Filter layer	If exposed, look for signs of subsidence, washout, or failure of the filter material	Filter layer should typically not be visible
Toe	Scour, washout, undermining, displaced or missing stone	

A.6 CAISSONS, COFFERDAMS, AND CELLULAR STRUCTURES

A.6.1 General

Caissons, cofferdams, and cellular structures are used to form wharves, piers, seawalls, quays, and freestanding mooring dolphins. These types of structures are also commonly encountered as exposed foundation members supporting the deck of a pier or wharf.

A caisson is a retaining structure constructed of reinforced concrete or a solid steel plate lowered or driven to form an enclosure filled with an engineered material. Caissons can be formed in many shapes but are often referred to as two main types: (1) box caissons, which are prefabricated with sides and a bottom; and (2) open caissons, which do not have a bottom.

Cofferdams form a water barrier most often used to provide temporary dry work areas. However, the reinforced concrete, steel plate, or steel sheet piles used to form the barrier may be left in place and can be integrated as a permanent member.

Cellular structures are usually formed by interlocking steel sheet piles, creating independent or connected closed or open cells (Figure A-26). Differing from a bulkhead, a cellular structure gains its strength and stability from the shear strength and weight of the contained infill or backfill material and does not depend completely on wall embedment or an anchor system. Closed cellular systems depend on their weight to resist applied loads, functioning as gravity structures. Open cellular systems

Figure A-26. Typical steel sheet pile cellular structure.
Source: Courtesy of Marine Solutions Inc., reproduced with permission.

have diaphragm walls extending landward, utilizing the lateral pressure of the wall backfill to anchor the cell.

A.6.2 Typical Components and Problem Areas

Many of the components and common problem areas associated with concrete and steel sheet pile bulkheads are similar for caissons, cofferdams, and cellular structures. However, a few considerations are specific to these types of elements. Some of the typical problem areas are described in the following sections. Table A-10 summarizes what to look for when inspecting the condition of these structures.

Table A-10. Caissons, Cofferdams, and Cellular Structures: Checklist for Inspections.

Section or part	What to look for	Comments
Foundation	Misalignment, bowing, plumbness	Settlement and lateral movement should be considered by observing and monitoring the overall structure's position
Steel sheet piling or plates	Corrosion, deformation, interlock separation, splitting, cracking, impacts, dents	Identification of the nominal steel section and thickness, interlock separation, and corrosion or damage to interlock is critical
Deck or backfill	Depressions, sink holes, cracking parallel to wall, joint separation, drainage conditions	Causes of backfill problems may not be evident by visual inspections of the wall; excavations, geotechnical borings, or ground-penetrating radar may be utilized
Cell cap	Settlement, cracking, drainage, edge spalling	If a hatch in the cap exists, it should be opened to observe fill level and composition; cap cracking can indicate settlement or loss of infill Edge spalling and deterioration can compromise the cap connection to the supporting wall

(Continued)

Table A-10. (*Continued*).

Section or part	What to look for	Comments
Reinforced concrete walls	Cracks, common precast concrete defects, reinforcement corrosion, joint displacement	Caisson or cofferdam walls may exhibit evidence of differential settlement
Mudline	Fill loss, scour, undermining, heaving	The depths and general contours of the mudline should be measured considering lateral stability if embedment is decreased

A.6.2.1 Settlement. Gravity structures not founded directly on competent bedrock are particularly susceptible to settlement and foundation failures, as they inherently apply a large dead load onto the bearing stratum. Settlement of the infill or backfill material can also occur independently of the foundation soils, especially if nonengineered fill is used (Figure A-27). As such, elevations and the structure's verticality, or plumbness, should be checked with each inspection. The alignment and configuration of the structure should also be recorded for monitoring through successive inspections. Differential settlements between the cell infill and concrete or steel walls should also be considered.

Figure A-27. Collapse of the concrete cap due to settlement of infill material. Source: Courtesy of Marine Solutions Inc., reproduced with permission.

A.6.2.2 Scour and Lateral Stability. Deepening of the mudline at the structure from scour, dredging, or propeller wash can undermine these types of structures; however, this is uncommon as the loss of lateral support provided by embedment usually leads to leaning (Figure A-28) or failure before the structure is undermined. Therefore, the design embedment, intended design mudline depth, and critical mudline elevations should be determined and evaluated during an inspection.

Figure A-28. Leaning of a cellular structure due to scour and slope instability. Source: Courtesy of Marine Solutions Inc., reproduced with permission.

A.6.2.3 Steel Sheet Pile Interlock Separation. Cellular structures depend on the interlock of the steel sheet piles forming the skin of the system to carry the lateral pressures applied by the infill or backfill material. Failure of the interlocks reduces the load-bearing capacity of the entire cell system by reducing the available tensile ring strength. If fill material is released, the load-bearing capacity is further reduced because of the reduction in shear capacity of the confined fill (Figure A-29). Failure of the interlocks is often the result of corrosion, section loss, impacts, increases in loading or fill pressure, or some combination of these occurrences. Often, once the interlock is separated, the length of separation will gradually increase over time, commonly referred to as *unzipping*. Tensile strength testing can be performed on coupon samples of steel sheet pile interlocks after bracing the joint where the sample is taken.

Figure A-29. Broken and separated sheet pile interlock resulting in the loss of infill material.
Source: Courtesy of Marine Solutions Inc., reproduced with permission.

A.6.2.4 Corrosion. Corrosion of steel sheet piling can cause or contribute to several failure mechanisms (Figure A-30). It is important to recognize that if corrosion occurs on the seaward side of the sheet piling, some corrosion is likely also occurring on the back of the sheet piling, although corrosion on the back may be less because of lower oxygen levels. Steel ultrasonic thickness measurements are critical to the evaluation of steel components. For steel caisson, cofferdam, and cellular structures, thickness readings should be taken in a procedure that allows mapping of the corrosion by elevation and area such that the risk of interlock separation, splitting, or cracking due to impact or yielding can be determined.

Figure A-30. Underwater photograph showing a hole in a cellular structure caused by corrosion.
Source: Courtesy of Marine Solutions Inc., reproduced with permission.

A.6.2.5 Concrete Cap Deterioration. Reinforced concrete caps covering closed circular cells are often designed to be supported by the cell walls but may be supported by the cell infill as well. Cracking parallel to the cell walls may indicate shear failure of the concrete at the walls. Common nondestructive and destructive testing techniques for reinforced concrete can provide valuable information to evaluate the condition and thickness of reinforced concrete caps. Coring and geotechnical borings may be necessary to determine the thickness of cell caps and the level and composition of cell infill.

A.7 FLOATING STRUCTURES

A.7.1 General

Floating structures are defined as any structure intended to remain floating during its service life. They can be moored or self-propelled. Examples of floating structures include large storage containers, dry docks, ferry slip docks, guide walls for navigation locks, breakwaters, bridges, and piers. Typically, the material used to construct these structures is composed of concrete or steel.

Floating bridges, ferry slip docks, breakwaters, and guide walls for navigation locks typically consist of concrete pontoons at the water surface. Most concrete pontoons are designed and constructed as a series of long, rectangular barges, with interior egg-crate bulkheads oriented both longitudinally and transversely. The pontoons provide the superstructure for the bridge (or ferry slip dock) or serve to carry a bridge (or ferry slip dock) superstructure. In the case of the guide walls for navigation, the pontoons serve as the entry channel wall(s). The position of the pontoons is either maintained by anchor cables that inhibit movement by working against the buoyancy of the hollow design of the pontoons or are held in place by open-pile clusters with a concrete cap.

Floating piers and docks typically consist of a deck that comprises one or more concrete or steel pontoons; however, timber, aluminum, and polymer composite materials have been used as the pontoon material. Steel pontoons typically require double-walled construction throughout the structure. Pontoons are anchored to the mudline and connected to the shore via bridges or ramps. Typically, the spud piles of floating piers are made of steel, concrete, or timber, and the connection hardware generally comprises hot-dip galvanized steel, marine-grade stainless steel, or aluminum.

The anchor cables of floating structures are typically structural strands made up of small-diameter, stranded steel wires or large-diameter chains. An anchor cable may either consist of a single run between the pontoon and the anchor or two runs resulting from a cable that is looped through the anchor. The cables are either attached to the anchors by means of a cable socket along with various pins, eyebars, and pin plates or threaded through the anchor. The cables enter the pontoons through ports where cable tensioners are typically located.

A wide variety of anchor designs are used for floating structures, including drag-embedment anchors, pile anchors, deadweight anchors, suction pile anchors, and direct-embedment anchors. A comprehensive description of these anchor types can be found in Section A.9.

A.7.2 Typical Components and Problem Areas

Tables A-11 and A-12 summarize components of floating structures that are typically inspected and the typical items of concern.

Table A-11. Floating Bridges: Checklist for Inspections.

Component	Section or part	What to look for	Comments
Pontoon	Submerged surfaces	Cracks, spalls, and loss of section, waterlogged filler material between steel double walls	Cracks or deep section loss may allow water infiltration
	Joints	Torn, loose, or bulging rubber membrane, Exposed and/or deteriorated grout, Excessive pontoon misalignment	
Anchor cable	Pontoon port	Misalignment and cable abrasion	
Anchor cable	Cable	Coating condition, corrosion, wire section loss, broken, and/or braided wires, and potential sources of cable abrasion	Exterior wire breaks related to stress may suggest comparable numbers of interior broken wires; stress breaks may also indicate the end of the cable's useful life

(*Continued*)

Table A-11. (*Continued*).

Component	Section or part	What to look for	Comments
	Anchor attachment assembly	Corrosion, misalignment, looseness, and cable abrasion or strain	
Anchor	Anchor assembly	Misalignment or movement instability, undermining, inadequate embedment, or ballast quantity	

Table A-12. Floating Piers: Checklist for Inspections.

Component	Section or part	What to look for	Comments
Pontoon	Submerged surfaces	Material deterioration, Waterlogged Styrofoam filler	Cracks or deep section loss may allow water infiltration
	Joints	Damaged and/or deteriorated joint filler and connections, Excessive pontoon misalignment	
Spud pile	Submerged surfaces	Ordinary pile considerations, Wear or abrasion related to misalignment	

(*Continued*)

Table A-12. (*Continued*).

Component	Section or part	What to look for	Comments
Tension line	Pontoon or anchor attachments	Corrosion, misalignment, looseness, and line abrasion or strain	
	Cable or chain	Corrosion, cathodic protection anode consumption (chains), breaks and/or abrasion, and potential sources of abrasion	Catenary chains may be subject to wear at mudline
Anchor	Anchor assembly	Misalignment or movement, instability, inadequate embedment	

A.7.2.1 Floating Concrete Pontoons. Floating pontoons are normally constructed of segmental precast, prestressed concrete, utilizing either standard-weight or lightweight concrete. The pontoon's interior and exterior surfaces need to be inspected for typical concrete deficiencies. Of particular concern is any cracking or deep section loss that could allow water to infiltrate the concrete and, thus, the inner compartment of the pontoon. Where a series of pontoons is employed, the rubber membranes and/or grout that are typically used in the joints between pontoons should be inspected for indications of overstressing, embrittlement, and general damage. Pontoon alignment across the joints should also be examined for indications of excessive differential movement.

Inspect the access hatch or point of entry to the pontoon for degradation of the sealing material and degradation of the surrounding concrete or steel material. When in the closed position, the hatch door should be watertight. Check for gaps around the edges. If the pontoon has multiple compartments, the bulkhead portal doors should be inspected in a manner similar to the access hatch. These types of inspections are most likely confined space inspections and require proper training, specialized equipment, and compliance with local and federal regulations. Refer to USACE EM 385-1-1 (USACE 2024), Section 34.B, or 29 CFR 1910.146 (OSHA 2011) for guidance on confined space inspections.

In some cases, the pontoons may have utility conduits, plumbing, duct banks, and/or cathodic protection systems. The overall condition of these systems should be inspected to identify general deterioration and to determine whether the utility conduits and plumbing are broken or detached. Beyond the general condition inspection of these systems, a specialized inspection may be required.

A.7.2.2 Floating Pontoon Anchor Cables and Chains. Anchoring cables should be inspected for the condition of protective coatings, extent of corrosion, and amount of individual wire section loss. In a strong current flow environment (2 to 3 + knots), vibration of the anchor cable can occur and further cause cyclic dynamic oscillations that can lead to fatigue, both in the cable and at the connection hardware. Cable misalignment and wire abrasion should also be checked at the pontoon ports. At the anchors, the attachment assemblies should be examined for any deterioration, looseness, or misalignment and for any adverse effects on the anchor cable.

Identify broken individual wires along the cable. When possible, determine the source of wire breakage that is abrasion or stress-related because exterior stress breaks typically suggest a comparable number of interior broken wires and also indicate that the end of the cable's useful life is approaching. The inspection of each anchor cable should also identify any potential sources of cable abrasion such as items hung on the cable (netting, anchor ropes, etc.) or obstructions at the mudline (see Section A.9 for additional inspection considerations).

Anchor chains should be inspected routinely for corrosion and section loss of the individual links, the amount of wear between adjoining links, the possible stretch of the chain over five continuous links, and alignment with the anchoring system. The anchor chains should be routinely inspected at critical locations throughout the chain assembly, including the riser, catenary, dip, and ground sections (see Section A.9 for additional inspection considerations). In some cases, the anchor chain will have cathodic protection throughout the system and should be inspected as described in Section A.12.

A.7.2.3 Floating Steel Pontoons. The steel portions of the pontoons should be inspected for typical material deficiencies. Of particular importance is any cracking, holes, or deep section loss that may allow water to infiltrate the interior cavities of a pontoon. In some cases, the steel pontoons will have double walls along the exterior portion of the hull. Therefore, both interior and exterior examinations of the walls are necessary. Where Styrofoam filler is used in steel pontoons, any exposed filler should be examined for material integrity and waterlogging. The joints between pontoons and related connections and/or fillers should be inspected for deterioration, damage, missing or deficient items, and other considerations, as described in Section A.7.2, where applicable.

A.7.2.4 Floating Pier Spud Piles or Tension Lines. Floating pier spud piles should be examined for the conditions associated with piles, as described in Section A.2. In addition, the piles should be checked for any excessive wear or abrasion resulting from misalignment. The top of the spud piles should be checked for a protective cap or coating that minimizes deterioration and water accumulation. Where tension lines are involved, the lines and attachment assemblies at either end should be examined for deterioration, breaks, and abrasion.

A.7.2.5 Floating Dry Docks. This section covers the basic components of a floating dry dock system. For comprehensive inspection techniques related to floating dry docks, refer to ASCE 77-22 (ASCE 2023).

A floating dry dock consists of a pontoon with stabilizing wing wall(s) and mechanical and electrical equipment to permit the controlled flooding and emptying of the ballast tanks. A floating dry dock is most conveniently inspected by having it dry docked as a unit and by performing the inspection in the dry. Self-docking may be an alternative for some floating dry docks, whereby portions of the structure are removed and dry docked on the remaining structure. In some cases, access to submerged portions of the pontoon is possible by creating a controlled list or pitch of the dry dock through the flooding of selected ballast compartments.

Those portions of the floating dry dock that may require underwater inspection include the pontoon hull and water intakes.

A.8 MOORING HARDWARE AND FENDER SYSTEMS

A.8.1 General

Mooring hardware and fender systems are the primary components in the berthing of vessels at any waterfront facility. The fender system is vital to protecting the waterfront structure and the vessel. The fender system must be able to resist strong impacts during initial vessel contact in addition to continuous loading from environmental conditions such as wind, current, and waves. Once a vessel is alongside the structure, the vessel's mooring lines are connected to the mooring hardware that is mounted to the structure. Typically, onboard winches adjust the mooring lines to position the vessel. As with fender systems, mooring hardware must be able to withstand constant environmental forces in addition to strong gusts increasing tension on individual mooring hardware.

A.8.2 Typical Components and Problem Areas

Mooring hardware and fender systems are no different from typical structural components, with common problem areas stemming from fatigue (cyclic loading), sudden high impact (overloading), and prolonged environmental exposure. Tables A-13 and A-14 summarize what to look for when inspecting the condition of mooring and fender system components, respectively.

Table A-13. Mooring Hardware: Checklist for Inspections.

Section or part	What to look for	Comments
Bollards, bitts, and cleats	Coating loss, corrosion, abrasion, displacement, cracking	Check high-wear areas, such as the base of the hardware, where the mooring lines sit. These areas are often the first to experience coating loss and corrosion
Fasteners	Corrosion, deformity	Check both above and below deck and if fasteners can be accessed from both sides, especially for timber structures
Supporting structure	Cracks, spalls, displacement, fungal decay, splits, corrosion	Assess concrete decks for cracking or spalling stemming from anchor bolts. Assess timber decks for fungal decay and splits. Assess steel decks for corrosions. Check below the deck if fasteners go through the deck, especially for timber structures

Table A-14. Fender Systems: Checklist for Inspections.

Section or part	What to look for	Comments
Absorption element	Deformation, cracking, abrasion, tears, sagging of unit fenders (e.g., rubber cell or cone fenders)	
Fender panels	Deformation, deterioration, debris accumulation, blocked drainage, loose, or missing hardware	Connections to fenders or chains are vulnerable areas

(Continued)

Table A-14. (*Continued*).

Section or part	What to look for	Comments
Supporting piles	Buckling, impact damage, broken sections, material-specific deterioration	Not found on all fender systems
Rub surface	Abrasions, impact damage, protruding bolts, missing elements	
Hardware	Corrosion and deterioration of chains and brackets, insufficient tension in weight chains holding fender panels, missing hardware	

A.8.2.1 Mooring Hardware. Mooring hardware is mainly composed of bollards, bitts, cleats, and hooks, which have various functions on the facility to secure vessel mooring lines. A general overall inspection of the mooring system should be performed, noting the location and type of mooring components; typical use of the system; and obvious deterioration, damage, or missing components.

A.8.2.1.1 Bollards, Bitts, and Cleats. Most bollards, bitts, and cleats are of steel construction and are subject to the same deterioration causes and symptoms of steel structural elements, as shown in Figure A-31. Typically, the mooring hardware is composed of cast steel or iron and is secured to the structure with bolts that either are anchored into or bolted through the structure. The actual cast fitting will have various wall thicknesses according to its capacity. As noted in Table A-13, problems arise from abrasion of the mooring lines, as this can remove the coating and leave the areas open to the marine atmosphere, which ultimately leads to corrosion. Corrosion will lead to loss of section and, therefore, a reduction in the strength of the fitting. In addition, the fitting can be subject to cracking and deformation. Hardware should be inspected for overall condition. Welds should be inspected for cracks. Base plates should be inspected for bending or uplift, indicating a previous overload. Hardware coatings and surfaces should be examined for damage and/or wear, especially from mooring lines. Surfaces should be smooth and free of defects that could catch or tear mooring lines.

Figure A-31. Typical mooring hardware: (upper left) a single bitt bollard with ears, (upper right) a cleat, (lower left) an angled double-bit bollard, (lower right) a double-, low-bit bollard.
Source: Courtesy of Jacobs, reproduced with permission.

A.8.2.1.2 Quick-Release Hooks. Quick-release hooks (QRHs), as shown in Figure A-32, require a high degree of maintenance. Detailed maintenance efforts are required to maintain the operational life of these units. However, a visual inspection of the hook can determine whether further inspection is required. Hook coatings and surfaces should be examined for damage and/or wear, especially defects that could catch or tear mooring lines. All exterior surfaces should be coated, including base plates and anchor bolts. The inspection should verify that unloaded hooks freely rotate and that the release mechanism is functional. Check for evidence of excessive lubricant loss, especially at the pins, such as streaking or pooling on the hardware on the deck below. Many QRH units are equipped with motorized capstans. Capstans should also be inspected for overall condition. Signs of distress may require further review by a specialist experienced in these types of units.

Figure A-32. (Left) A double quick-release mooring hook with a manual quick-release mechanism and cleat, (right) a quadruple quick-release mooring hook with electric capstan and tension monitoring equipment.
Source: Courtesy of Simpson Gumpertz & Heger Inc., reproduced with permission.

A.8.2.1.3 Anchorage. Anchorage conditions of all mooring hardware should be carefully inspected and documented. Maintaining a continuous load path of the hardware is critical for safe use of these items. Hardware should not be loose-fitting. Common signs of anchorage damage include corrosion (Figure A-33), missing/loose bolts or nuts, bolts that have been partially pulled through timber structural elements, or cracking of the grout pad or concrete deck supporting the hardware. When observations such as these are noted, attempt to document the cause of the damage. Recessed bolt holes should be filled with mastic, silicone, or a similar material to prevent water ponding and bolt corrosion.

Figure A-33. Moderate anchorage corrosion of a bollard.
Source: Courtesy of Simpson Gumpertz & Heger Inc., reproduced with permission.

A.8.2.2 Fender Systems. Fender systems are typically composed of two to three critical elements:

- Absorption element absorbs the energy from the impact of a vessel and minimizes the forces transmitted directly to the structure.
- Supporting elements either support the absorption element or provide a contact/reaction surface—although not all systems have these.
- Facing or rub surface provides the area of direct contact with the vessel. Reducing friction between the fender system and the vessel is typically desirable, preventing unintended loading or damage to the absorption element. The facing typically consists of a low-friction facing such as ultra-high-molecular weight polyethylene (UHMW-PE) or timber, along with framing to secure the facing to the supporting piles or absorption elements.

A.8.2.2.1 Absorption Elements. Two main types of absorption elements include elastomeric fender units that are normally directly attached to a structure and pneumatic or foam-filled fender units that float in front of the structure.

A.8.2.2.2 Elastomeric Fender Units. Elastomeric fender units come in many shapes and sizes and are typically located between the structure and a rub surface, as shown in Figure A-34. Rubber compounds are subject to physical damage, such as abrasion, fatigue, prolonged stretching, overload, deterioration from heat aging, UV damage, and chemical degradation. Rubber reacts especially poorly to contact with chemicals or petroleum products, which can cause hardening and cracking.

Figure A-34. (Left) Leg/element-type elastomeric fender unit, (right) a cone-type unit (right).
Source: (Left) Courtesy of Childs Engineering Corp., reproduced with permission; (right) Courtesy of Jacobs, reproduced with permission.

Fender units should be checked for rubber deterioration in the form of cracking, hardening, abrasion, tears, and compression set. If fender panels are not properly supported, fender cracking may occur. This can be either circumferential around the fender at the connection with the supporting structure or longitudinal on top of the unit because of excessive tension associated with self-weight and/or the fender panel. The fender unit will often sag, and the panel will no longer be plumb. Radial cracking on the sides of the unit may indicate excessive angular loading. Look for dirt, sand, and debris accumulation.

A.8.2.2.3 Pneumatic or Foam-Filled Fender Units. Pneumatic and foam-filled fenders are located on the seaward side of pile-supported fender systems or connected directly to the side of the structure. They are typically cylindrical-shaped objects with large eyelets on either end that allow a chain to be attached to secure the fender to the fender system structure, as shown in Figure A-35.

As discussed in Table A-14, pneumatic and foam-filled fenders should be visually checked for overall signs of obvious deterioration and damage, such as disfiguration, cracking, tears, sagging, or permanent distortion. Urethane covers should be checked for signs of damage like tears, gouging, ribs, or holes. Fittings at the ends of these units should be checked for splitting or separation from the fender and for corrosion, deformation, or abrasion damage.

Figure A-35. (Left) Foam-filled fender unit with concrete backing piles, (right) similar units with timber and steel backing.
Source: Courtesy of Simpson Gumpertz & Heger Inc., reproduced with permission.

A.8.2.2.4 Steel Frames and Structural Supporting Members. Rubber fenders often support a steel frame, a fender pile system, or some other structural system that will be the contact point for the berthing vessels. These structural systems should be checked for deterioration on the basis of material composition.

Steel fender panels should be checked for corrosion, cracks (especially at the connections to the elastomeric fender units), deformation, and coating conditions (Figure A-36). Check all connection hardware (bolts, nuts, washers) for intactness and corrosion. Also, check whether all connection hardware is tight-fitting and in place. Some fender panel designs are prone to collecting debris and water, accelerating corrosion. Inspect these areas for signs of localized deterioration damage.

Timber piles sometimes serve only as the supporting element and have separate absorption and facing elements. In older structures, timber piles often supplement all three elements of a fender system – the absorption element, support element, and rub surface.

Timber pile systems should be checked for rot, checking, splitting, abrasion, displacement, and structural failure in a manner similar to structural timber piles as described in Section A.2.2.1. Also, check connection hardware for corrosion and damage (Figure A-37).

Concrete fender pile systems should be checked for spalling, cracking, displacement, and corrosion in a manner similar to structural piles, as described in Section A.2.3.1. Like all fender piles, connection hardware should also be checked for deterioration and damage.

Figure A-36. (Left) Moderately corroded steel fender framing, (center) typical chains supporting the fender framing and panel, (right) UHMW-PE fender facing typical wear.
Source: (Left) Courtesy of Simpson Gumpertz & Heger Inc., reproduced with permission; (center, right) Courtesy of Childs Engineering Corp., reproduced with permission.

Figure A-37. (Top left) Steel H-piles supporting a timber fender framing and rub strip, (top right) timber fender pile system, (bottom left) clusters of timber fender piles banded together with wire rope, (bottom right) polymer composite fender pile system with a broken wale.
Source: Courtesy of Childs Engineering Corp., reproduced with permission.

A.8.2.2.5 Fender Chains and Fittings. As a result of constant movement, chain links may wear considerably at the contact points between them, decreasing the chain's overall strength. Chains and fittings should be checked for corrosion, wear, permanent deformation, and signs of overloading. Check all connections to fenders, panels, and supporting structures for signs of overloading (see Section A.20 for a more comprehensive discussion of chain inspection).

For elastomeric fender systems with fender panels, weight chains, tension chains, and shear chains need to be properly tensioned to prevent unintended damage or loading to the fender units. Weight chains should effectively prevent sag of the fender system, tension chains should be able to prevent rotation from low berthings, and shear chains should prevent drift from lateral loading.

A.8.2.2.6 Synthetic Facing Materials. Fender panels and fender piles might be covered with a synthetic facing material to provide a low-friction surface against which the vessel hull can bear. This material is usually UHMW-PE or other low-friction material. Synthetic facing materials should be checked for normal wear and tear, chemical deterioration, and displacement. Check connecting bolts for corrosion and proper recess in the facing material to avoid damage to the vessel hulls from the bolts. Older installations should also be checked for deterioration associated with UV exposure, such as fading, cracking, or embrittlement of the material.

A.9 MOORING BUOY SYSTEMS

A.9.1 General

Mooring buoys are standard means of mooring vessels or floating structures to the sea floor by laterally leading lines to anchors. Different types of buoys used in moorings include the drum buoy, the peg top buoy, the drum nonriser buoy, and the cylindrical buoy (Figure A-38). Mooring buoys provide primary, temporary, or contingency berthing. A mooring buoy system usually consists of a floating buoy, anchor chain(s) or cables, anchor(s), and connecting hardware. Standard types of mooring buoy systems currently in use include the free-swinging mooring, the bow and

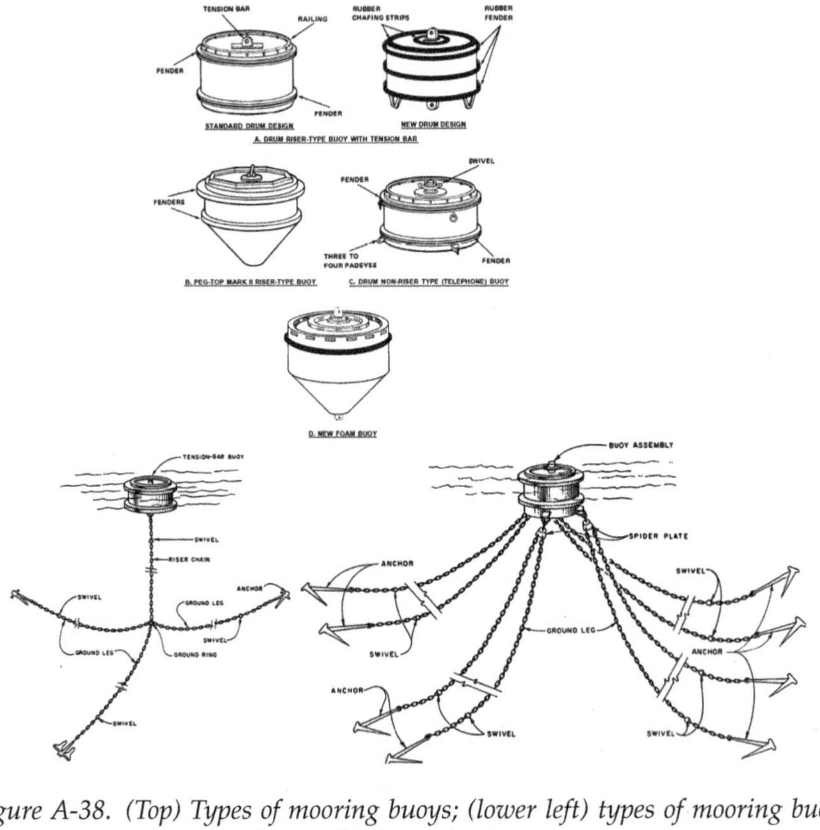

Figure A-38. (Top) Types of mooring buoys; (lower left) types of mooring buoy arrangements, riser type; (lower right) nonriser type.
Source: US Navy (1987).

stern mooring, the spread mooring, the Mediterranean mooring, and the buoy dolphin mooring. Specialized moorings are used in nonstandard configurations and are designed to satisfy unique mooring applications or specific operational requirements. Mooring buoy systems are classified by their respective holding power. The US Navy classification ranges from holding capacities of 50,000 to 300,000 lbs (222 to 1,340 kN).

A.9.2 Typical Components and Problem Areas

Unless a mooring can be completely lifted out of the water for assessment, the bulk of the inspection is performed underwater. Because mooring buoys are typically difficult to access and are located in less-than-ideal conditions, experienced personnel, and reliable equipment are required to accomplish these tasks. Table A-15 summarizes the specific components of common mooring types and the most common areas of concern for each component. The inspection of mechanical and electrical equipment that may form part of the mooring system is beyond the scope of this manual.

Consistent measurements and inspection of the accessible portions of a mooring usually provide a good indication of the overall condition of the mooring. However, if portions of the mooring are buried, such as anchors and chain anchor legs, they are usually not uncovered during the inspection. Pop-up buoys may be attached to chain anchor legs where they disappear into the mud to permit survey crews to determine their location.

Table A-15. Mooring Buoy Systems (Riser and Nonriser Type Moorings): Checklist for Inspections.

Component	Section or part	What to look for	Comments
Buoy, upper portion	General	Size, freeboard, physical damage (dents, holes, list), coating condition	
	Fenders, chafing rails, or strips	General condition	
	Top Hardware	General condition of tension bar, hawse pipe, manhole covers, etc., as applicable	Check for hardware wear and corrosion, damage, and so on

(Continued)

Table A-15. (*Continued*).

Component	Section or part	What to look for	Comments
Buoy, lower portion	Hull	General condition, physical damage (dents, holes, etc.)	Check the condition of coatings
	Bottom Hardware	Condition of tension bar, hawse pipe, etc., as applicable; condition of chain pad eyes (for nonriser moorings)	Tension bar: Check eye and retaining plate for wear/distortion. Hawse pipe: Check chain for wear/corrosion, rubbing casting for wear
Riser chain assembly	Chain links, connecting hardware, swivel, ground ring	Type and general condition, wear/corrosion, distortion	Measure chain links at three locations for each shot of chain, that is, each end and midway between; at each location, take single and double link measurements
Anchor chain subassembly	Chain links, connecting hardware, swivel (if applicable)	Type and general condition, wear/corrosion, distortion, compass bearing of each chain	Check for pitting corrosion
Anchor subassembly (if visible)	Anchor, connecting hardware	Type and general condition	Note the orientation of anchor flukes if visible
Cathodic protection system, buoy, and chain			See Section A.12

A.9.2.1 Buoys. Most mooring buoys have steel hulls protected with several coats of paint or with fiberglass coating. To protect the coating from damage during the underwater inspection, not attempting to remove any marine growth from the buoy may be advisable (Figure A-39).

The buoy should be examined to determine its overall condition. Figure A-40 shows a typical drum buoy with a steel hull and topside hardware. The upper hardware should be inspected for excessive wear and corrosion. In addition, any excess top hardware or wire rope cables attached to the buoy should be reported. Physical damage to the hull of the buoy, such as holes, dents, metal distortion, or listing, should be documented. The buoy's freeboard should be measured and documented. The condition of fiberglass on fiberglass-coated buoys should be inspected, and any cracks, wear, peeling, or rust bleeding should be identified. The condition of the coating should be checked for cracking, chipping, and/or peeling. The condition of water drains should be examined for broken parts, surface rust, and surface pitting. The condition of fenders and chafing strips should be checked for physical integrity, and their connections should be secure to the buoy's surface. Fender/chafing strip brackets or studs should be inspected for corrosion and/or cracks.

Figure A-39. Underside of buoy with marine growth intact.
Source: Courtesy of Appledore Marine Engineering, LLC., reproduced with permission.

Figure A-40. Topside hardware of typical drum buoy with Minor to Moderate corrosion and coating loss.
Source: Courtesy of Appledore Marine Engineering, LLC., reproduced with permission.

A.9.2.2 Riser Chain Subassembly. The riser chain subassembly normally consists of a chain, swivel, ground ring, and connecting hardware such as shackles and special links. The swivel permits the buoy to turn 360 degrees without twisting the anchor chains. The ground ring connects the riser chain to the anchor chain legs.

The mooring chain is susceptible to two basic forms of corrosion: uniform and fretting. Uniform corrosion occurs over the entire chain link. The links initially corrode at a somewhat fast, uniform rate, which decreases with time. Therefore, the diameter of the chain should be measured at several locations throughout the assembly, typically accomplished using calipers and a tape measure. Fretting corrosion, which is more damaging and difficult to prevent, occurs at the crossing of two adjoining links (called the grip area). It results when the movement of the chain links grinds away the outer corroded layer of steel in the grip area (Figure A-41). This process continuously exposes new, noncorroded surfaces of the steel, which are then corroded at progressively faster rates. Loss of chain diameter is accelerated in the grip area, and the useful life of

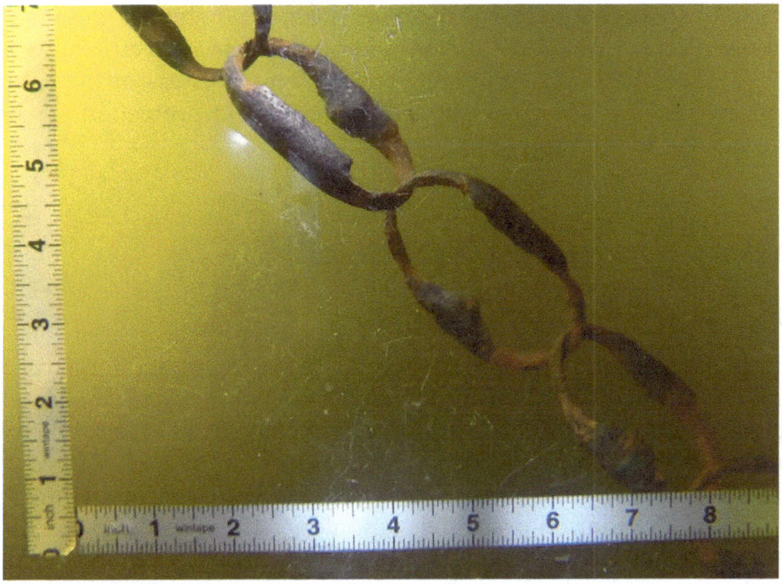

Figure A-41. Severe fretting corrosion on chain.
Source: Courtesy of Appledore Marine Engineering, LLC., reproduced with permission.

the chain is reduced. The amount of corrosion at the grip area can be determined by measuring the overall length of the two adjoining links, called a double link measurement.

Additional areas of concern of the chain assembly are overall alignment and possible stretch. Stretching can be determined by comparing a measurement over five continuous links to the original construction measurement over five links. A tolerance of 2.5% variation in the five-link measurement is typical. The alignment and orientation of the chain legs can be checked using an underwater compass. Catenary measurements can be made using an inclinometer at predetermined locations along the chain.

A.9.2.3 Anchors and Anchor Chains. See Section A.20 for a discussion of anchors and anchor chains.

A.9.2.4 Cathodic Protection System. Many buoys are cathodically protected by sacrificial zinc anodes attached to the hull. The anodes' condition, dimensions, and connection need to be checked during an underwater investigation. To enhance chain protection, continuity cables are attached to the anode, woven through the chain, and connected at links by means of clamps or U-bolts. In this fashion, a shot of chain can be protected by a single anode and should be inspected. Electrical potential readings should be taken to determine the effectiveness of the cathodic protection system. The inspection of the cathodic protection system is further covered in Section A.12.

A.10 WAVE SCREENS AND ATTENUATORS

A.10.1 General

Many marinas and small harbors include wave attenuation structures to reduce the effects of wave action and wake on moored vessels. These structures can be fixed or floating and are typically installed where a rubble mound breakwater is too costly or has too much of an environmental impact to be practical. These structures can be integrated into other components of a marina or can be standalone structures. The orientation of these structures is designed to block waves from a specified direction, typically dictated by the prevailing wind direction.

The typical construction of a fixed wave screen includes piles that support wave screen panels or vertical planks. The panels or planks extend a specified depth below sea level to disrupt the natural formation of waves as they travel toward the facility. The panels are connected with beams or wales that span between the support piles. The panels or planks typically have gaps between one another to allow water to flow through the structure.

The typical construction of a floating wave attenuator is similar to a floating dock. The floats are designed to extend a specified depth below sea level to disrupt the formation of waves. The floating units are either fixed in place by guide piles or by an anchorage system. The floating units are typically connected to each other with hardware specifically designed by the wave attenuator manufacturer.

A.10.2 Typical Components and Problem Areas

Because wave screens and wave attenuation structures are designed to block waves and wakes, they are typically oriented in the path of waves. As such, they are subject to impact from debris and heavy loading during storm conditions. If these structures are not integrated into other elements of the marina or harbor, such as an access dock or mooring float, their structural integrity is not as critical as other structures that support pedestrian or vessel loads. Table A-16 summarizes what to look for when inspecting the condition of these systems.

Table A-16. Wave Screens and Attenuators: Checklist of Inspections.

Component	Section or part	What to look for	Comments
Fixed wave screens	Support piles	Damaged or missing piles, deterioration, alignment (straightness) of piles from top to bottom, exposed length	Check for ice jacking in colder climates
	Beams and wales	Damaged or missing elements, deterioration, connecting hardware condition	
	Panels or planks	Damaged or missing elements, deterioration, connecting hardware condition, length of element below water	Check for fouling from floating debris
Floating wave attenuators	Guide piles	Damaged or missing piles, deterioration, alignment (straightness) of piles from top to bottom, exposed length	Check abrasion at pile guides
	Floats	General alignment, condition of attached wave attenuation devices, connection between floats, bottom surface condition	See Section A.7 for general float inspection
	Anchor chains or cables	General condition, wear/corrosion, distortion, connection to float, connection to anchor	
	Anchors	Location of anchor, embedment, general condition	

A.10.2.1 Fixed Wave Screens

A.10.2.1.1 Piles. The piles that support fixed wave screens should be inspected for damage and deterioration similar to other piles in the marine environment (refer to Section A.2 for additional information). The alignment of the piles should be verified. Some wave screens are designed with a batter, and some are designed with vertical piles. Design and as-built drawings should be reviewed prior to the inspection. If the piles were originally designed to be vertical but are observed to be leaning, this could indicate that the allowable lateral loads on the piles are being exceeded. The exposed length of the pile should be measured during the inspection of these structures to verify that the seabed or mudline has not changed significantly. A greater exposed pile length can significantly affect the lateral capacity of the piles supporting the wave screen.

A.10.2.1.2 Beams and Wales. The beams and wales that span the support piles and support the wave screen panels or planks should be inspected for damage and deterioration as described in Section A.2. Attention should be paid to the connection between the beams and the piles and between the beams and the panels or planks. These structures are exposed to significant lateral loads, and connections may be the first element to show signs of damage or deterioration.

A.10.2.1.3 Panels or Planks. The panels or planks that provide the wave screening should be inspected for damage and deterioration. The length that these elements extend below water should be verified against the design or as-built drawings. The depth these elements extend below water is critical in their effectiveness in attenuating waves. In addition, the wave screens should be inspected for debris that may be trapped between the panels or planks. A buildup of debris adds load to the overall system by preventing flow through the system.

A.10.2.2 Floating Wave Attenuators

A.10.2.2.1 Guide Piles. Guide piles can be of timber, steel, or concrete and should be inspected for damage and deterioration similar to other piles in the marine environment. In floating wave attenuation structures, the guide piles are subject to significant lateral loads and wear from the vertical travel of the floats they support. Abrasion in the tidal zone should be inspected. Guide piles are always installed vertically, and plumbness should be verified in the field. If the piles are not vertical, it may be an indication that the allowable lateral load of the pile has been exceeded.

A.10.2.2.2 Floats. Similar to marina floats, the float units that make up a wave attenuation structure should be inspected for alignment, damage, and deterioration. Listing or partial submersion may indicate that the flotation system is damaged. The floats should be inspected with consideration of their material composition and their design details. These floats may be either timber or steel with flotation units integrated into the structure or concrete box construction. Flotation units may be constructed of Styrofoam, closed-cell foam, polystyrene units, or various other flotation systems. The integrity of the flotation should be visually inspected for soundness. Concrete box construction may rely on the buoyancy of the box to maintain flotation and should be inspected for leakage, cracks, or spalls.

The pile guides on float systems may be cast into the units or externally mounted on the face of the units. These should be inspected to verify they are operating as designed. Many have rollers or elastomeric units to provide protection to the guides. These elements should be inspected to verify they are operating as designed.

A.10.2.2.3 Anchor Chains or Cables. For systems that are secured by anchor chains or cables, these elements should be inspected to verify the integrity of the anchorage. Chains should be visually inspected, and links should be measured to check for corrosion. Cables should be inspected for frays and integrity of the strands making up the cable. Connections between the anchor chain or cable and the float unit should be inspected. If a winch system is used to maintain tension on the anchor's lines, it should be inspected for function and condition. Winch stands that are bent out of alignment may indicate impact damage or overloading. The connection between the anchor line and the anchor block at the mudline should be inspected for damage and deterioration.

A.10.2.2.4 Anchor. For anchored systems, the location of anchor blocks should be verified. Movement of the anchor blocks may indicate that the design loads have been exceeded. A change in the orientation of the overall system may also indicate this. Anchors may be constructed of concrete blocks, helical piles, mushroom anchors, or drag anchors. These anchors may be buried and cannot be visually inspected. In this case, the orientation of the overall system should be verified to investigate possible movement of the system. Proper alignment of the wave attenuation system is a key factor in the effectiveness of the system.

A.11 WATERFRONT SECURITY BARRIERS

A.11.1 General

Waterfront security barriers are becoming common for Department of Defense installations and major ports worldwide. Typical waterfront security barriers (WSBs) use a continuous floating fence that protects maritime assets located within its perimeter. It also serves as a line of demarcation to separate controlled areas from commercial and public water traffic. The WSB fence is very much like a common chain link fence on land, except that it floats on the water. Instead of fence posts, it has moored buoys that anchor it into position. Instead of a chain link, the WSB has a series of connected modular floats that hold up a net.

Early designs utilized entirely steel construction, whereas more recent upgrades have transitioned to fiberglass and high-density polyethylene (HDPE) components. Regardless of construction materials, the WSB comprises the same primary components: buoys, floats (pontoons and beams), connectors, and netting.

The WSB's moored (anchored) buoys are used to hold the barrier strings (multiple floats connected together) in place on the water at most sites. Some shallow-draft locations have anchors attached directly to WSB floats. Other locations have only the end of the barrier attached to a vertical wire rope instead of a buoy. The function and construction of buoys may vary.

A.11.2 Typical Components and Problem Areas

Several different WSB float designs are used at various barrier sites. Older WSB floats typically have steel pontoons and round-pipe trusses, whereas newer floats have plastic HDPE pontoons and square-tube trusses. The WSB connectors are used to attach the WSB floats to each other and the buoys. Most of these are static connectors; once installed, they are not disconnected from the floats and buoys. However, some connectors are designed for frequent opening and closing at barrier gates, allowing ships and other vessels to pass.

Conditions that degrade the ability of any portion of the barrier to stop a craft should be reported immediately to harbor control so that they can arrange backup security. These conditions include floats or buoys that are close to sinking, the failure of both primary and secondary connectors, and significant damage to the netting. Table A-17 summarizes what to look for when inspecting the condition of these barriers.

Table A-17. Waterfront Security Barriers: Checklist for Inspections.

Section or part	What to look for	Comments
WSB floats	Verify floats are intact, not showing signs of advanced corrosion, and not taking on water internally.	The freeboard of multiple floats should be compared to determine whether a buoy is significantly lower in the water, indicating possible water intrusion issues
WSB buoys	Verify buoys are intact, not showing signs of advanced corrosion, and not taking on water internally; verify buoy locations have not changed from the original configuration and mooring chains are not showing greater than 80% wear.	Compare the freeboard of multiple buoys to determine whether a buoy is significantly lower in the water, indicating possible water intrusion issues. Typically, diving will be required to inspect mooring chains. See Section A.9 for additional information
Nets	Verify all nets are intact, properly connected to structural members, and hung properly	The life span of the WSB nets may be decreased with exposure to natural outdoor environments
Connectors	Verify all connections, both primary and backup (if applicable), are intact and operating properly; verify connections on both ends are secure and properly tightened	Periodically check urethane connectors for cracking and interior deterioration
Kayak guards	Verify all kayak guards are intact and properly connected to the WSB	Kayak guards should be positioned to prevent small vessels from transiting under the WSB
Navigational lights	Verify all navigation lights are properly working and connected securely to the WSB; verify the power supply, if solar, is not covered by bird guano	Navigational lights may need to be periodically cleaned to prolong life
Warning signs	Verify warning signs are legible, properly mounted to the WSB, and appropriately positioned	Warning signs may need to be periodically cleaned to ensure legibility

A.12 COATINGS AND CATHODIC PROTECTION SYSTEMS

A.12.1 General

Marine structures constructed with steel or reinforced concrete components are subject to corrosion. These structures can be fitted with a cathodic protection system to prolong their service life. Cathodic protection is simply a means of structuring the cathode in an external electrochemical cell. This transfers the uncontrolled corrosion of the structure to controlled corrosion of external anodes, which can be replaced.

Waterfront structures are typically divided into corrosion zones. These include the atmospheric zone, the splash zone, the tidal zone, the immersed zone, and the embedded zone. Corrosion rates are typically the highest in the splash zone and at the lower end of the tidal zone.

Three types of typical cathodic protection systems exist: coatings, the galvanic anode system (also known as the passive or sacrificial system), and the impressed current system, also known as the active system.

All submerged steel members and components must be bonded together metallically to ensure electrical continuity for proper cathodic protection. For instance, on pile structures, each bent of a pier would typically be tied together electrically by means of welded cross bracing. Electrical continuity may be achieved by welding across interlocks for steel sheet piling. Continuity may be achieved on concrete structures by bonding the reinforcing steel or by providing supplemental wire connections. See Chapter 4, Sections 4.4 and 4.6, for more details on corrosion zones' corrosion rates.

A.12.2 Typical Components and Problem Areas

Cathodic protection systems for waterfront structures often include both above- and below-water components. The below-water portion of the structure may experience the largest demand for the cathodic protection system. Therefore, an underwater investigation may be required to verify system performance. Table A-18 summarizes what to look for when inspecting the condition of these systems.

Table A-18. Cathodic Protection Systems: Checklist for Inspections.

Element	Component	What to look for	Comments
Coating	Overall	General condition, amount of coverage, amount of bare steel exposed, dry film thickness, adhesion/bonding	
	Zones	Atmospheric zone condition	
		Splash zone condition	
		Tidal zone condition	Include a description of marine growth
		Submerged zone condition	Include a description of marine growth
	Edges	Condition of coating at edges of flanges and other members	
	Welds and connections	Condition of coating at welds and other connections	Look for coating unbonding at welds, bolt heads, and other connections
Galvanizing	Overall	General condition, layer thickness	
	Zones	Atmospheric zone condition	
		Splash zone condition	
		Tidal zone condition	Include a description of marine growth
		Submerged zone condition	Include a description of marine growth
	Welds and connections	Condition of galvanizing at welds and other connections	

(*Continued*)

Table A-18. (*Continued*).

Element	Component	What to look for	Comments
Galvanic cathodic protection system	Anodes	Size	Record dimensions and % remaining
		General condition	Describe appearance
	Connection	Welded	
		Bolted	
		Hanging	
		Sled	
		Cables	
Impressed current cathodic protection system	Rectifier	AC power supply to the unit	
		AC power across the transformer	
		DC power across the rectifier	
		DC power at output taps	
		Voltage and current to anodes	Record voltage and current levels
	Cables	General condition	
		Connections	
	Anodes	General condition	
		Size	

A.12.2.1 Coatings. Inspection of coatings could be considered a specialty discipline. The coating industry trains and certifies coating inspectors. The reader is referred to the National Association of Corrosion Engineers, the Society for Protective Coatings, and the American Galvanizers Association for information related to certified coating inspectors and important standards and reference materials. It is beyond the scope of this manual to summarize the large body of knowledge surrounding this topic. This manual is concerned with some of the basic concepts associated with coating inspections.

A visual inspection of coatings should include a description of the coatings at various zones of the structure and an estimate of the amount of bare steel exposed and/or of the remaining coating. For example, "80% of the coating appears to be intact." Check the edges of members, the area around fasteners and connections, and areas that retain water for coating unbonding. The inspector should look for cracking, peeling, or blistering of the paint, and report chalking (deterioration from UV light).

Coatings can be tested for dry film thickness and adhesion. Magnetic pulloff gauges and ultrasonic instruments can accommodate this task. Typically, a disk is glued to the surface of the coating, and a special hydraulic instrument is used to pull a divot of coating from the base metal. This instrument is calibrated to measure coating adhesion.

A visual inspection of galvanizing should include a description of the coatings at various zones of the structure, thickness, appearance, and level of marine growth. It should also include an examination and description of whether the coating is continuous or has bare steel patches exposed. Fresh galvanizing typically has a complete and continuous thick silver coating and has little or no marine growth. As the galvanizing is consumed, the coating becomes thinner. At some point, marine growth starts to adhere to the coating. Once the coating nears the end of its service life, gaps in the galvanizing will appear, and bare steel will be seen in patches. Observing different conditions of galvanizing in different zones on the same structure occurs frequently. For example, galvanizing may be nearly consumed in the intertidal zone but be in good condition in the atmospheric zone.

A.12.2.2 Galvanic Cathodic Protection System. Galvanic anode systems are subject to various types of deterioration. The three most common types are consumed anodes, broken connection wires, and coating failure (see previous section). Impact forces, abrasion, and environmental factors can cause these types of damage.

A.12.2.2.1 Anodes. Anodes should be visually inspected for even consumption. No evidence of consumption indicates that the system is not effective. Observe and record the physical condition of the anodes, including the general dimensions, such as cross-sectional dimensions and length. The percentage of remaining material can be estimated if the original dimensions are known. The presence or absence of marine growth should be reported. Typically, an active anode has little or no marine growth attached and will have a chalky, flaky exterior. An anode with marine growth and a hard exterior may be passivated and may not be working.

Anodes should be replaced before they are completely consumed to ensure continued effective functioning of the system.

A.12.2.2.2 Connections. The connections and attachments of the anodes should be inspected, and the condition reported. Attachments often provide an electrical connection to the structure. This connection is critical to the operation of the anode and should be identified, described, and reported. If the electrical connection is made by a mechanical clamp, the clamp must be in firm contact with the structure. Galvanic anodes typically have very low driving voltages, often less than 1 V, so any resistance at the connection can have a large effect on the performance of the anode. All connections in the system should likewise be examined to ensure that they are sound (e.g., welded cross-bracing, connection wires, support wires, straps, and conduits). These components ensure the electrical continuity of the structure. Breaches can affect the performance of the system. Wave action, debris, ice, severe weather, and impact can damage these components.

A.12.2.2.3 Potential Survey. A potential survey can be conducted to verify cathodic protection is adequate by conducting a potential survey using a reference cell and a voltmeter to take readings at regular intervals along members. The readings are also used to verify that all the members in the structure are continuous and protected. See the following section for more information.

A.12.2.3 Impressed Current Cathodic Protection System

A.12.2.3.1 Rectifier. Significant voltage and current may be present at the rectifier, and detailed inspection should be done only by an electrician or trained individual. Having an electrician turn the power off for a visual inspection may be appropriate in some situations. Some rectifiers have an indicator light to signify that the power is on. In this case, the condition of the indicator light should be noted. Some rectifiers have voltage and amperage meters located on the front panel. If so equipped, document the voltage and amperage. These items may indicate that the power is on and the current is flowing to the anodes. Rectifiers are subject to electrical failure and broken wires, which can result in maladjustment of the system and lead to under- or overprotection of the structure. Insufficient current results in underprotection and corrosion. Excessive current leads to overprotection, failure of protective coatings, and the potential for both hydrogen embrittlement of high-strength steel and stray current corrosion on adjacent structures.

Anodes, connections, and protective coatings should be examined for damage and deterioration as described previously for the galvanic system. A typical mode of failure for an impressed current anode system is cable failure. If the anodes are visible, the size and condition should be noted. Also, perform a potential survey to verify the cathodic protection adequacy.

A.12.2.3.2 Potential Survey. One of the most common methods of evaluating cathodic protection is to measure the potential or voltage of the structure with respect to a reference electrode. The reference electrode is referred to as a half-cell because it acts as one pole. The structure acts as the other pole in the cell. Various types of standard reference electrodes exist. For field investigations, the two most common are the copper–copper sulfate (Cu-$CUSO_4$) and the silver–silver chloride (Ag-AgCl) reference half-cells. Cu-$CUSO_4$ electrodes are commonly used in above-water surveys to take potential readings on decks and the ground over buried structures. Ag-AgCl electrodes are commonly used to take potentials on submerged structures.

Ag-AgCl readings of −0.800 to −1.2 V typically indicate adequate cathodic protection in seawater. This range can vary depending on factors such as the water salinity and type of half-cell. Readings of about −1.1 V are typical near active sacrificial anodes. Readings of up to about −2.0 V sometimes indicate active impressed current anodes. Because the most important thing is not the voltage at the anode but the voltage at the structure, readings should be taken in areas distant from the anodes. Impressed current systems should be turned off just prior to taking readings. In this case, the polarized or residual potential of the structure will be taken. Corroding steel in seawater with no cathodic protection typically produces readings between −0.5 and −0.7 V.

A.12.2.3.3 Stray Currents. Stray electrical currents promote corrosion by accelerating the electrochemical process. These currents can originate from a variety of sources, including welding activity, faulty grounds from vessels or other structures, cathodic protection systems for neighboring structures or pipelines, transit systems, DC industrial generators, power stations, or substations. This type of accelerated corrosion and member deterioration also results in damage similar to that encountered in steel and concrete structures.

A.13 MARINA AND SMALL CRAFT HARBOR COMPONENTS

A.13.1 General

Marinas and small craft harbors may comprise various marine structures, many of which are common to other types of waterfront facilities. Open-piled piers, bulkheads, timber cribs, wave attenuators, breakwaters, and gangways are all structures that may form part of a marina and are described elsewhere in this appendix. This section deals specifically with those marine components that are parts of marinas, such as floats, anchor systems, fenders, roofs, gangways, appurtenances, and utilities.

Table A-19 summarizes common problem areas associated with the specific above-water components for marinas.

Table A-19. Marinas and Small Craft Harbors: Checklist for Inspections.

Component	Section or part	What to look for	Comments
Floats	General	Excessive float misalignment, tilt, reduced freeboard	Misalignment may suggest anchor slippage; tilt or loss of freeboard could be caused by leakage, excessive marine growth, and so on
	Decking or surface	Worn, uneven, spalled, loss of nonslip surfacing, corrosion, loose deck members	May result in a tripping or slipping safety hazard
	Shell	Surface deterioration; physical damage (dents, holes, etc.); loose or leaking access hatch covers	Check for a waterlogged filler
	Joints between float units	Damaged, loose, and/or deteriorated connecting hardware; excessive float misalignment	
Anchor systems and piles	Anchor chain and connecting hardware	General condition, wear/corrosion, distortion	
	Anchor cable winches	Secure attachment to the deck, and verify whether cables are not binding within the winc	
	Pile system	Damaged or missing piles, alignment (straightness) of piles from top to bottom	Check for ice jacking in colder climates

(Continued)

Table A-19. (*Continued*).

Component	Section or part	What to look for	Comments
	Pile surfaces	Deterioration, wear/corrosion, mechanical damage	
	Pile guides	Deterioration, damage, wear/corrosion, missing components, binding	Integrity and condition of pile/float connection
Fenders and appurtenances	Fenders or bumpers	Deterioration, missing/loose members, condition of attachment hardware	Check corner bumpers
	Cleats	Damage, missing connecting hardware	
	Ladders	Damage, missing connecting hardware	
Roofs and covers	Roof panels	Holes, tears, missing connectors	
	Trusses and columns	General condition, missing or bent members, missing or damaged connecting hardware, column attachments	

A.13.2 Typical Components and Problem Areas

A.13.2.1 Floats. Floating docks at marinas typically consist of floating units anchored by piles or chains and anchors. Floats come in various materials, such as timber, concrete, aluminum, metal, and polymer composites. Other materials used in float construction include foam and rubber; these materials are also used in conjunction with some of the other materials noted previously. Foam billets may need to be coated to protect against deterioration caused by the environment or exposure to fuel and oil in the water. Plastic tubs are often used in lightweight applications such as floats for residential recreational craft. Hardware or rubber sheets are often used to connect the float modules. Where finger floats attach to main floats, steel or aluminum knee braces or triangles are used to provide connectivity and stability. Hardware used for floats is usually hot-dip galvanized. Figure A-42 depicts a typical marine float with tall guide piles to accommodate a wide range of water-level elevations.

Individuals familiar with these types of structures can inspect floating dock components above the water. Divers must be used to inspect the undersides of floats. Typical problems encountered on floats usually relate to the deterioration of a specific float material.

Floats should not have a visible list and must float high enough to keep walers out of the splash zone, typically no less than 12 in. for small craft. In its simplest form, the timber float consists of a timber deck supported on a log or timber framing. Loss of buoyancy from waterlogging of the timbers can become a problem. Modern timber floats may incorporate polystyrene flotation billets to improve buoyancy. However, these systems may lose buoyancy because of leakage associated with shell cracking or deterioration, impact damage (particularly for concrete and fiberglass floats, which may be more susceptible to cracking), utility hole openings, accumulation of marine growth on floats and anchor chains, and deterioration of foam buoyancy units. Float hardware should be inspected for wear, stress/fatigue damage to float connecting components, loosened float waler nuts, and missing hardware components.

Figure A-42. Timber float on polyethylene flotation billets.
Source: Courtesy of Reid Middleton Inc., reproduced with permission.

A.13.2.2 Anchor Systems. Anchors for marina floats secured by chains or lines usually consist of precast concrete blocks or mushroom anchors placed directly on the bottom. They are most suitable for soft bottom conditions where they can achieve improved resistance against sliding. Screwed-in helical piles may also be used as anchors. Chains, cables, or lines are normally used to secure the floats to their anchors.

Guide piles consisting of concrete, steel, timber, or polymer composite piling are another type of float anchorage system. Float and pile connections are critical systems subjected to heavy wear and repeated load reversals. Metal or timber guides attached to the float encompass the piles, allowing the piles to move vertically with the changing tides and laterally with wind and wave forces acting on the float. Guides may be internal or external to the float modules. Antifriction blocks or wheels are used to permit the pile to move freely within the guide. Figure A-43 shows a typical steel pile and guide.

If the anchor systems consist of an anchor chain assembly, any above-water connecting hardware at the float should be inspected for general condition. Cable winches should be inspected for loose hardware, corrosion, and freedom of movement. For floats that utilize a pile and guide system, guides should be inspected for condition, wear, missing sliding hardware such as blocks or wheels, and binding. Float and deck material adjacent to the guides should be inspected for stress cracking.

Figure A-43. Galvanized steel pile with UHMW antifriction guides. Source: Courtesy of WSP USA Inc., reproduced with permission.

A.13.2.3 Fenders (Bumpers). Timber walers that are part of the structural unit sometimes serve as fender protection on concrete and timber floats. Timber rub boards (treated or untreated), low-friction UHMW-PE, or vinyl bumpers are often added to protect both the floats and the vessels. Attachment to the floats may be accomplished with bolts, screws, or nails. Corner bumpers often consist of vinyl bumpers or wheeled assemblies.

Fenders are made mainly of timber, vinyl, or rubber. More recently, UHMW-PE has been used, but this protects the dock more than the vessel. Other materials sometimes used include fire hoses, tires, and rubber belts. Depending on the material, fenders should be inspected for missing sections, material degradation, rot, cracks, tears, and missing attachment hardware.

Vinyl bumpers are particularly susceptible to attachment hardware pullout because galvanized nails are often used, which do not stand up well to heavy vessel impact accompanied by shear forces. Corner bumpers that utilize wheels should be inspected for alignment and freedom of movement.

A.13.2.4 Covers and Roofs. Rigid covers and roofs provide weather protection for moored vessels. Roofing material is sloped to provide rain runoff and consists primarily of metal sheeting (panels), although wood or fiberglass are also used. The sheeting is attached to supporting roof trusses with rivets or bolts. Posts or column trusses extend from the roof and attach to the float deck with bolts. In some cases, roof-covered slips may be partially or fully enclosed by walls.

Covers and roofs should be inspected for missing or loose roof panels, missing roof connecting hardware, holes, and general condition. Trusses, columns, and posts should be inspected for missing elements, deterioration or corrosion, missing connectors, and attachment to the float. Walls, if present, should be inspected for general condition.

A.13.2.5 Appurtenances. Primary appurtenances consist of cleats and ladders. Secondary appurtenances include float-mounted safety items and amenities such as fire extinguishers, fire hoses and cabinets, float throw rings on stands, telephones, utility pedestals, and dock storage boxes. Some float-mounted appurtenances are shown in Figure A-44.

Appurtenance condition is determined by age and wear. Appurtenances should be checked for missing connecting hardware, attachment condition, damage, deterioration, and safety. Ladders should be inspected for

Figure A-44. Typical float-mounted appurtenances: throw ring, fire hose, and fire pull box.
Source: Courtesy of Reid Middleton Inc., reproduced with permission.

attachment to the float, deterioration, and missing rungs. Ladder rungs should extend at least two rungs into the water to ease climbing out of the water. If the rungs telescope downward when weight is applied, they should be checked for freedom of movement. Secondary appurtenances should be inspected for attachment to the float and general condition.

A.13.2.6 Utilities. Utilities, such as potable water, electrical cables (including cable TV and telephone), fuel stations, piping for sewage pumpouts, and storm sewer outfalls for upland drainage, are normally found at marinas. Utilities transition from the shore to the floats through tide loops lying on the embankment or mounted on the gangway. Once on the floats, utilities may be run outside or inside the floats.

The above-water components of these utilities should be inspected as outlined in Section A.19. Internal utilities within junction boxes are not usually checked. Utility tide loops and utilities running under gangways and along the outside edges of floats should be checked, if accessible. Figure A-45 shows a utility tide loop covered in marine growth. Marine growth adds stress to the cables, couplings, and supports and often hides visible damage.

Figure A-45. Marine growth on a utility cable tide loop.
Source: Courtesy of WSP USA Inc., reproduced with permission.

A.14 GANGWAYS

A.14.1 General

Gangways are transition structures providing access between land and marina floats or between fixed structures and vessels. They are typically hinged on the landward end and are free to move (roll, slide, or pivot) on the float end. Skids or rollers that travel over threshold plates are attached to the gangway's free end.

Vessel gangways are those that are carried, maintained, and operated by the vessels themselves. Vessel gangways are often fixed to the vessel and free on the structure side. Unless the vessel is permanently moored to the structure, inspection of vessel gangways is typically beyond the scope of this manual.

Table A-20 summarizes common problem areas associated with the specific above-water components for marina gangways.

Table A-20. Gangways: Checklist for Inspections.

Section or part	What to look for	Comments
Ramp structural members	Missing or bent members, weld cracks, and damaged coatings	See Section A.19 for utilities
Ramp deck, walking surface	General condition and antislip elements	

(Continued)

Table A-20. (*Continued*).

Section or part	What to look for	Comments
Skids, rollers	General condition, freedom of movement	
Guardrails	General condition, height, intermediate rail, attachment to deck	Pay particular attention to splinters (if wood) or burrs (if metal), which may cause hand injury
Connections	Deterioration, signs of overstress and fatigue	

Gangways are typically constructed of aluminum to be lightweight but may also be constructed of steel. Guardrails and antislip decking are provided for safety. Gangway slopes vary with the water level but should always slope down toward the float end. Gangway width and length vary. Several manufacturers have standardized designs that can meet the needs of most marina and float owners. Handicapped accessible ramps consisting of electrically operated lifts can also be provided. An 80 ft long heavy-duty marina gangway is shown in Figure A-46.

Figure A-46. 80 ft Long heavy-duty gangway.
Source: Courtesy of Reid Middleton Inc., reproduced with permission.

Larger gangways, such as cruise vessel mobile passenger gangways, are often used to provide access between buildings above the first story and higher decks of vessels. They are typically fabricated from structural steel and are wheel-based mechanical devices that use a combination of electrical and hydraulic power systems. These gangways can be positioned in a particular location to interface with a ship and facilitate the embarkation/disembarkation of passengers and supplies. Structural steel elements should be inspected and assessed in accordance with the recommendations for steel structures described in this manual. Inspection and assessment of the electrical systems, hydraulic elements, and appurtenances are beyond the scope of this document. A mobile passenger gangway is shown in Figure A-47.

Figure A-47. Cruise vessel mobile passenger gangway.
Source: Courtesy of Tampa Port Authority, reproduced with permission.

A.14.2 Typical Components and Problem Areas

A.14.2.1 Ramp. Discussion of typical components focuses primarily on smaller marina-type gangways. The ramp consists of a travel surface and its supporting structure. Walking surfaces, which should be provided with antislip features or materials, may consist of grating, diamond plating, timber elements, or plywood.

While lighter-duty ramps may have bolted components, heavier-duty ramps typically consist of welded aluminum or galvanized steel. Welds should be visually inspected for cracking or other signs of stress and fatigue, and connections should be inspected for corrosion. Inaccessible welds should be noted. Steel connections used on aluminum elements should be inspected for accelerated corrosion caused by contact between dissimilar metals. Steel ramps should be inspected for weld cracks, tight connections, and corrosion. Galvanizing and coatings should be inspected for proper coverage and adhesion. Skids and rollers at the gangway ends wear over time. Skids' ability to move freely should be verified, and rollers should be inspected for binding and flat spots. The surface on which the skids or rollers ride should be attached firmly to the float and not exhibit signs of excessive wear.

Because the travel surface gets frequent use, the condition of the antislip features should be evaluated for effectiveness. If grating is used, pay special attention to missing or corroded attachment clips. Inspect fiberglass grating for breaks or holes. Inspect timber and plywood deck surfaces for damage, deterioration, and rot. Because timber is not very slip-resistant when wet, antislip materials or coatings are often added to the top surface. This antislip material should be inspected for wear.

A.14.2.2 Guardrails. Guardrails should be inspected for secure attachment to the ramp, including welds and connecting hardware, such as connecting pins. Verify the height of the top railing and the presence of an intermediate rail. Coatings should be inspected for general condition. Take note of any misaligned or detached components of the guardrail.

Handrails and guardrails are important to provide accessibility and safety along waterfronts. In the United States, the Americans with Disabilities Act regulations provide additional dimensional requirements for railings, such as height and length beyond the ends of ramps. However, jurisdictional regulations may vary from site to site. Although checking the design of railings is beyond the scope of typical condition assessments, inspectors should note areas that are clearly out of compliance, such as leaning, loose, or broken railing elements.

A.14.2.3 Connections. Connections from the gangway to the supporting structures include hinges, transition plates, and deck clips. A transition plate and rollers are shown in Figure A-48. Inspection should focus on missing, loose, or corroded connecting hardware and clips. Hinges at the top of the ramp should be free to move.

Figure A-48. Gangway transition plate and gangway rollers.
Source: Courtesy of Reid Middleton Inc., reproduced with permission.

A.14.2.4 Utilities. Utilities provided to marina floats are often routed alongside or beneath gangways. Hangers are used to attach the utilities to the structure. Deck lighting may be provided on handrails to increase safety during nighttime. These components should be inspected as outlined in Section A.19.

A.15 BOAT RAMPS

A.15.1 General

A boat ramp is an inclined surface that allows for launching into and removing a boat from the water. This section covers ramps used to launch boats using hydraulic and vehicle trailers. Ramps may consist of one or more lanes. The elements of a ramp include the toe, ramp, and transition curve.

Boat ramps are constructed primarily of cast-in-place concrete or precast concrete panels. Some ramps may also use the natural material in the area, provided it has sufficient traction. Traction for concrete ramps is often provided using roughened surfaces or V-shaped grooves. In freshwater areas, ramps can be constructed of heavy-duty timber planks with metal chain links or other metal fabric attached to the top surface to enhance traction. Ramps may have side curbs and end stops to prevent trailers from leaving the ramp. Material for the curbs and stops can include concrete, timber, rubber, or vinyl.

Boarding floats or piers are frequently provided to assist in the launch, retrieval, and queuing of boats. A transition plate spanning between the float and an abutment at the top of the ramp makes the float more accessible. In locations where water levels fluctuate, grounding skids or blocks are provided on the underside of the floats to prevent damage during lower tides.

A.15.2 Typical Components and Problem Areas

Because boat ramps are located at the water's edge and consist of in-water and out-of-water elements, above-water components are best inspected during low water conditions. Table A-21 summarizes common problem areas associated with boat ramps.

Table A-21. Boat Ramps: Checklist for Inspections.

Type	Component	What to look for	Comments
Boarding floats	Grounding skids or blocks	Presence, condition	See Sections A.7 for floats and A.19 for utilities
Ramps	Ramp surface	Concrete condition, undermining, cracking, surface traction	
	Abutments	Concrete condition, undermining, cracking	
	Curbs	General condition, spalls, alignment	

A.15.2.1 Ramps. Ramp surfaces are the most important component of a boat ramp. If the surface is in good condition, a boat could be launched or retrieved even if it had damaged curbs or boarding floats. Therefore, inspection should focus on the general condition of the ramp. For ramps on natural ground, surfaces should be checked for potholes, buildup, or drifting of aggregate, and loose deep soil. Concrete ramps should be inspected for undermining, spalls, cracks, reinforcing steel corrosion, and excessive aggregate exposure. Evaluate traction features, such as V grooves and surface roughing. Asphalt ramps should be inspected for potholes and cracking. Document evidence of sediment accretion. Figure A-49 shows a boat ramp and boarding floats at low tide, exposing sediment buildup at the lower reaches of the ramp. Curbs should be inspected for damage.

Concrete abutments rarely suffer damage but should be examined for evidence of vehicular impact.

Propeller wash can scour the soil below the base of the ramp if it is not properly protected. This type of damage can typically be observed only by underwater inspection.

Figure A-49. Boat ramp with sediment covering the lower extents of the ramp. Source: Courtesy of WSP USA Inc., reproduced with permission.

A.15.2.2 Boarding Floats. Boarding floats should be examined in accordance with float inspections outlined in Section A.7. One advantage of observing floats out of the water is that all sides of each out-of-water float module can be examined. Damaged concrete, torn flotation material, holes and punctures, and deformations will be readily visible. In addition, the presence of grounding skids or blocks should be verified, and their conditions noted. Inspect float connections to the abutment and transition plates.

A.16 MARINE RAILWAYS

A.16.1 General

A marine railway is a cradle-and-track system used to lift vessels out of the water. They can be used at dams, locks, portages, shipyards, and marinas. Typical operation consists of positioning the vessel over the cradle, raising the cradle to allow the floating vessel to rest within the cradle, and then winching the cradle along tracks to launch or extract the vessel from the water.

A.16.2 Typical Components and Problem Areas

The main components of the marine railway are the track structure, the cradle, and the winching system. The track structure typically consists of two or more pile-supported track beams that are connected transversely to prevent the track gauge from changing. The cradle rests on either a system of free rollers or a wheel system built into the cradle. The cradle is typically raised and lowered by a cable winch system.

The piles, transverse beams, and track beams can be fabricated from timber, concrete, or steel members and should be inspected as described for open-piled structures, with additional attention paid to the rail-to-track beam fastening system.

The cradle may also be fabricated from concrete, timber, or steel members. However, steel and timber are more commonly used. The inspection of the cradle system should be performed as described for typical above-water structures, with particular attention paid to the connections. The cable connection at the cradle should be thoroughly inspected for damage and corrosion.

The winch and braking system may require inspection by a mechanical engineer experienced in winch design and inspection.

A.17 BULL RAILS, LADDERS, AND SAFETY FEATURES

A.17.1 General

Bull rails are the nosings or vertical projections at the waterward edge of a pier or wharf deck. They may be permanent, as in the case of a cast-in-place concrete structure, or removable, as in the case of a heavy timber cross section. Bull rails are installed to reduce the likelihood of a person or piece of moving equipment inadvertently going over the edge of the structure. In some instances, bull rails are also used to provide smooth, rounded edges where mooring lines would otherwise drape over the sides of marine structures and potentially snag.

Ladders are typically placed intermittently along the wharf or pier face to reduce swimming distance. To work as intended, a wharf-side ladder must extend several feet below the lowest anticipated water surface. Constant wetting and drying of the ladder and connections by tidal action can accelerate corrosion if the materials are not treated to account for this.

In addition to bull rails and ladders, safety features on a waterfront facility may include Type IV (throwable) personal flotation devices, pedestrian refuges, hazard and pedestrian area striping, and audible alarms for moving equipment.

A.17.2 Typical Components and Problem Areas

A.17.2.1 Bull Rails. A visual inspection of the bull rails should focus on corroded connections, chipped or spalling elements, and areas damaged by equipment (Figures A-50 and A-51). Some bull rails are considered sacrificial, which should be considered when evaluating the extent of damage sustained. Take note of defects or elements that could inhibit the bull rail from serving in its capacity as a safety feature.

Figure A-50. Split timber bull rail.
Source: Courtesy of Simpson Gumpertz & Heger Inc., reproduced with permission.

Figure A-51. Concrete bull rail with integrated mooring points.
Source: Courtesy of Port of Tacoma, reproduced with permission.

A.17.2.2 Ladders. Wharf-side ladders are a means of emergency egress from the water (Figure A-52). Wharf-side ladders typically do not have standard fall protection features. An inspection of a wharf-side ladder should include climbing the ladder, assuming that the ladder appears safe. Field personnel should follow all appropriate safety precautions when climbing up the ladder.

Figure A-52. Wharf-side ladder.
Source: Courtesy of Port of Tacoma, reproduced with permission.

A.17.2.3 Miscellaneous Safety Features. Safe operation of a waterfront facility depends on the users of the facility. Examples of items to look for include improper stowage of dockside materials, missing or damaged safety equipment, hindered access to pedestrian refuge, and tangled or frayed mooring lines.

A.18 CRANE RAILS, TRENCHING, AND CABLES

A.18.1 Crane Rails

Crane rails are typically oriented longitudinally to the structure below or parallel to the shoreline in the case of landside crane rails. Contemporary crane rails are installed using a continuous welded rail section, whereas older installations may utilize jointed rail. They may bear on shim plates and flowable grout or on cross ties on a ballasted substrate.

Crane rails are attached to the structure by mechanical anchors. In the case of concrete ties or a concrete substrate, these are typically metallic clips or clamps that hold the rail base in place by applying a compression force. In the case of timber cross ties, standard rail spikes are used. Crane rail anchorages are often encased in asphalt or low-strength cementitious grout, concealing them from visual inspection.

A.18.2 Trenching and Cables

Power is supplied to shore side cranes via an electrical cable or a collector bar system. Electrical cables can be either in an open trench (Figure A-53) or in a covered trench (Figure A-54).

Cable trenches run parallel to the wharf face, typically near the waterside crane rail. Crane cables are typically stored on the crane and played out from the crane on a self-return cable reel. Cables are laid out in the trench and connected to the power supply in a subsurface vault on the wharf or uplands. Cables feed through the wharf via cable horns that allow the cables to lie in the trench in either direction along the face of the wharf. Cable horns allow the cables to penetrate the wharf structure and then run under the structure transverse to the crane rails and back to the power supply at an upland transfer station or substation. Cables may be visually inspected from the structure deck from a splice to the cable horn. In the case of an open-face wharf or pier, any other

Figure A-53. Cable trench (to the right of the crane rail).
Source: Courtesy of Port of Tacoma, reproduced with permission.

Figure A-54. Cable trench protection system.
Source: Courtesy of Liftech Consultants Inc., reproduced with permission.

visual inspection of cable or conduit must occur below the deck structure. In the case of a closed-face wharf, power cables are either run in an accessible utility trench or buried and, therefore, unavailable for visual inspection.

Collector bar systems utilize live electrical buss bars in a hinged plate-covered trench parallel to the crane rails. As the crane moves along the rails, a wedge-shaped projection from the crane frame lifts the plates open and exposes the collector bars to the electrical connections feeding the crane.

A.18.3 Typical Components and Problem Areas

A visual inspection of crane rails may not reveal internal structural distress or connection/support issues that warrant attention. However, visual indications of rail or connection fatigue, delamination, and signs that the rail may be near the end of its useful life are often apparent. A visual inspection of electrical cables only reveals cracks in the insulation casing or pinched/kinked cables. Shorts in the cable must be investigated by a qualified electrical equipment inspector. Table A-22 summarizes common problem areas associated with crane rails and associated components.

Table A-22. Crane Rails, Trenching, and Cables: Checklist for Inspections.

Element	Component	What to look for	Comments
Crane rail	Head	Mushrooming or plastic deformation of running surface	Thin layers of steel may be seen as sliding off the rail head
		Uneven wear	
		Lateral, out-of-plane deformation	
	Web	Cracking at connection to the head	
		Rotation	Indicates rail is insufficient size for loads applied or crane wheels are misaligned
	Joints	Worn connection plates	Look for widened bolt holes
		Loose bolts	
	Support	Cracking or spalling in grout under crane rail base	
	Connection to base	Loose clips or clamps	Rail attachments to concrete
		Loose or missing spikes	Rail attachments to timber ties
Trench, cable, and collector bars	Cable	Cracked insulation	Refer to facility operator immediately
		Burn marks or other signs of an electrical short	STOP! Initiate lockout/tagout process
		Kinked or bent cable	
		Cable in water (on deck or under deck)	

(Continued)

Table A-22. (*Continued*).

Element	Component	What to look for	Comments
	Trench	Corrosion, concrete spalling	
	Trench	Operation of cover plates	
		Cover plates broken, buckled, or corroded	
		Debris in trench	
	Collector bars	Burn marks and other signs of an electrical short	STOP! Initiate lockout/ tagout process
		Burn marks and other signs of an electrical short	
		Burn marks and other signs of an electrical short	

A.19 WATERFRONT UTILITY SYSTEMS

A.19.1 General

Utility systems are often essential components of military, industrial, commercial, or recreational waterfront facilities. They provide safe lighting, electrical power, water, sewer, and fire protection for people and equipment utilizing the structure.

The most common waterfront utility systems are as follows:

- Potable water: Fresh water for drinking and washing.
- Nonpotable water: Fresh water or seawater for washing, flushing, or fire protection.
- Fire protection: Typically, nonpotable water is used to contain fire-related emergencies. For flammable liquid handling facilities, fire protection may consist of foam or other medium in conjunction with water or dry chemical systems.
- Electrical power: Power transmission from source (power plant, generator, or local utility feed) to the destination on the pier (shore power station, disconnect, electrical panel, etc.). The final destination

may provide power to vessels (cold ironing) or other large equipment (gantry cranes) or provide feeds for local services such as lighting, aids to navigation, security, or cathodic protection systems.

- Grounding system: Provides grounding for operational and personnel safety for all electrical systems, apparatuses, machinery, electrical conduit, and all accessories that are a part of the power distribution system. Grounding systems are also provided for structural supports, frames, towers, safety fencing, hardware, equipment enclosures, system neutrals, and buried ground cable networks and counter poses used in substations and similar areas. Grounding systems also provide lightning protection.
- Communications: Provides speech or data transmission via wire or cable over distances (telephone, data transmission, coded transmission, cable TV, and signal or alarm circuits).
- Steam service: Provides steam along piers and other waterfront structures typically used for active military berthing and ship repair and at the perimeter of graving dry docks.
- Compressed air: Provides pressurized air for control or operation of pneumatic systems.
- Sewer/sanitary waste: Removes sewer/sanitary waste from the facility and/or from ship connections.
- Oily waste/waste oil: Removal of bilge or ballast water from vessels.
- Product lines: Provides petroleum, oil, lubricants, or chemicals that may be part of ship supply or part of commercial storage and transport operations.

Utility systems can be located directly on decks, within utility trenches, alongside structures, or underneath structures on pipe hangers. The specific location depends on the quantity and type of utility lines and the installation time. For example, new container facilities will likely be designed to accommodate the utilities in preformed conduits or trenches during construction. However, utility retrofits or upgrades for facilities with inadequate space may require new runs under or alongside the structure.

Inspectors should comply with all safety precautions, including lock-out/tag-out for relevant systems. These precautions are particularly important for grounding systems because grounding wires may or may not be in conduits and have relatively low currents (approximately 5 mA) that can be difficult to detect but still dangerous.

This section addresses the physical condition of visible utility components such as conduits, pipes, and hangers. Although testing the performance or code compliance of these systems is also important, it is beyond the scope of this manual.

A.19.2 Typical Components and Problem Areas

Table A-23 summarizes some common problem areas associated with various utility systems.

Table A-23. Waterfront Utilities: Checklist for Inspections.

Type	Component	What to look for	Comments
Utility Support	Pipe or conduit hangers	Missing, broken, or structurally failed supports; corrosion; bending or distortion; loose hardware or other signs of connection failure	Sagging pipes or conduits are a sign of missing hangers
	Trenches	Water intrusion or ponding, failed drainage, or pump systems	
	Cover plates or utility holes	Improper seating on the frame, corrosion, leakage, distortion, elongation, distortion of holding clamps, and misalignment	
Utility Runs	Piping	Obvious leakage; strain or torsion; excessive corrosion; failed coating; misalignment; lack of support; excessive vibration; broken, loose, deteriorated, or strained connections; and cracks or other breaches	Immediately notify owner of active leaks
	Conduits	Breaks, visible wires, corroded or mechanically damaged conduit, loose or missing attachments to structures, failed conduit inspection plates	Initiate lockout/tagout process for any signs of electrical shorting or exposed wiring
	Fittings	Discontinuity from loose, missing, or broken connections; signs of burning or overheating; corrosion	Initiate lockout/tagout process for any signs of electrical shorting or exposed wiring

A.19.2.1 Utility Supports

A.19.2.1.1 Pipe/Conduit Hangers. Hangers, brackets, bolts, and specially fabricated supports and braces should be inspected for corrosion, damage, loose hardware, or other signs of connection failure. In addition, missing, broken, or structurally failed supports should be documented, as the conduits or piping may not be designed for increased spans. Sagging or deflection of these lines is an obvious sign of locations where hangers have completely failed. In addition, burning or charring wood at hanger contact points could indicate a grounding deficiency.

A.19.2.1.2 Trenches. Utility trenches should be inspected for water intrusion or standing water, specifically at valves and connections. Trenches designed to be self-draining should be inspected for blocked drain outlets. Trenches with automatic pumps for drainage should be tested for proper operation by testing the float switch, if accessible.

A.19.2.1.3 Cover Plates, Access Hatches, and Manholes. Plates, hatches, and utility holes in decks should be inspected for flush-mounted covers to reduce tripping hazards. These elements can also snag and damage mooring lines during line handling if they are damaged. In addition, corrosion, leakage, distortion, elongation, distorted holding clamps, and misalignment should be noted during the inspection.

A.19.2.2 Utility Systems

A.19.2.2.1 Piping. Inspection of piping systems should usually be limited to visual inspection. Defects include obvious leakage; strain or torsion; excessive corrosion; failed coating; misalignment; lack of support; excessive vibration; broken, loose, deteriorated, or strained connections; and cracks or other breaches. In addition, inspectors should note worn or illegible identification markings and failed piping enclosures, such as insulation or heat tracing. Note broken valves or gauges.

A.19.2.2.2 Conduit. A visual inspection of conduit is similar to that of piping systems. Inspectors should note breaks, visible wires, corroded or mechanically damaged conduit, loose or missing attachments to structures, and failed conduit inspection plates.

A.19.2.2.3 Grounding Systems. Grounding systems for lighting protection or electrical service may or may not be in conduits. In general, an inspection of these systems should be limited to a visual review for discontinuity from loose, missing, or broken connections and signs of burning, overheating, or corrosion.

A.20 ANCHORS AND CHAINS

A.20.1 General

Anchors and chain assemblies are often used for mooring purposes. They can be deployed for both permanent and temporary use. Mooring buoys (Section A.9) are the most common application of anchors and anchor chains. Anchor systems work differently depending on their applications and the seafloor conditions. Inspectors of anchors and anchor chains should have a general understanding of the anchor system being inspected.

A.20.2 Typical Components and Problem Areas

Table A-24 summarizes some common problem areas associated with anchors and chains.

Table A-24. Anchors and Chains: Checklist for Inspections.

Component	Section or part	What to look for	Comments
Anchors	Drag embedment anchors	Anchor positioning relative to structure/ object being anchored; steep chain angles	Drag-embedment anchors are intended for horizontal holding power. Chain angle to the sea floor should be shallow
	Pile anchors	Corrosion; section loss, displacement/tilt from overload	
	Dead-weight anchors	Signs of dragging along the sea floor, section loss, concrete spalling	
	Suction pile anchors	Corrosion; section loss, displacement/tilt from overload	

(*Continued*)

Table A-24. (*Continued*).

Component	Section or part	What to look for	Comments
	Direct-embedment anchors	Signs of soil displacement	Anchors will likely be fully embedded below the sea floor and not directly accessible for visual inspection
Chains	Chain links, connecting hardware, swivel (if applicable)	Type and general condition, wear/corrosion, distortion, section loss	

A.20.2.1 Anchor Chain Subassembly. Anchor chain assemblies consist of the multiple legs of chain and connecting hardware that connect the ground ring (in the case of riser moorings) or the buoy (in the case of nonriser moorings) to the anchors. The chain should be checked for wear and corrosion as described previously for the riser chain subassembly and should be inspected at critical locations throughout the chain assemblies, such as the catenary section, the dip section, and the ground section (Figure A-55).

The catenary section is the portion of the chain that maintains constant curvature and is located between the riser section and the touchdown point of the chain. The dip section is the portion of the chain leg that initially comes in contact with the seabed. This section of the chain leg receives the most wear, abrading the seabed as it moves. The ground section is the portion of the static chain leg on the sea floor. Vertical movement of the mooring element does not move the chain leg at this section.

The Kenter link, which is a special link used to connect two lengths of chain, should be inspected for the connection stud, keeper pins, and lead pellets (plugs). A Kenter link that connects different size diameter chain lengths may have a studless link to allow for the connection; however, the keeper pins and lead pellets are still required.

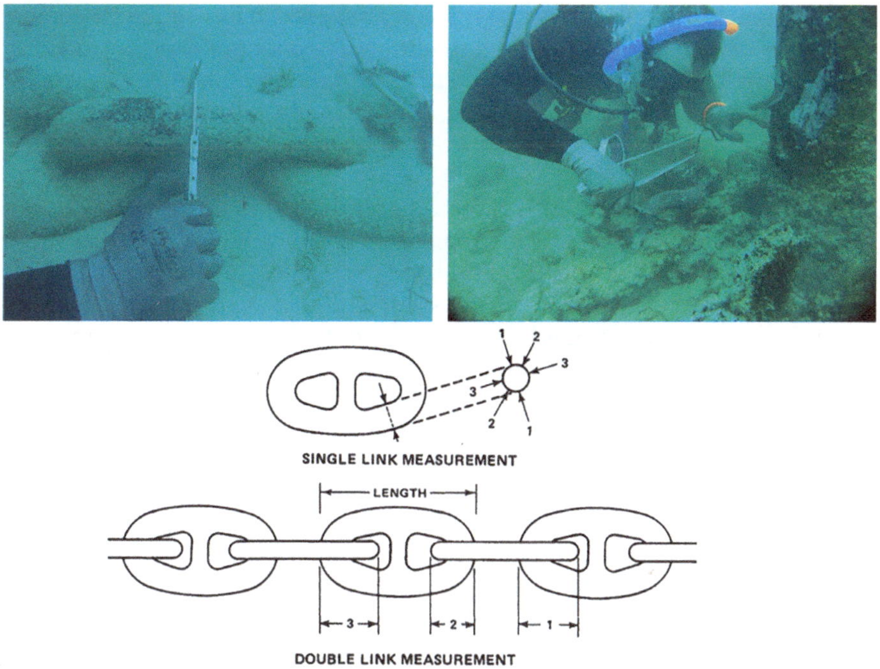

Figure A-55. (Upper left) Measurement of chain hardware, single link; (upper right) double link.
Source: (Top) Courtesy of Moffatt & Nichol Inc., reproduced with permission; (bottom) US Navy (1987).

A.20.2.2 Anchor Subassembly. A pop-up float may be attached to an anchor so that the position of the anchor relative to the mooring buoy can be observed from the surface. Buried anchors are normally not uncovered for inspection. However, wherever possible, the anchor and anchor chain connection should be inspected for wear, corrosion, and alignment. If possible, the type of anchor and connecting hardware should be identified. If anchors are found to be uncovered, investigate the cause of uncovering. Some mooring designs have two anchors in tandem separated by one chain shot. The anchor closest to the buoy is called the *buoy anchor*, and the one farthest from the buoy is called the *tandem anchor*. The buoy anchor is connected to the ground section at the shank end and to the chain at the crown end of the anchor. The chain at the crown end of the anchor is located at the opposite end of the shank end and also connects the flukes.

A.20.2.3 Anchor Types. Several types of anchors can be used in moorings, including drag-embedment (conventional) anchors, pile anchors, deadweight anchors, suction pile anchors, and direct-embedment anchors. Each type of anchor works using different mechanisms. Inspect the anchors' general condition and whether they are properly deployed.

A.20.2.3.1 Drag-Embedment Anchors. Drag-embedment anchors are the most commonly used anchors in moorings (Figure A-56). Drag-embedment anchors have an anchor shank that is used to transfer the mooring line load to the anchor flukes, which have large surface areas to mobilize soil resistance. The leading edge of a fluke, called the fluke tip, is sharp so that the fluke will penetrate the sea floor. Tripping palms, located at the trailing edge of the flukes, cause the flukes to open and penetrate the sea floor (Figure A-57). The shank-fluke connection region is called the crown of the anchor. Some anchors have stabilizer bars located at the anchor crown oriented perpendicularly to the shank. Stabilizers resist rotational instability of the anchor under load.

Drag-embedment anchor performance is sensitive to sea-floor soil type and drag angle. Drag-embedment anchors are designed to resist horizontal loading. A near-zero angle between the anchor shank and the sea floor (shank angle) is required to ensure horizontal loading at the anchor.

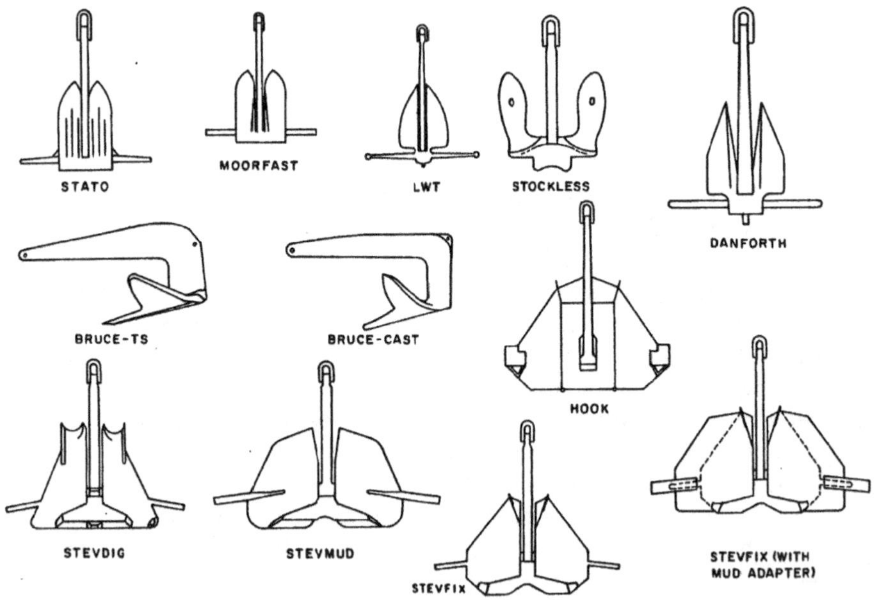

Figure A-56. Types of drag embedment anchors.
Source: US Navy (1985).

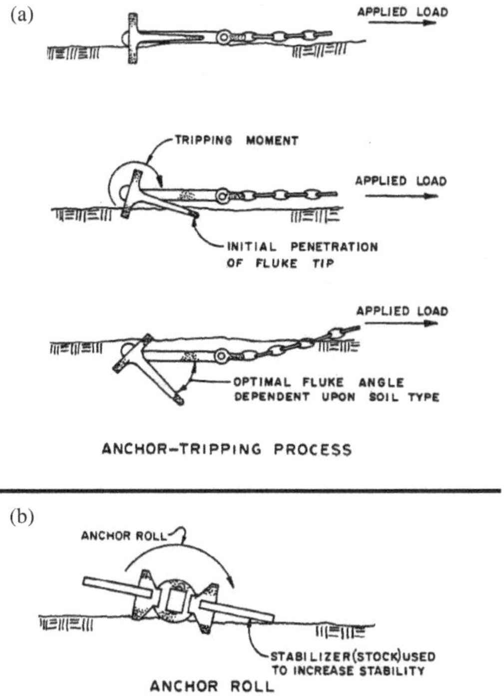

Figure A-57. Performance of drag embedment anchors under loading. Source: US Navy (1985).

Sufficient scope in the mooring line will result in a near-zero shank angle. (Scope is defined as the ratio of the mooring line length from the mooring buoy to the anchor to the water depth.) As the shank angle increases from zero, the vertical load on the anchor increases, and the anchor's holding power decreases.

A weight called a sinker, usually made of concrete, may be placed on a mooring leg in a location to ensure horizontal loading at the anchor end and to aid in energy absorption. A steel rod (called a hairpin) is cast into the sinker to provide the connection to a mooring chain. The connection between the mooring chain and the sinker is critical. If this connection fails, the sinker will be lost, and the entire mooring may fail. Therefore, certain precautions must be observed. First, the connection must allow free movement of the chain links to avoid distortion and failure of the links. Second, a sinker must not be cast around the chain itself.

A.20.2.3.2 Pile Anchors. A pile anchor consists of a structural member, driven vertically into the sea floor, designed to withstand lateral (horizontal) and axial (vertical) loading (Figure A-58). Pile anchors are usually

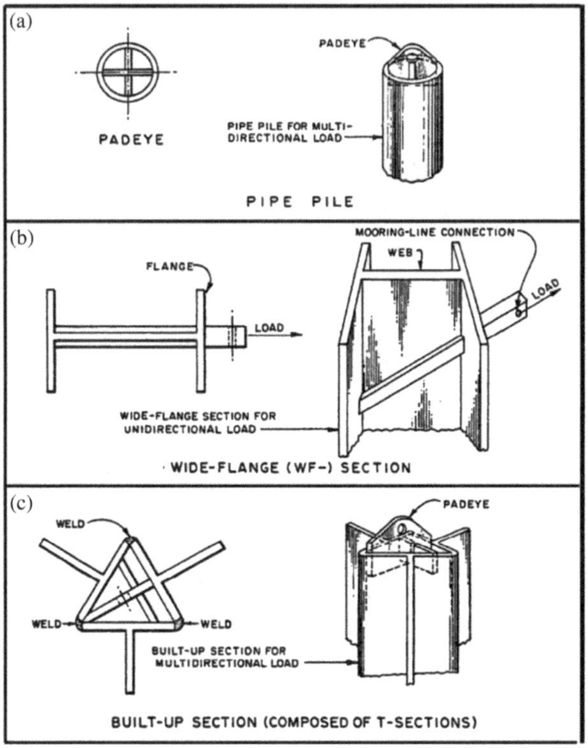

Figure A-58. Types of pile anchors.
Source: US Navy (1985).

standard structural steel shapes fitted with a mooring-line connection; however, they may be composed of concrete, timber, polymer composites, or steel helical shapes screwed into the ground. Piles achieve their lateral and axial holding capacity by mobilizing the strength of the surrounding sea-floor soil. The lateral strength of a pile anchor derives from lateral earth pressure and its axial strength derives from skin friction.

Pile anchors can fail in three ways: by pulling out of the sea floor, by excessive deflection, or by structural failure. In the first, the anchor pile may pull out of the sea floor when uplift loads exceed the axial capacity offered by skin friction. In the second, lateral loads applied at the upper end of the pile cause the pile head and surrounding soil to deflect. Excessive and repeated deflections of the pile head and surrounding soil will cause a reduction in soil confining stress and may result in failure of the pile anchor. Finally, large lateral loads on a pile may result in stresses in the pile that exceed its structural strength. Inspect the anchor for structural defects or failures, especially at connection points, and note lateral deflections indicative of previous overloading.

A.20.2.3.3 Deadweight Anchors. A deadweight anchor is a large mass of concrete or steel that relies on its own weight to resist lateral and uplift loading (Figure A-59). Deadweight-anchor construction may vary from simple concrete clumps to specially manufactured concrete and steel anchors with shear keys. Deadweight anchors are designed to withstand uplift and lateral loads and overturning moments. soil strength primarily through friction. In cohesive soils, deadweight anchors may also have shear keys to take advantage of passive resistance from the soils. A deadweight anchor will drag when the applied load exceeds the resistance offered by static friction. Look for signs of dragging along the seafloor and connection failures at the anchor attachment points.

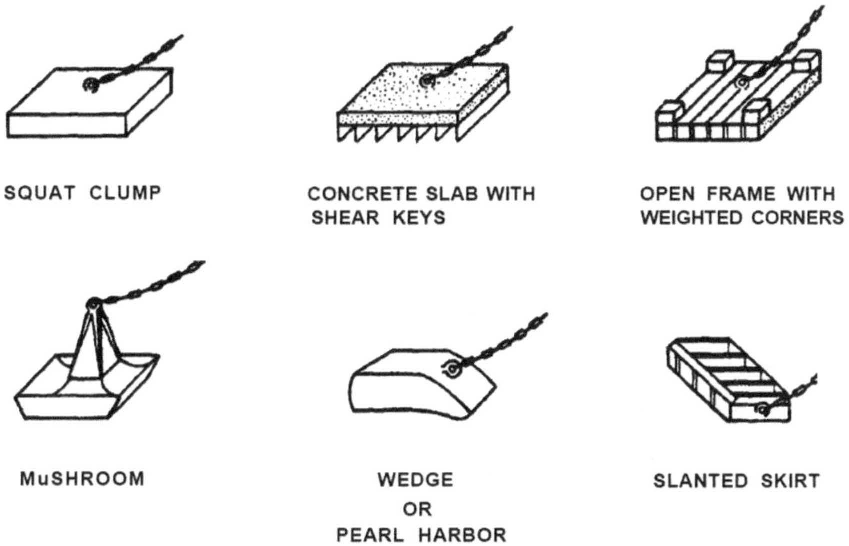

Figure A-59. Types of deadweight anchors.
Source: US Navy (1985).

A.20.2.3.4 Suction Pile Anchors. Suction anchors are typically large-diameter, cylindrical steel piles with a sealed cap and valve at the top. Once suction anchors are lowered to the sea floor, the top valve is closed, and water is pumped out from the inside of the pipe. The net differential hydrostatic pressure suctions the anchor down and drives the anchor into the sea floor.

Suction anchors gain their vertical capacity through skin friction along the outside of the anchor from surrounding soils and along the inside of the anchor from the soil plug. An inspection of suction pile anchors should be performed similarly to other structural steel piles, checking for damage to the pile, lateral deflections, and damage at the connection points.

A.20.2.3.5 Direct-Embedment Anchors. A direct-embedment anchor is driven, vibrated, propelled, or screwed into the sea floor, after which the anchor fluke is expanded or reoriented to increase pullout resistance (Figure A-60). Direct-embedment anchors are capable of withstanding both uplift and lateral loading. Direct-embedment anchors achieve their holding capacity by mobilizing soil-bearing strength. Shallow anchor failure is characterized by the removal of the soil plug overlying the anchor fluke as the anchor is displaced under loading. A deep anchor failure occurs when soil flows from above to below the anchor as the anchor is displaced under load. The tendency toward the shallow or deep anchor-failure mode depends on the anchor fluke size and the embedment depth. Because direct embedment anchors are typically below the sea floor, the inspection is limited to observations of the surrounding soils. Look for signs of soil displacement.

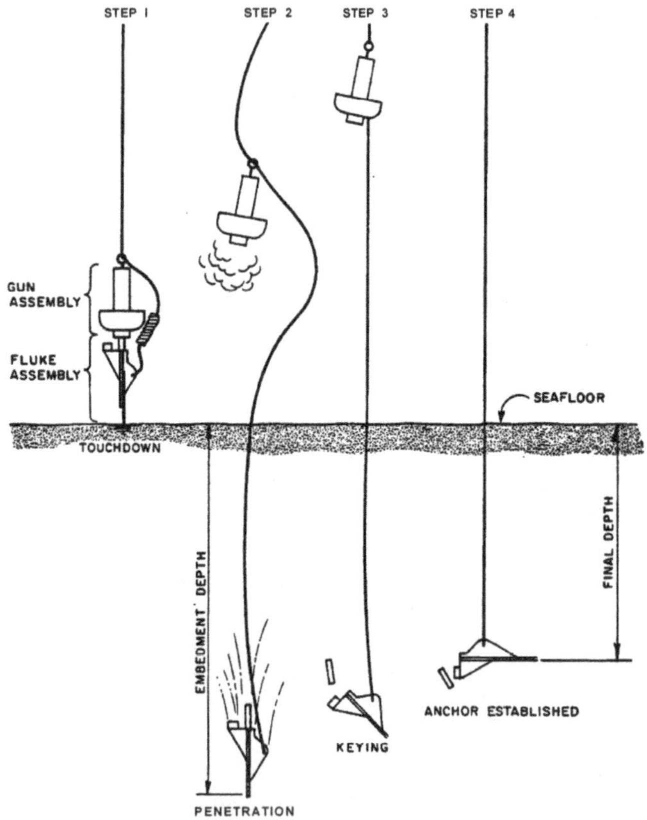

Figure A-60. Propellant-embedded anchor.
Source: US Navy (1985).

REFERENCES

ASCE. 2023. *Dry dock standard*, ASCE/COPRI 77-22. Reston, VA: ASCE.

OSHA (Occupational Safety and Health Administration). 2011. *Permit-required confined spaces*, 29 CFR 1910.146, Subpart J. Washington, DC: OSHA.

USACE (US Army Corps of Engineers). 2024. *Safety and health (SOH) requirements manual*, EM-385-1-1. Washington, DC: USACE.

US Navy. 1985. *Fleet moorings basic criteria and planning guidelines*, Design Manual 26.5. Washington, DC: Naval Facilities Engineering Command.

US Navy. 1987. *Mooring maintenance manual*, NAVFAC MO-124. Washington, DC: Naval Facilities Engineering Command.

APPENDIX B
TYPES AND CAUSES OF DEFECTS

B.1 INTRODUCTION

This appendix provides general information on various types and causes of material damage and deterioration, collectively referred to as *defects*. The types of defects described are those most encountered when inspecting marine structures. Some of the defects discussed may be found in many other types of construction (buildings, roadways, etc.); however, all types and causes of defects discussed stem from practices commonly used in marine construction, exposure to the marine environment, or applied forces specific to waterfront facilities.

This appendix intends to overview the most encountered defects. Although some defects discussed are complex, the subjects are only briefly discussed, providing a short summary of the information useful in identifying, assessing, and addressing these conditions. The reader is encouraged to pursue additional sources to further explore these conditions' scientific and engineering aspects.

B.2 CONCRETE STRUCTURES

Most types of defects that are found on concrete waterfront structures are common to all concrete structures. However, certain types of defects are more commonly found on waterfront structures and can lead to severe consequences in the marine environment. Note that these defects typically apply to the various types of concrete members, including unreinforced, reinforced, precast, prestressed, and fiber-reinforced concrete. Although not specifically addressed in this manual, it is important to consider the differing effects the various defects discussed may have on the capabilities of the affected member.

B.2.1 Cracking

Cracking is a linear separation of concrete into two or more parts caused by induced tensile stresses in the concrete that are greater than the tensile strength of the concrete. The cracking may extend partially or completely through the member. Cracking is often classified as structural, nonstructural, and degradation type cracking. The type of cracking is identified by the position of the crack on the member as well as an understanding of load paths and deterioration mechanisms.

For waterfront structures, the forces that induce cracking can be extreme when compared to other applications and environments. Live loads from cargo, ship berthing and mooring, cargo-handling equipment, and vehicles often apply large and concentrated forces to the structure. In the marine environment, members are submerged, wetted, and dried by tidal or flood variations or splash, and they are exposed to high levels of humidity and large variations in temperatures. Perhaps most detrimental is the introduction of chlorides, sulfates, and other chemical constituents from salt or fresh water and the marine atmosphere that can have deleterious effects on concrete members (Figure B-1).

Figure B-1. Cracking of a reinforced concrete pile cap under a pier. Source: Courtesy of Marine Solutions Inc., reproduced with permission.

Structural cracks caused by stresses from imposed loads can often be identified as flexural or shear cracks. Flexural cracks often begin at the maximum tension zone and progress toward the compression zone, commonly occurring within the areas of maximum positive or negative moment of force. Shear cracks are often diagonal or vertical cracks that normally occur in a member's web or near a point of bearing where maximum shear stress is applied. They often begin at one end of the member and extend diagonally toward the center, but may be vertical (Figure B-2).

Figure B-2. Structural crack on a reinforced concrete pile cap.
Source: Courtesy of Marine Solutions Inc., reproduced with permission.

Nonstructural cracking includes cracks caused by improper placement, curing, or finishing, temperature cracks caused by thermal expansion and contraction of the concrete, shrinkage cracks caused by contraction of the concrete during curing, and mass concrete cracks created by thermal gradients in large concrete sections (Figures B-3 and B-4). Shrinkage cracking is common in the marine environment where highly variable temperatures often occur.

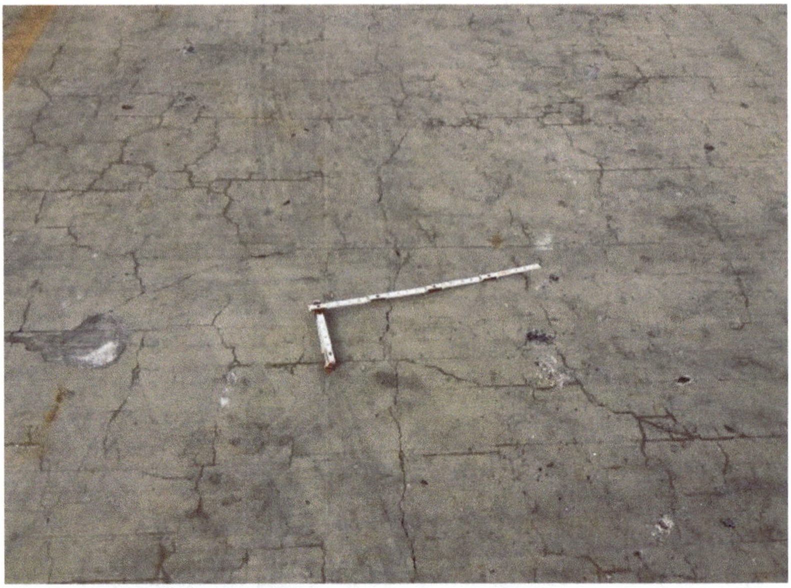

Figure B-3. Nonstructural cracking on the surface of a reinforced concrete deck. Source: Courtesy of Marine Solutions Inc., reproduced with permission.

Figure B-4. Nonstructural cracking on the face of a concrete bulkhead. Source: Courtesy of Marine Solutions Inc., reproduced with permission.

Various types of shrinkage include plastic shrinkage, drying shrinkage, autogenous shrinkage, thermal shrinkage, and carbonation shrinkage. Although nonstructural cracks are not attributed to insufficient load-carrying capacity, they can be problematic because they provide openings for the intrusion of water, chlorides, sulfates, and other contaminants into the concrete matrix or to internal steel reinforcement.

Degradation cracking of concrete is caused by a form of deterioration such as delamination of the concrete surfaces, corrosion of reinforcing steel, or chemical reactions within the concrete matrix. In the marine environment, where many degrading agents exist, degradation cracking often leads to progressive deterioration much faster than cracks formed in other environments. Additional discussion of the various types of degradation that induce this type of cracking can be found in subsequent sections of this appendix.

B.2.2 Corrosion of Reinforcing Steel

Corrosion of internal reinforcing steel is often the limiting factor in the service life of reinforced concrete elements placed in the marine environment. The corrosion of reinforcing steel in concrete largely occurs because of a change in the chemical composition of the concrete substrate from exposure to the environment. Corrosion is an electrochemical process requiring an anode, a cathode, and an electrolyte. A moist concrete matrix provides an acceptable electrolyte, and the steel reinforcement provides the anode and cathode. Electrical current flows between the cathode- and the anode-transferring electrons and, with the electrons, mass in the form of steel particles.

At an early age, concrete has a high level of alkalinity (pH) that results in the transformation of a surface film of ferric hydroxide covering the reinforcing steel. If this passivation film is not disturbed, it will keep the steel passive and protected from corrosion. The corrosion process begins with the penetration of deleterious elements, such as chlorides or carbon dioxide, into the porous concrete. Over time, these elements penetrate the reinforcing steel, at which point the protective film is destroyed, and the steel begins to corrode when oxygen and moisture are present (Figure B-5).

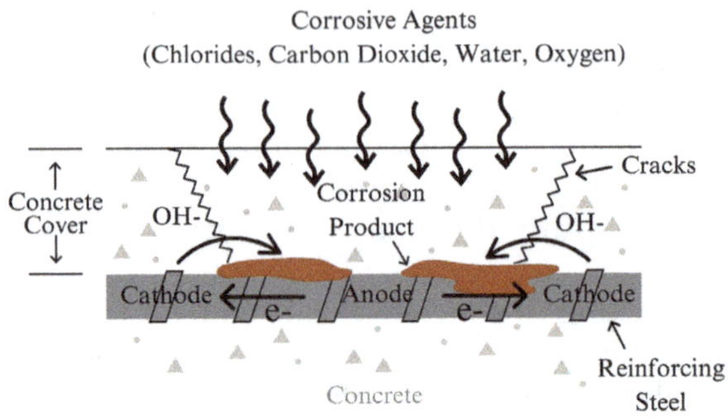

Figure B-5. Illustration of the corrosion process of reinforcement in concrete. Source: Courtesy of Vector Corrosion Technologies Inc., reproduced with permission.

The rate of corrosion can be very slow in good-quality concrete. Accelerated corrosion will occur if the alkalinity is lowered, aggressive chemicals are present, or dissimilar metals are introduced into the concrete. Stray electrical currents and concentration cells caused by an uneven chemical environment can also speed up corrosion. In addition, concrete structures can also be vulnerable to corrosion of reinforcing steel by the action of sulfate-reducing bacteria found in sediments, spilled industrial products, or industrial waste discharge.

This action produces a porous and disintegrating matrix. In each case, the amount of concrete cover over the reinforcing steel is a critical factor in the induction of corrosion of the reinforcing steel. As such, mechanical damage can also reduce the layer of concrete over the reinforcing steel, initiating and accelerating internal corrosion.

Damage from corrosion of reinforcing steel is often first observed within the tidal or splash zone where oxygen and moisture are both abundant. Shallow concrete cover over the reinforcing steel can be a primary instigator of the corrosion process. Further, damage from corrosion of reinforcing steel is often the most severe at the corners of the members where there is increased exposure to seawater and oxygen, two exposed concrete faces, a concentration of longitudinal and transverse reinforcement, and often less concrete cover over the reinforcing steel.

Cracking and spalling caused by the expansive forces allow increased access by oxygen, chlorides, and moisture, intensifying corrosion and accelerating steel loss. The intrusion of chlorides into the concrete from rainwater, saltwater spray, and deicing agents also contribute to the corrosion process. These agents penetrate the concrete either through pores, cracks, and holes, or by diffusion, alldue oxygen and water to attack the reinforcing steel. Subsequently, the concrete cracking and spalling, the loss of bonding of the steel reinforcement to the concrete, and the loss of the steel reinforcement bar section, threaten the structural integrity and load-bearing capacity of the concrete member.

Structural implications of internal corrosion are more acute in prestressed concrete than in conventionally reinforced concrete. This is caused by the difference in the quantity of steel in the two types of construction, with the steel strands used in prestressed construction being considerably smaller than the cross-sectional area of reinforcing steel used in non-prestressed construction. The concrete requires the same amount of steel corrosion by-products to crack the concrete surface, but a larger percentage of the prestressed strand is consumed before the concrete cracks. This reduction in steel area is such that little of the strand's cross-sectional area may be left by the time the initial crack forms and becomes visible.

B.2.2.1 Corrosion Cracking. A corrosion crack is a splitting crack that occurs in concrete because of the expansion of the chemical products generated by the corrosion of steel reinforcement. Corrosion cracks are formed when the internal stresses produced by the expansion of corrosion products exceed the tensile strength of the concrete. Corrosion cracks appear close to the corroding steel and propagate along the bars. The main characteristic of these cracks is their location, which is adjacent and parallel to corroding reinforcing.

Corrosion cracks can vary in width, with the maximum width at the point of most intense corrosion; leading edges with an opening that gradually reduces until the crack tapers out; and red, orange, or brown corrosion products (rust staining), which often bleed from the crack. Corrosion cracking often first appears at the corner of a concrete member because the concrete has less restraint and less resistance to the internal pressure of the expansion products (Figures B-6 and B-7).

Figure B-6. Typical corrosion crack on a reinforced concrete pile.
Source: Courtesy of Marine Solutions Inc., reproduced with permission.

Figure B-7. Typical corrosion crack on a reinforced concrete pile cap. Source: Courtesy of Marine Solutions Inc., reproduced with permission.

In marine structures, members subjected to seawater spray, such as piling, pile caps, beams, and the underside of deck slabs, are more susceptible to corrosion cracking because of the accelerated increase in chlorides caused by wave action against the structure (i.e., splash, spray, and wicking). The initiation of corrosion cracks accelerates the corrosion process, alldue direct access of oxygen, moisture, and chlorides to the reinforcing steel.

B.2.2.2 Closed Corrosion Spall. As corrosion cracking continues, the defect may degrade into a closed corrosion spall. A closed corrosion spall is defined as a partially separated section of concrete completely or partially surrounded by cracks induced by corrosion of the underlying reinforcing steel. When it is struck with a hammer or steel rod, the spall gives off a hollow sound, indicating the existence of a fracture plane below the surface.

It is the intermediate stage in the process of complete separation of a fragment of concrete cover. As such, a closed corrosion spall is often referred to as an *impending spall*. A closed corrosion spall often indicates advanced corrosion, potentially significant section loss of the affected reinforcing steel bar, and a loss of bonding between the concrete and the steel bar (Figure B-8).

Figure B-8. A closed corrosion spall on a reinforced concrete beam.
Source: Courtesy of Marine Solutions Inc., reproduced with permission.

Figure B-9. Open corrosion spall on the corner of a reinforced concrete pile cap.
Source: Courtesy of Marine Solutions Inc., reproduced with permission.

B.2.2.3 Open Corrosion Spall. An open corrosion spall is an area of broken off and missing concrete section caused by corrosion of the underlying reinforcing steel. The steel is usually covered with corrosion by-products at this stage (Figure B-9).

In open corrosion spalls, the structural significance of the damage depends on the extent of the loss of the concrete- and steel-reinforcing bar section and the loss of bonding between the steel bar and the concrete. Open spalls can also increase the rate of corrosion of further internal reinforcement through the newly exposed, unfinished concrete surfaces (Figures B-10 and B-11).

Figure B-10. Open corrosion spall on the underside of a reinforced concrete deck. Source: Courtesy of Marine Solutions Inc., reproduced with permission.

Figure B-11. Open corrosion spall on the underside of a reinforced concrete beam. Source: Courtesy of Marine Solutions Inc., reproduced with permission.

B.2.2.4 Delamination. Delamination is a separation of layers of concrete along a plane parallel to, and near, the surface of the concrete. Delamination is commonly caused by premature or improper finishing or curing causing a thin layer of entrapped air or water within a few inches of the concrete surface. Delamination can also be caused by corrosion of the reinforcing steel near the concrete surface or poor bond between two-course placements. Concrete placed in the marine environment is especially susceptible to delamination due to variable conditions during placement, finishing, and curing (i.e., wind, sun, temperature, and humidity). Nontypical configurations and challenging access result in reinforcing steel placement with insufficient embedment and concrete cover.

The delamination may range in size from several square feet to very large areas. The concrete surface may exhibit cracking and color differences before separating and breaking away because of traffic, freezing, or moisture. Delamination can be detected by a hollow sound when tapped with a hammer or steel bar, a roller percussion tool, heavy chain drag, or impact echo and ground-penetrating radar technologies (Figure B-12).

Delamination can create an undesirable working surface and lead to further deterioration of the exposed unfinished concrete areas, spalling, and loss of concrete section. If reinforcing steel is exposed, the structural capacity of the member may be affected (Figure B-13).

Figure B-12. Delaminated area marked after sounding the concrete on a reinforced concrete pier deck.
Source: Courtesy of Marine Solutions Inc., reproduced with permission.

Figure B-13. Delaminated area on a reinforced concrete pier deck.
Source: Courtesy of Marine Solutions Inc., reproduced with permission.

B.2.3 Chemical Attack

A chemical attack occurs because of the constituents of the concrete mixture, curing methods, and, at times, the ambient environment. The most common types of chemical attack found in marine structures are alkali–aggregate reaction, chloride contamination, sulfate attack, delayed ettringite formation, carbonation, and acid attack.

B.2.3.1 Alkali–Aggregate Reaction. Alkali–aggregate reaction (AAR) can cause expansion and cracking in concrete. The two most common forms are alkali–silica reaction (ASR) and alkali–carbonate reaction (ACR). Both are caused by a reaction between chemical ions in the alkaline cement solution in concrete and reactive forms of aggregate. An expansive gel is produced as a product of the reaction. The resulting increase in volume exerts an expansive pressure-induced cracking. The moisture-rich marine environment can accelerate this reaction. Reactive silica aggregates are much more common than reactive carbonate reactions; therefore, ASR is the more common reaction.

Alkali–aggregate reaction may be typically first observed as patterned or map cracking reflecting the expansive cracks forming around aggregate and the internal pressure generated. Over time, bulging and distortion of the concrete member can occur. Cracking due to chemical attack can be differentiated from cracking caused by finishing, curing, or volumetric changes by the depth of cracking and by petrographic examination to detect chemical reactions. The reaction may create a small popout if the reactive aggregate is close to the surface.

B.2.3.1.1 Alkali–Silica Reaction. Alkali–silica reaction occurs when certain aggregates, such as reactive forms of silica, react with alkalis, typically potassium and sodium, and calcium hydroxide from the cement. When sufficient moisture is present, a gel forms around the reacting aggregates. As the gel develops, it expands, causing microtension cracks to form around the aggregate. The gel continues to form until the reactive aggregate or alkalis are consumed. The reaction can continue almost indefinitely in the marine environment, where an abundance of alkali occurs.

Alkali–aggregate reaction causes microcracking within the concrete matrix. Damage is often first noticeable when longitudinal cracks develop in areas with the least restraint. The damage may progress with macrocracking, pattern cracking, and delamination (Figure B-14). Alkali–aggregate reaction may be found above water, but because of the presence of moisture and alkalis found in seawater, the greatest amount of damage from ASR may be found below water. Corrosion of the reinforcing steel exposed by the degradation resulting from the chemical attack is a secondary effect and can sometimes be misdiagnosed as the primary cause of damage.

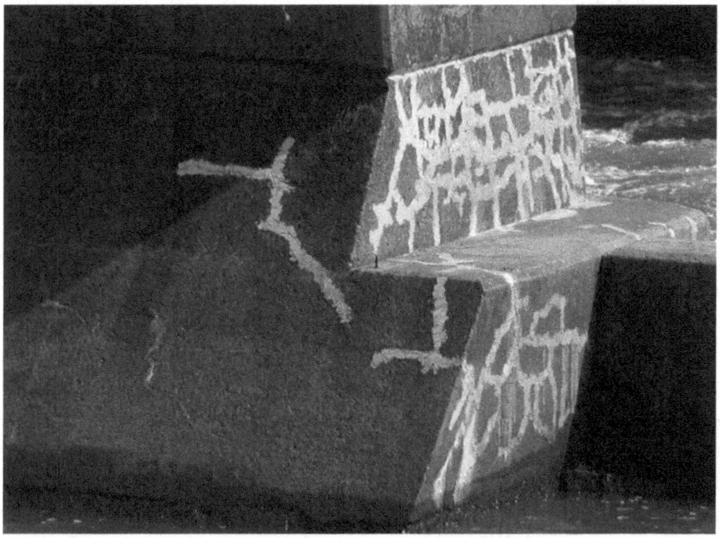

Figure B-14. Repaired epoxy-covered cracks formed from ASR.
Source: Courtesy of Marine Solutions Inc., reproduced with permission.

B.2.3.1.2 Alkali–Carbonate Reaction. Alkali–carbonate reaction can cause damage like ASR. Alkali–carbonate reaction occurs between some of the cement's dolomitic limestone aggregates and alkalis. The reaction forms a gel, which swells, causing cracking around the reacting aggregates when sufficient moisture is present. As the gel swells, it expands, causing micro-tension cracks to form around the aggregate. The gel continues to form until the reactive aggregate or alkalis are consumed or the moisture is depleted.

Like ASR, ACR cracks are typically oriented vertically and are often found in a member's submerged zone. Deterioration is believed to take place underwater because the seawater provides additional alkalis for the reaction and intensifies it.

Alkali–carbonate reaction cracks do not significantly reduce a member's capacity, as long as they remain shallow. Deeper ACR cracks have the potential to expose the reinforcing steel to corrosion and, thus, may seriously affect the integrity of prestressed members.

B.2.3.2 Chloride Contamination. Chloride contamination is initiated by the presence of chlorides within the concrete mix and can naturally occur in certain aggregates. Concrete made with beach sand or mixed with seawater has an increased level of chlorides from the outset. Water-soluble chlorides are the most damaging because they readily recrystallize within the concrete. The newly formed salt crystals cause the swelling of capillary cavities in the concrete, and the expansive force of the increase in volume can cause microcracking, leading to the disintegration of the concrete. Salt water and other external contaminates will also introduce chloride contamination to the concrete.

B.2.3.3 Sulfate Attack. Sulfate attack is caused by exposure of the concrete to sulfates; the most common sulfates are sodium, potassium, magnesium, and calcium. In the presence of moisture, the sodium, potassium, and calcium sulfates react with various hydration products to form ettringite. The magnesium sulfate reacts with the hydration products to break down the calcium silicate hydrate present in cement. Sulfate attack often occurs on members exposed to industrial chemicals, often ferilizers, spilled onto the structure.

Sulfate attack often leads to softening of the concrete. Damage is often first noticeable at the corners of the members where longitudinal cracks may develop. The damage will progress to degradation of the corners and eventual rounding of the member (Figure B-15). Often, the concrete will degrade to a soft, chalk-like texture that crumbles and can be removed and broken by hand. Accelerated by moisture and sulfates found in seawater, the greatest amount of damage from sulfate attack is found below water. Corrosion of the reinforcing steel exposed by the degradation resulting from the chemical attack is a secondary effect and can sometimes be misdiagnosed as the primary cause of damage.

Figure B-15. Underwater photograph of a reinforced concrete pile with softening and disintegration of the concrete and exposure of the steel reinforcement. Source: Courtesy of Marine Solutions Inc., reproduced with permission.

B.2.3.4 Delayed Ettringite Formation. Delayed ettringite formation (DEF) is a unique type of chemical attack affecting marine structures. Many as-built factors, such as concrete composition, curing conditions, and exposure, influence the potential for sulfate attack; however, DEF is believed to be a result of improper heat curing of the concrete.

DEF occurs when the concrete is subjected to elevated temperatures during hydration. Ettringite is not stable at elevated temperatures, resulting in the formation of monosulfates. When the concrete returns to normal temperatures, the monosulfates are not stable, thus ettringite is formed; however, this delayed ettringite is formed while the concrete is in a hardened state and with the presence of moisture. The ettringite forms and, if space is insufficient, the ettringite formation will cause expansion. Because of this expansion, microcracks form around the aggregates and are filled with ettringite. The formation of the ettringite can lead to extensive damage (Figure B-16).

Damage is often first noticeable when longitudinal cracks develop in areas with the least restraint. The damage may progress with macrocracking, pattern cracking, and delamination. DEF may be found above water in moist or humid conditions, but the greatest amount of damage from DEF is typically found below water. Corrosion of the reinforcing steel exposed by the degradation resulting from the chemical attack is a secondary effect and can sometimes be misdiagnosed as the primary cause of damage (Figure B-16).

Figure B-16. Reinforced concrete pile with degradation below the high tide level due to DEF.
Source: Courtesy of Marine Solutions Inc., reproduced with permission.

B.2.3.5 Carbonation. Carbonation is the reaction between acidic gases (usually carbon dioxide) in the atmosphere and cement hydration products. This can occur in industrial areas where the level of carbon dioxide in the air is usually higher than normal. Carbon dioxide enters through pores in the concrete by diffusion, forms a mild carbonic acid, and reacts with the calcium hydroxide dissolved in the pore water. The alkalinity of the concrete is reduced, which, when the carbonation reaches the level of the reinforcing steel, removes the passivation film, resulting in the onset of corrosion if oxygen and moisture are present. Carbonation also causes an increased concentration of chloride ions along the carbonation front, thereby increasing the corrosion.

The process of carbonation requires continuous wet–dry cycles, carbon dioxide, and moisture. At normal levels, such as in rural environments, the reaction is minimal; however, in areas with higher concentrations of carbon dioxide, such as in large cities and industrial ports, the rate of carbonation increases with the increase in carbon dioxide. Carbon dioxide amounts are normally very small in seawater. However, concentrations become higher in the presence of decaying organic matter, causing carbonation to occur. Carbonation can also take place in concrete exposed to water emanating from underground. Carbonation does not occur when concrete is constantly submerged.

B.2.3.6 Acid Attack. Concrete is highly alkaline and typically not used in the presence of acids without protection. However, acid attack of concrete is observed in a few instances. The first is in sewers, where sulfuric acid can form. Sulfuric acid creates multiple reactions, including sulfate attack and an attack of the calcium hydroxide and calcium silicate hydrates in the concrete.

The second reaction is a reaction between acidic waters and carbonate aggregates. The carbonate aggregates, typically limestone and dolostone, dissolve in the acidic water. The acidic waters are usually found in swamps and sloughs but may also be encountered in industrial areas.

B.2.4 Mechanical and Other Types of Damage

In the harsh marine environment, mechanical and other types of damage common to most concrete structures can be amplified by extreme temperature changes, moisture conditions, and exposure to chlorides and sulfates. Somewhat minor cracking or spalling may provide exposure to the concrete substrate and pathways to steel reinforcement, accelerating secondary damage from steel reinforcement corrosion or chemical attack. As such, the secondary effects of even minor damage from somewhat typical defects should be considered.

B.2.4.1 Volumetric Changes. Volumetric changes in concrete are caused by thermal expansion and contraction, plastic shrinkage, freezing and thawing, and changes in moisture content. Tensile stresses are often generated in the concrete, resulting in cracking and spalling. Significant damage can particularly be caused when the concrete member is confined.

B.2.4.2 Impact and Overload. Impact damage is caused when ships, boats, or other objects strike a concrete member. The extent of damage depends on the mass and velocity of the object and can range from superficial damage to fracture or failure of the member.

Usually, the consequence of impact is localized structural damage ranging from cracks, voids, chipped corners, and local spalling to major structural distress. This damage is usually located on a marine structure's berthing face and in the piles' intertidal zone. Impact damage can cause complete failure of a structural element or can accelerate corrosion of reinforcing steel by causing cracks or broken sections of concrete reducing or removing the concrete cover (Figure B-17).

Overload damage is caused when loads are applied to the structures in excess of their capacity, also referred to as *overstressing*. When applied loads result in stresses exceeding the concrete's compressive or tensile stress capacity, or when the yield stress of the reinforcing steel is exceeded, damage occurs. As with impact damage, the extent of the damage can range from superficial to failure of the member (Figure B-18).

Overstressing damage does not necessarily occur at the point of load application but at locations where the resulting stress meets or exceeds

Figure B-17. Reinforced concrete pile damaged from a vessel impacting the pile. Source: Courtesy of Marine Solutions Inc., reproduced with permission.

Figure B-18. Overloaded and broken reinforced concrete pile. Source: Courtesy of Marine Solutions Inc., reproduced with permission.

the structural capacity of the structural section. The result is a deformation of the member or an overstress crack, which is often characterized by sharp edges and wedges of broken concrete along the length of the crack (Figure B-19).

Figure B-19. Overstress cracks and broken sections of a reinforced concrete pile cap. Source: Courtesy of Marine Solutions Inc., reproduced with permission.

B.2.4.3 Scaling. Scaling is the gradual and constant loss of surface mortar and aggregates from an area of concrete. It is most often found near the waterline and underwater because of a breakdown of the finished surface of the concrete (Figure B-20). Improper finishing and curing can often contribute to the early onset of scaling. Scaling can lead to cracking and spalling and the eventual depletion of the concrete cover.

Figure B-20. Concrete scaling at the base of a seawall.
Source: Courtesy of Marine Solutions Inc., reproduced with permission.

B.2.4.4 Freeze–Thaw Damage. Freeze–thaw damage takes place when freezing and thawing cycles act on porous concrete that has absorbed water. Water contained in the pores expands as it freezes, causing expansion forces that break the surrounding concrete, which is comparatively weak in tension. The disintegration occurs in small pieces, working from the outer surfaces inward. Freeze–thaw deterioration typically occurs on vertical surfaces that are near the waterline where they are exposed to wave action and spray (Figure B-21).

Figure B-21. Freeze–thaw damage along the face of a reinforced concrete bulkhead. Source: Courtesy of Marine Solutions Inc., reproduced with permission.

B.2.4.5 Honeycombing. Honeycombs are voids or hollows in the concrete and are a construction deficiency caused by inadequate concrete consolidation. The lack of vibration segregates the coarse aggregates from the fine aggregates and the cement paste, creating small voids and pockets within the concrete (Figure B-22). Honeycombing can occur on the interior or the surface of a concrete component. Honeycombing is often an as-built condition that can lead to further deterioration of the affected area.

The impact of honeycombing on the structural capacity of concrete components depends on the size and depth of the area affected. These areas will have high permeability and will be more susceptible to chemical attack. Surface honeycombing has the additional potential of alldue the penetration of corrosion agents into the reinforcing steel and eventual cracking and spalling (Figure B-23).

Figure B-22. Honeycombing on the side of a reinforced concrete bulkhead. Source: Courtesy of Marine Solutions Inc., reproduced with permission.

Figure B-23. Honeycombing on the underside of a reinforced concrete pier deck. Source: Courtesy of Marine Solutions Inc., reproduced with permission.

B.2.4.6 Popouts. Popouts are shallow, cone-shaped holes in the surface of the concrete. They are formed when conical fragments break away from the concrete surface. A shattered aggregate particle is usually found at the bottom of the hole, and another piece of this particle may still be attached to the small end of the cone that popped out, if present.

The presence of reactive aggregates, high alkali cement, or salt water can cause popouts. They can occur when aggregates that expand with moisture, such as shale, are in the makeup of the concrete or when the ends of internal reinforcing steel corrode, developing localized expansive forces. They can also occur when concrete surfaces are frequently wetted by salt water, resulting in an increased concentration of chloride ions fueling alkali reactions (Figure B-24).

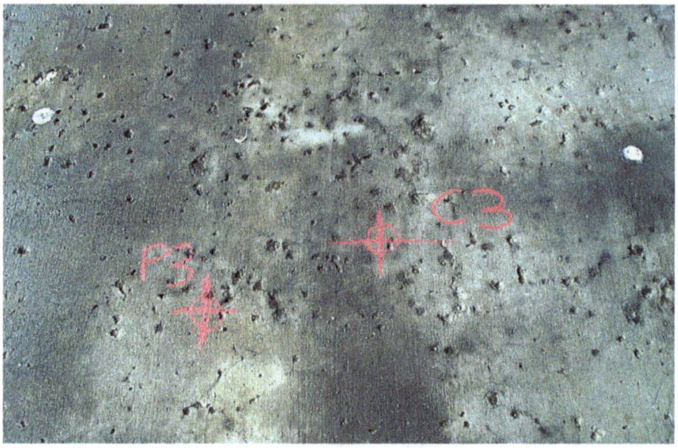

Figure B-24. Close-up of concrete pier deck affected by alkali silica reaction causing pop-outs of aggregate from the deck surface.
Source: Courtesy of Moffatt & Nichol, reproduced with permission.

B.2.4.7 Abrasion or Erosion. Abrasion of the surface of concrete structures is the result of external forces acting on the concrete. Major causes of abrasion or erosion are wave action, waterborne solids (debris), friction from floating components, propeller wash, and water currents (Figure B-25).

Figure B-25. Severe erosion of concrete due to chemical attack and wave action on degraded concrete sheet piles.
Source: Courtesy of Marine Solutions Inc., reproduced with permission.

B.2.4.7.1 Waterborne Solids. Floating debris, detached fender components, sand, rocks, and other debris carried in wave or current action can cause abrasion damage at the water line or the tidal zone. Ice floes can also abrade at this elevation. Abrasion can also occur at the mudline caused by the action of abrasive material carried in the swift current of some tidal areas and rivers (Figure B-26).

Figure B-26. Damage to a prestressed reinforced concrete pile from floating debris. Source: Courtesy of Marine Solutions Inc., reproduced with permission.

B.2.4.7.2 Friction. Floating or moving components such as docks, fenders, chains, mooring lines, or other moving components can be a source of continuous friction and can damage concrete components (Figure B-27).

Figure B-27. Damage to a concrete pile from friction against a floating dock. Source: Courtesy of Marine Solutions Inc., reproduced with permission.

B.2.4.7.3 Propeller Wash and Water Currents. Wash from vessels repeatedly and quickly starting and reversing their propellers (e.g., tug or ferry boats) can act like an abrasive on underwater components. Heavy currents from rivers and large tide cycles can degrade concrete surfaces by erosion and scaling. Often cosmetic, severe scaling, erosion, gouges, and cavities can occur (Figure B-28).

Figure B-28. Diver measuring severe concrete erosion damage. Source: Courtesy of Marine Solutions Inc., reproduced with permission.

B.2.4.8 Biological Deterioration. Organisms that grow on concrete can affect its condition. In fresh and brackish water, algae and hydroids can grow on a structure's continuously moist or submerged zones. They form a dense covering, which tends to seal the concrete and decrease gas permeability, consequently reducing carbonation and the availability of oxygen (Figure B-29). Depending on the type of marine growth, accumulation, and density, marine growth may promote or deter other causes of deterioration.

Plants of higher orders, such as seaweed, in seawater, are potentially aggressive agents. They will not grow until carbonation lowers the pH of the concrete. When they do become established, the root systems can break down the concrete. In addition, the release of carbon dioxide during daylight hours may increase carbonation, and the action of sulfur developed during the decomposition of seaweed can degrade the concrete.

Figure B-29. Heavy marine growth including barnacles, muscles, and algae located on the underside of a pier.
Source: Courtesy of Marine Solutions Inc., reproduced with permission.

B.2.4.9 Contamination. Certain chemical solutions will attack marine structure components. Various acids, alkaline solutions, and salt solutions are all examples of these aggressive chemicals. Chemical reactions on concrete involve the reaction between the acid and the calcium hydroxide of hydrated portland cement. The reaction produces water-soluble calcium compounds, which are leached away. When limestone or dolomite aggregates are used, the acid may dissolve them.

B.2.4.10 Weathering. Gradual weathering by sun, wind, and water is usually indicated by erosion or shallow, fine cracking of the concrete surfaces (i.e., erosion, scaling, aggregate popouts, rounded corners, or shallow pattern cracking). The location and severity of the weathering often varies by the contributing environmental condition and may be affected by the finishing and curing practices used. For example, weathering may be more severe in areas subject to heavy water currents or underwater, where unfinished surfaces are present.

B.3 STEEL STRUCTURES

B.3.1 Steel Deterioration

Deterioration of steel structures in the marine environment is typically caused by corrosion, fatigue cracking, and impact or overload damage. Two or more of these agents often work collectively to cause the degradation.

B.3.1.1 Corrosion. Corrosion of steel is the deterioration and eventual destruction of the metal due to its reaction with the environment. Chemically, it is the transformation of a metal to its oxide through a reaction involving oxygen, water, and other agents. Corrosion is most advanced slightly above high water (splash zone) and slightly below low water but may also be advanced in the submerged zone of the member, particularly near the mudline. The rate of deterioration in these zones may be significantly affected by tide ranges and other site-specific characteristics.

B.3.1.1.1 Progression of Corrosion. Numerous factors determine the rate and progression of corrosion. These include environmental conditions, type of steel, surface protection, and other parameters.

Environmental effects include temperature, humidity, and the exposure of the metal. Warm water and higher ambient air temperature increase the rate of corrosion, as does high humidity. Exposure to the drying effects of wind and sun decreases corrosion rates, whereas sheltered areas retain moisture and can accelerate corrosion. Impurities like salt can make water a more efficient electrolyte and speed up corrosion.

The presence of organisms in swamps, bogs, heavy clay, stagnant or brackish water, and contaminated water may cause bacteriological corrosion. Movement of water in the splash zone caused by tides, waves, and high-velocity currents also affects corrosion. This type of water movement allows for a greater number of wet–dry cycles, increasing the supply of oxygen to the metal. Water movement can also facilitate the removal of the prime layer of corrosion, which normally provides some protection and helps reduce the rate and progression of corrosion.

In addition, the presence of abrading elements in the moving water can also remove the buildup of corrosion by-products and increase the rate of deterioration. Repeated marine growth removal to facilitate inspection of the same steel member or component at the same location may also increase corrosion rates in these areas.

Other factors affecting corrosion include atmospheric pollutants, animal deposits, stray electric currents, galvanic action, and surface growth. Atmospheric pollutants can act similarly to water salt, and acids from atmospheric gases can directly attack steel. Bacteria often destroy the protective film or coating on metals, forming deposits and occasionally attacking the steel itself.

Bird droppings retain moisture and form deposits, which chemically attack the steel. Stray electrical currents from adjacent sources may promote corrosion by speeding the rate of the electrochemical process. Galvanic action occurs when other metals are in contact with steel and causes corrosion similar to rust. Marine growth on steel located in seawater can occasionally deter corrosion, but it can also hide areas of damage.

B.3.1.1.2 Characteristics of Corrosion. Different types of corrosion are classified according to how the corrosion attacks the metal. Some of the more common types of corrosion encountered during waterfront facilities investigation include the folldue.

- **Uniform Corrosion.** Uniform corrosion (rusting) is the general thinning of metal in an overall manner. It occurs when bare metal is exposed to a corrosive environment and identified by uniform rust or section loss over the entire surface (Figure B-30). It comprises many small pits joined together. The corrosion by-product reduces corrosion by forming a barrier between the metal and the environment.
- **Microbial-Induced Corrosion.** Microbial-induced corrosion (MIC) is also known as accelerated low-water corrosion (ALWC). Microbial-induced corrosion is a form of localized and aggressive corrosion that typically occurs around low water and is evident by bright orange oxide (Figure B-31). Left untreated, MIC may corrode the steel at a rate of around 1/32 in. per year. Further discussion of MIC and ALWC is included in Section B.3.1.2.

Figure B-30. Uniform corrosion on steel pipe piles.
Source: Courtesy of Marine Solutions Inc., reproduced with permission.

Figure B-31. Corrosion of steel sheet piles with bright orange oxide indicating possible MIC.
Source: Courtesy of Marine Solutions Inc., reproduced with permission.

- **Crevice Corrosion.** Crevice corrosion occurs at confined locations that entrap water or moisture, alldue a stagnant solution that is trapped and not renewed. Concentrations of oxygen cells or metal ion cells in these confined areas create an environment conducive to corrosion. Chloride ions are also often trapped in crevices. Crevice corrosion is typically found between mating surfaces or between back-to-back members (Figure B-32).

Figure B-32. Concentrated crevice corrosion at the intersection of two steel members.
Source: Courtesy of Marine Solutions Inc., reproduced with permission.

- **Pitting.** Pitting is localized corrosion that causes the formation of isolated penetrations into steel surfaces. It forms when chemical or physical differences occur such as imperfections in the steel under the paint or debris deposits. Pitting can act as a stress riser and cause failure by cracking (Figure B-33).

Figure B-33. Underwater photograph of pitting on a steel pile.
Source: Courtesy of Appledore Marine Engineering Inc., reproduced with permission.

- **Galvanic Corrosion.** Galvanic corrosion occurs when metals of different types or differing corrosion protection are in contact in the presence of an electrolyte. Combination type walls where steel sheetpiles and steel piles are used in comination and connected are one example. The difference in their corrosive potential produces an electron flow, with one metal becoming the cathode and the other the anode. This results in an area of thinning and perforation of the steel. Galvanic corrosion can occur in a single piece of steel because of differing potentials within the material. Table B-1 illustrates the corrosion potential between common metals.

Table B-1. Galvanic Series in Seawater.

Material	Electrical potential (V)
Titanium	0.00
Stainless steel, Types 316, 317	−0.06
Nickel-copper alloy 400 (Monel)	−0.07
Stainless steel, Types 302, 304, 321, 347	−0.08
Nickel–chromium alloy 600 (Inconel)	−0.17
Lead	−0.22
Copper	−0.37
Naval brass, yellow brass, red brass	−0.38
Low alloy steel	−0.61
Mild steel, cast iron	−0.66
Cadmium	−0.71
Aluminium alloys	−0.88
Zinc	−1.02
Magnesium	−1.61

- **Stress Corrosion.** Stress corrosion occurs when member stresses and a corrosive environment coexists. Areas of high stress can lead to accelerated corrosion causing localized areas of section loss. Corrosion causes the initiation of discontinuities in the metal that act as stress risers, leading to section loss and possible cracks (Figure B-34).

Figure B-34. Concentrated severe corrosion at an area of high stress.
Source: Courtesy of Marine Solutions Inc., reproduced with permission.

- **Erosion Corrosion.** Erosion corrosion is the attack on a metal caused by the flow of liquid over its surface with sufficient velocity to remove adhering surface corrosion products. It is caused by particle erosion, where particles in water abrade the metal surface, wearing away the surface coating of corrosion protection products. This allows corrosion to continually attack bare metal and, consequently, speeds the rate of deterioration. It is found in areas where river currents, tidal flow, or propeller wash carry particulate matter such as silt and sand. This type of corrosion is usually identified as damage at a particular elevation or band of deterioration on the member (Figure B-35).

Figure B-35. Heavy rust scale and corrosion products eroded by wave action and water current.
Source: Courtesy of Marine Solutions Inc., reproduced with permission.

B.3.1.1.3 Effects of Corrosion. Steel components used in waterfront construction that are subject to corrosion are predominantly H piles, pipe piles, sheet piling, and bracing. Connections (e.g., bolt, rivets, and welds) are also susceptible to corrosion. Corrosion has five primary effects on the structural integrity of these components.

1. **Loss of Section**. Reducing member capacity leads to lower bending, axial, and shear capacity. Typically, thinning, knife edging, areas of missing section, and localized buckles identify this type of deterioration (Figure B-36).
2. **Creation of Stress Risers**. The formation of holes and notches by corrosion causes stress concentrations, providing locations for initiating cracks. Pitting, corrosion nodules, or other localized imperfections can identify this type of deterioration.
3. **Changes in Section Properties**. Corrosion in a defined area of a member or on one side of a member can change the properties of the member and structural section including the moment of inertia. Corrosion in a specific area of a member such as areas of maximum moment or other high stress areas may significanty affect the performance of the member.
4. **Introduction of Unintended Fixity**. When corrosion freezes moving parts of a structure, such as expansion devices or fender systems, the structure behaves differently than originally designed. As a result, members can be subjected to unexpected high stresses and subsequent damage.
5. **Introduction of Unintended Movement**. Corrosion buildup in constricted areas can generate pressure that bends or moves components with damaging effects.

Figure B-36. Underwater photograph of severe corrosion, loss of section, and knife edging on the flange of a steel H pile.
Source: Courtesy of Marine Solutions Inc., reproduced with permission.

B.3.1.2 Biological Deterioration. Microbial-induced corrosion and ALWC, a severe form of MIC, is caused by the presence of microbes whose metabolism produces acids and sulfides. These microbes and their byproducts in the presence of metals can produce a film conducive to accelerated local corrosion. Microbiological involvement in corrosion provides additional means by which aggressive ions can be formed, and it facilitates the acceleration of the corrosion process. Microbial-induced corrosion can attack ductile iron, stainless steel, galvanized steel, and copper.

Microbial-induced corrosion can form large, unusually shaped pits, or a corrosion product film, depending on the bacteria involved. It has been observed in recirculating cooling water systems, on the interior of stainless-steel storage tanks and aluminum alloy fuel tanks, and in buried pipes. It may also be found on marine structures in certain environments (Figures B-37 and B-38). Microbial-induced corrosion can be controlled by pH adjustment in certain situations, cathodic protection, or a barrier system such as a coating. An effective prevention and maintenance program for MIC must combine microbiology and corrosion science expertise.

Figure B-37. Steel sheet pile with bright orange oxide indicating possible MIC. Source: Courtesy of Marine Solutions Inc., reproduced with permission.

Figure B-38. Steel sheet pile with deep pitting caused by MIC.
Source: Courtesy of Appledore Marine Engineeing, LLC, reproduced with permission.

B.3.2 Impact or Overload Damage

Loads that exceed the capacity of a member or structure may cause deformation or failure. Overloading or impact damage may result in deformation or partial breakage of a component (Figure B-39). Deformation of tension members can be identified by elongation and a decrease in cross section (necking). In compression members, the symptoms of plastic deformation are single bow buckling or S-buckling and double-bow buckling, where the component under compression is fixed at the center point. Deformation of flexural members can also be identified by local buckling and elongation and by locations of concentrated corrosion in areas of high stress.

Signs of a one-time overload failure are a fibrous appearance at the point of separation, gross distortion at the point of failure, necking down under tension, and buckling under compression or bending. Partial breakage of a component occurs when a member is not completely broken but a portion has been severed or is missing. The member may be functional but will have reduced structural capacity. In severe cases, overloading or impact damage may result in complete breakage of a component. The portions of the member near the break are discontinuous, failing the member.

Figure B-39. Impact damage on a steel brace.
Source: Courtesy of Marine Solutions Inc., reproduced with permission.

B.4 TIMBER STRUCTURES

B.4.1 Timber Deterioration

Biological factors, mechanical means, or chemical agents can cause the deterioration of timber structures in the marine environment. Various factors affect the rate of deterioration.

B.4.1.1 Biological Damage

B.4.1.1.1 Marine Borers. The two types of marine borers responsible for most damage to structures in the seawater environment are crustacean borers and mollusk borers. The Limnoria is a crustacean borer. The Teredo, Bankia, and Pholad are mollusk borers.

- **Limnoria.** The Limnoria is a waterborne, surface-boring crustacean. It is also known as the wood louse or gribble. It is the only free marine borer that swims and can move from member to member. The Limnoria bores into wood, preferably untreated wood, as soon as it hatches.

Adults reach a length of 1/8 to 1/4 in. Limnoria bore a tunnel to a depth of 1/4 in. and then burrow parallel to the surface of the wood. As the tunnel length increases, auxiliary tunnels are bored to the surface to provide access to water for respiration. The end result of Limnoria infestation is a seriously weakened honeycomb-like surface. Wave action and debris then break down this fragile timber lattice. As this occurs, the Limnoria can burrow deeper into the member.

The continuous burrdue of Limnoria causes the progressive deterioration of a member's cross section, typically resulting in an hourglass shape to piles in the tidal zone, where sufficient oxygen is present to support the organisms (Figure B-40). Damage can also extend to the mudline in shallow waters but usually does not occur below the mudline where the organisms' oxygen and access are limited. Infestation can also occur through construction damage caused by overdriving, open bolt holes, or cut member ends left untreated after field cutting. This type of infestation is typically identified as internal cavities in the damaged member and may be

Figure B-40. Typical hourglass shape formed in a timber pile due to damage from Limnoria.
Source: Courtesy of Marine Solutions Inc., reproduced with permission.

very difficult to locate and assess. Some species of Limnoria are tolerant to creosote at milder levels.

- **Teredines.** Teredo and Bankia are both mollusk borers and members of the teredinideae family of internal marine borers. They are also known by the common name of *shipworms*. Teredo and Bankia are clam-like mollusks that burrow by rasping with a pair of finely serrated shells on the head of a worm-like body.

  The shipworm begins its life cycle as a free-swimming larva that attaches to timber, preferably untreated and starts boring. It can only spread from one member to another when it is in this free-swimming larval stage. Once it bores into timber, the timber imprisons it. A small opening is maintained at the surface of the timber to provide fresh water for respiration and to possibly obtain nourishment from seawater. Once the larva enters the wood, metamorphosis occurs, and the shipworm acquires its adult form.

  Shipworm larvae do not settle on timber that is well treated with creosote or waterborne toxic salts. However, they do attain access to a member through areas where the protective layer is defective or damaged, such as open bolt holes, untreated field cuts, and splits or cracks sustained during construction or service life. Adult shipworms have been found to penetrate the creosote layer of a pile via a firmly attached piece of untreated timber.

  The loss of timber volume caused by the substantial diameter and length of the shipworm can be extensive, and a small number of shipworms can destroy a pile in as little as 6 to 9 months. Damage may occur in the intertidal zone but most commonly is found throughout the submerged and mudline zones. Because shipworm damage is restricted to the interior of a timber member, areas of damage are difficult to visually identify.

  When alive and actively boring, the only visible sign is the two slender posterior siphons, which extend beyond the surface at the entry hole. When the animal is dead, the only external sign of damage is the original pinhole point of entry (1/16 in. diameter or less). The presence of large internal cavities can sometimes identify timber members with a heavy infestation of shipworms. Often, these cavities are caused by a combination of Limnoria and shipworm infestation (Figure B-41).

Figure B-41. Underwater photograph shdue holldue of the interior and subsequent peeling of the exterior of a timber pile from Teredo.
Source: Courtesy of Marine Solutions Inc., reproduced with permission.

- **Pholads.** Pholads are rock-burrdue clams related to shipworms. These borers have also been found to burrow in timber. Their bodies are entirely enclosed in a pair of shells. Like shipworms, they become imprisoned within the timber. They bore into the surface of the timber, making a pear-shaped burrow that grow larger as it grows. The Martesia striata is the species most commonly associated with timber attack. Adults range in size from 2 to 2-1/2 in. long and up to 1 in. diameter.

  As pholads grow, they enlarge the entrance hole to 1/4 in., making it more readily detectable than shipworms. Pholad attack is most severe in the tropical waters of Hawaii and Mexico, but sightings have also occurred along the western coast of the United States.

B.4.1.1.2 Fungi. Fungal decay is primarily a problem in the above-water portions of waterfront structures. Fungal decay can often occur wherever moisture is collected on a timber member (Figure B-42). It can occur when moisture is collected from water runoff or regularly fluctuating water levels.

Fungi require four conditions to grow:

1. Oxygen: atmospheric air;
2. Favorable temperature range: 21 to 30 °C, although some species grow slowly at temperatures as low as 0 °C and as high as 45 °C;

Figure B-42. Rot at the top of timber fender piles.
Source: Courtesy of Marine Solutions Inc., reproduced with permission.

3. Adequate food supply: timber; and
4. Adequate supply of moisture: 20% to 50%.

Decay is usually found in areas with consistent wet and dry cycles (Figure B-43). Moisture is often retained at member interfaces and in the vicinity of steel fasteners, spikes, bolts, or drift pins, which, when temperatures are favorable, provide optimal conditions for fungal attack.

Figure B-43. Severe rot caused by moisture located on a timber beam.
Source: Courtesy of Marine Solutions Inc., reproduced with permission.

Three groups of fungi are found grdue on timber: wood-destroying fungi, soft rot fungi, and stains and molds.

1. Wood-destroying fungi: Two types of these fungi exist: brown rot and white rot. Brown rot is common in softwoods. The wood becomes brittle, is brown, and displays distinct cross-grain checks. Brown rot significantly reduces timber strength. White rot attacks hardwoods. The wood takes on a whitish or tan color, flecked with dark pencil-like lines. The wood is not checked and may be soft or punky. White rot causes less strength loss than brown rot.
2. Soft rot: Soft rot causes the gradual softening and degradation of the timber surface. It occurs at high moisture levels and significantly reduces strength properties (Figure B-44).
3. Wood stains and molds: Stains are spots, streaks, or patches of varying colors, which penetrate the sapwood. They are a superficial phenomenon. Molds are powdery circular growths of varying colors. They can be brushed off when the wood is dry. Molds do not cause decay.

Figure B-44. Splitting and internal decay of a timber pile due to soft rot. Source: Courtesy of Marine Solutions Inc., reproduced with permission.

Neither molds nor stains cause decay, but they do indicate conditions favorable to the growth of fungi.

B.4.1.1.3 Insects. Insect damage is common in timber structures. The inspector should have a basic knowledge of the variety of common wood-destroying insects including termites, carpenter ants, and buprestid beetles. Like their marine borer counterparts, these insects can cause significant cross section and structural integrity loss to affected members.

Specific to the marine environment, the caddis fly is an insect that can damage timber members. It is typically found in freshwater but can also tolerate brackish water. The caddis fly has been known to burrow into creosote-treated timber. Bacterial and fungal infections in the timber attract the caddis fly. Timber that caddis flies have damaged is characterized by the appearance of many, smallpox-like pits.

The caddis fly is closely related to moths and butterflies. As a larva in water, it digs small holes in the timber for protection. It prepares a shelter for the pupa stage by enlarging, deepening, and strengthening the hole. At the end of the pupa stage, the pupa cuts its way out of the hole, swims to the surface, and the adult emerges. The cycle then begins again.

The next generation may use and enlarge the existing holes. A high density of caddis flies on a member, combined with bacterial and fungal decay, and the abrasive action of river or tidal currents, can reduce the cross section of the member and affect its structural strength.

B.4.1.1.4 Cellular Degradation. Microscopic organisms are present in timber, particularly in the marine environment. These bacteria are known to attack the cell wall, detoxify preservatives, and increase the permeability of timber. They are highly resistant to many timber preservatives and may aid in the infestation of timber by marine borers. At present, little of their destructive process is known. The continuing study of timber-inhabiting microbes may further define the role of bacteria in the breakdown of timber.

B.4.1.2 Mechanical Damage

B.4.1.2.1 Impact and Overload. Timber components will split, shear, or fracture under high-velocity impact or from overloading. Impact at lower velocities can sometimes be absorbed without significant structural damage. However, any impact can result in compression or tearing of the timber surface. This damage can compromise the protective coating and expose untreated timber to attack by wood-destroying organisms. Damage due to overload can be caused by applied loads greater than those for which the facility was designed for, but is often precipitated by some form of deterioration weakening the pile (Figures B-45 and B-46).

Figure B-45. Impact damage on a timber fender pile.
Source: Courtesy of Marine Solutions Inc., reproduced with permission.

Figure B-46. Overloaded and failed timber pile.
Source: Courtesy of Marine Solutions Inc., reproduced with permission.

Abrasion. Floating debris, marine traffic, floating docks, anchoring systems, and waterborne materials such as ice and sand can cause abrasion of the surface of timber members. Areas of timber damaged by abrasion appear worn and smooth. The effects of abrasion damage are twofold: (1) gradual reduction in the diameter of the member, and (2) damage to the protective coating that exposes untreated timber to attack by wood-destroying organisms.

B.4.1.2.2 Construction Damage. Timber members can be damaged during construction. Splits or cracks in members caused by improper handling, pile damage as a result of overdriving, scarring caused by machines used to place members, and any drilling or field cutting that does not receive field treatment to restore the integrity of the protective coating may leave the member vulnerable to attack by wood-destroying organisms. Depending on the damage's specific nature, the member's load capacity may be reduced.

B.4.1.2.3 Chemical Damage. Chemicals commonly encountered in a marine environment, such as chlorides and sulfates, typically do not cause the degradation of timber. However, animal waste can initiate damage to timber. The presence of bird droppings increases the amount of nitrogen in wood, which normally contains very little nitrogen. The added nitrogen can stimulate fungal decay. In addition, the buildup of droppings can act to retain moisture and thereby also promote fungal decay.

B.5 MASONRY STRUCTURES

Stone masonry, used today more for ornamentation or historical restoration, is often found as the primary structural building material at many older facilities, typically as mass-gravity-retaining elements such as quay walls, drydocks, abutments, and other similar structures. Granite, limestone, and sandstone are the most common types of stone found in masonry construction. Traditionally, stone masonry was used in the marine environment because of its increased durability in the tidal zone, particularly when subjected to freeze–thaw cycles or abrasion (Figure B-47).

Cyclopean structures typically consist of graded rock ranging from 100 lbs to 7 tons (45 kg to 6.35 mtons) or more, stacked below water. After placement of the rock in lifts, the voids between rocks are infused with concrete to create a solid conglomerate. Like their more modern counterparts, masonry structures in aquatic or marine environments are susceptible to various destructive processes. These can affect either the rock or the mortar and can lead to significant damage or destruction of the structure.

Figure B-47. Typical masonry structure under construction.
Source: Courtesy of Appledore Marine Engineering Inc., reproduced with permission.

B.5.1 Erosion and Movement

By nature, stone walls rely on individual stones' geometry for structure stability. The movement of individual stones due to erosion of fill or mortar can affect the overall global stability of the structure. In granite quay walls and other dry-laid structures, sinkholes are common as fines wash out from behind the wall. Over time, unless laid directly on bedrock or a pile foundation, the wall may settle or move, differentially affecting the overall batter (Figure B-48).

Figure B-48. Erosion and movement of a masonry block structure.
Source: Courtesy of Appledore Marine Engineering Inc., reproduced with permission.

B.5.2 Spalling

Spalling occurs when small pieces of the stone break away from the surface leaving a depression in the stone. A common cause of spalling is repeated freeze–thaw cycles where expansive forces generated by water freezing in the fissures and pores in the rock break the rock apart. Spalling of stone structures can also occur because of high temperatures caused by fire or other sources.

B.5.3 Splitting

Splitting of the stones used in masonry occurs when cracks open up in the rocks, eventually breaking the stone into small pieces. This can be caused by volume changes such as seasonal expansion and contractions of the rock and freeze–thaw cycles (Figure B-49). Plant growth can also generate and increase the size of cracks in the stone. Roots and stems in the crevices of stone can exert a wedging force, which can break up the stone.

Figure B-49. Splitting of masonry blocks.
Source: Courtesy of Marine Solutions Inc., reproduced with permission.

B.5.4 Abrasion

Abrasion and weathering cause the hard surface of the stone to degenerate into small granules, giving the stone a smooth, rounded appearance. It is caused by waterborne materials such as sand, debris, and ice. This type of deterioration can also be caused by chemicals (gases and solids) dissolved in the water. Oxidation and hydration of some compounds found in the stone can also cause damage. In addition, lichens and ivy can chemically attack the surface of the stone, resulting in degradation of the stone surface.

B.5.5 Degradation of Mortar

The mortar used in masonry construction is also subject to deterioration. Frequently, this deterioration occurs more rapidly than that of the stones themselves. The mechanisms of mortar deterioration are similar to that of concrete. Mortar is particularly susceptible to degradation at the waterline caused by freeze–thaw cycles (Figures B-50 and B-51). Finding the mortar essentially gone from this area is not uncommon. Abrasion and chemical damage are also agents of mortar deterioration.

Figure B-50. Mortar degradation in the tidal zone.
Source: Courtesy of Appledore Marine Engineering Inc., reproduced with permission.

Figure B-51. Closeup view of loss of mortar between masonry blocks.
Source: Courtesy of Appledore Marine Engineering Inc., reproduced with permission.

B.5.6 Marine Borers

Marine borers can attack stone. Stone-burrdue clams, called Pholads, use chemical secretions to bore into the stone. They make pear-shaped burrows that enlarge as they grow, up to 2.5 in. in length and 1 in. in depth. This results in loss of cross section in the stone under attack.

B.6 POLYMER COMPOSITE STRUCTURAL COMPONENTS

The use of polymer composite materials for marine and waterfront construction is a somewhat new development when compared to conventional materials. Consequently, few polymer composite structures can provide long-term information about the performance and durability of the various materials and member types.

Fiber-reinforced polymer (FRP) polymer composites or recycled plastics are typically less prone to deterioration than traditional materials. Polymer composite materials are typically composed of glass, aramid fibers, or carbon. Glass fibers are economical but have limited structural value because of their low strength. Aramid fibers are usually limited to specialty applications of high stress and vibration because of their high cost and low modulus of elasticity.

Carbon fibers offer significant advantages for marine applications that require high tensile strength, high modulus of elasticity, and low susceptibility to deterioration. Although the cost of carbon fiber is somewhat high, the material has proven to be versatile and cost-effective in many applications when used in combination with other materials and when life-cycle costs are considered.

Common applications of polymer composite materials in marine construction include piles, decking, and strengthening of members using carbon fiber laminates or wraps. Polymer composite piles are currently available in many proprietary configurations, including

- Recycled plastic with steel reinforcing or recycled plastic with FRP reinforcing.
- Hollow steel core with a recycled plastic shell (the core may be filled with concrete or sand).
- Hollow FRP core with a recycled shell (the core may be filled with concrete or sand).

B.6.1 Ultraviolet Deterioration

Early applications of polymer composite piling exhibited deterioration such as cracking and splitting as a result of ultraviolet (UV) degradation, thermal stresses, and quality problems with constituent materials. Deterioration and degradation of protective coatings can expose fiber reinforcement (Figure B-52).

Figure B-52. Deteriorated coating and degraded fiberglass-reinforced pile casing shdue abraded fibers.
Source: Courtesy of Marine Solutions Inc., reproduced with permission.

B.6.2 Material Incompatibility

Problems may be associated with the quality of the polymer materials used, manufacturing process, and application. Damage can be caused by differential thermal responses of the polymer composite materials that can cause distress to the component during the polymer curing process or when encountered post installation (Figures B-53 and B-54). The relatively low stiffness of polymer composite members used in combination with higher stiffness members may also cause excessive bending and overstress. Adhesives used in the polymer composite members may also deteriorate over time when exposed to the marine environment.

Figure B-53. Damage to an FRP pile from heat exposure.
Source: Courtesy of Jacobs, reproduced with permission.

Figure B-54. Cracked FRP pile likely caused by shrinkage of plastic material. Source: Courtesy of Jacobs, reproduced with permission.

B.6.3 Corrosion Damage

Carbon fiber materials are susceptible to galvanic corrosion when used adjacent to metal. A barrier material, such as epoxy resin, must be used to separate the materials. Corrosion may also be a concern for polymer composite components that rely on steel reinforcing or a steel core for strength. Cracks in the outer protective covering may leave the steel susceptible to corrosive deterioration.

B.6.4 Swelling

Some of the fibers used in manufacturing polymer composite elements are susceptible to swelling in the presence of water. The fibers must be encased in a protective layer, typically resin.

B.6.5 Overstress Damage

Polymer composite materials often undergo significant deflection as a result of heavy loads. Large deflections can result in cracking of the outer shell, potentially compromising the member's structural integrity. Such deflections may also result in the separation of the polymer composite materials, as in the case of polymer composite piles with a concrete-filled core. Such overstress damage is often encountered in fender piling (Figures B-55 and B-56). Permanent deflections can also result from the hysteresis effect of repeated large deflections. These deflections can render the component ineffective and may lead to the damage of adjacent members. Long-term deflection due to creep can also be a concern with polymer composite materials.

Figure B-55. Cracked FRP pile from vessel impact.
Source: Courtesy of Marine Solutions Inc., reproduced with permission.

Figure B-56. Cracked and broken FRP piles of a mooring dolphin.
Source: Courtesy of Jacobs, reproduced with permission.

B.7 COATING AND WRAP SYSTEMS

Several different types of coatings and wrap materials are used to protect steel against the effects of seawater and corrosion. Asphalt enamels, coal-tar enamels, coal-tar epoxies, polyurethane materials, inorganic zincs, and other coatings are available. In addition, various types of wraps including polyvinyl chloride (PVC) or fiberglass jackets and heat-shrink-applied barriers are common to protect marine piles. Concrete and other materials are also commonly used to encase structural members in the marine environment. The basis of these systems is usually to isolate the structural member from the corrosive and abrasive marine environment.

B.7.1 Coatings

Coatings can fail for several reasons. However, disbonding and breakdown of coatings can often be traced back to poor surface preparation. Shop-applied coatings provide greater control over conditions such as cleaning and moisture. However, damage to the coating can occur during transport or installation. Structural steel shapes are prone to coating problems because of sharp corners and angles. When coatings are applied to these areas, internal forces in the coating draw it away from the edge, leaving the steel member either exposed or only partially covered (Figure B-57).

Operational causes of coating deterioration include impact damage or a buildup of marine fouling, which can cause breaches of the coating and allow corrosion of the steel. Isolated areas of coating loss or holidays in the coating can cause accelerated corrosion at the uncoated location where an electrochemical differential occurs between the coated and the uncoated

Figure B-57. Coating failure on steel bracing.
Source: Courtesy of Marine Solutions Inc., reproduced with permission.

Figure B-58. Deteriorated coal–tar epoxy polyamide coating resulting in corrosion of a steel fender pile.
Source: Courtesy of Marine Solutions Inc., reproduced with permission.

areas (Figure B-58). A poorly adjusted cathodic protection system can result in blistering of the coating caused by electro-osmosis or hydrogen gas evolution.

B.7.2 Wraps

Several wrap materials are also available for either shop application or field installation. These products are usually composed of plastic, PVC sheet material, or mastic-coated tapes that are installed over structural members. As with coatings, these products are used as a preventative maintenance technique to provide protection against corrosion, marine organisms, floating debris, and moored vessels.

Various factors cause deterioration of wrap material, including impact damage, UV radiation, punctures or tears from floating debris and vessels, or heavy accumulation of marine growth. A disbanded, damaged, torn, or incomplete wrap system can cause accelerated corrosion or concentrated areas of marine organism attack (Figure B-59). Further, if not providing sufficient protection, wraps may hide ongoing deterioration. Dissolved oxygen testing may be performed to indicate the conditions beneath the wraps to evaluate corrosion potentials or conditions supporting marine organisms. Alternatively, selective removal and reinstallation of wrap systems are often necessary to correctly evaluate the effectiveness of the system and detect hidden deterioration.

Figure B-59. Torn PVC wrap exposing a timber fender pile.
Source: Courtesy of Marine Solutions Inc., reproduced with permission.

B.8 LOAD ISOLATORS AND BEARINGS

Base isolation devices are being introduced for use with waterfront facilities. This is typically done as a means of retrofit for improved performance of a structure undergoing lateral forces, usually from earthquakes. These devices usually consist of a combination lead-rubber bearing assembly introduced between the supporting piling and the pile cap. This retrofit approach has been used for several decades with the foundations of buildings as a means of improving the structural performance of the building's response to earthquakes. Slide bearings are commonly used to improve performance at the interface between bridges and supporting elements like abutments.

The typical lead-rubber base isolation unit consists of a cylindrical lead core surrounded by layers of rubber and steel laminated in such a fashion to make a larger cylindrically shaped element that is sandwiched between a top and a bottom steel-mounting plate. Diameters of the isolators can range from 12 to 60 in. (0.3 to 1.5 m). The anchor bolts and steel retention brackets mate with the top- and bottom-mounting plates of the isolator and provide a means of connection to the substructure and superstructure (Figure B-60).

The rubber in the isolator unit acts as a soft spring that deforms laterally but is very stiff in the vertical direction. The lead core provides a damping mechanism by deforming under lateral loads. Slide bearings are of similar construction but typically incorporate a low-friction disk substituted for the flange plate.

Figure B-60. View of a lead–rubber base isolation unit installed below the pier deck. Source: Courtesy of Moffatt & Nichol Inc., reproduced with permission.

Bearings are often located close to the splash zone (the intertidal area) beneath a marine structure. Placement of these units in locations of severe exposure promotes corrosion of the exposed steel elements. Typically, the steel fabrications of these assemblies are made from galvanized or stainless steel or covered with an epoxy coating. An inspection of these elements should address the folldue:

- **Corrosion.** The steel-mounting plate and associated connection hardware should be inspected for corrosion. Excessive corrosion could be deleterious to the service life and performance of the bearings.
- **Debris.** Because the bearings are usually placed within a narrow space, finding birds' nests or other biofouling or the accumulation of trash or other deleterious material is not uncommon.
- **Damage to Concrete.** Because of the possibility of entrapped moisture, the surrounding concrete element should be carefully inspected for obvious signs of cracks and/or spalls that are related to the corrosion of embedded reinforcing steel. Refer to Section B.2 for discussion on the corrosion-related concrete damage mechanism.
- **Alignment.** Because the bearings are designed to facilitate movement of the superstructure in relation to the substructure, the possibility of permanent deformation after significant lateral loading exists. The bearings should be inspected relative to the alignment of the top- and bottom-mounting plates to preclude permanent distortion.

B.9 MUDLINE CHANGES

Waterways are dynamic and often cause changes in the mudline that can affect the performance of waterfront structures. The waterway and mudline can change significantly because of a wide variety of occurrences such as storms, floods, and changes in facility use or environment. These changes can result in aggregation causing inadequate water depths for vessels or scour causing loss of foundation support and differing conditions from which the structure was designed.

B.9.1 Aggradation

Aggradation is the buildup of the mudline due to sediment deposition. Aggradation or the term *general aggradation* is often used to describe widespread changes in the mudline along a considerable marine area or river. However, the term *local aggradation* can be used to describe smaller areas of sediment deposition. Aggradation can be caused by long-term environmental processes, storms, or changes in the waterway. Local aggradation may be associated with scouring of an adjacent area.

B.9.2 Scour

Scour is the movement of seabed or riverbed material below its given elevation by the action of moving water resulting from accelerated currents, wave energy, vessel propulsion, or other turbulence. The loss of bottom material due to scour may cause a loss of foundation support by undermining footings or removing supporting soils around piles. Even weathered rock can be eroded if the currents are able to move soft or loose bedrock during storms or floods. The three forms of scour are characterized as general scour, contraction scour, and local scour.

B.9.2.1 General Scour. General scour is the general degradation or loss of the bed material along a considerable length of a marine area or river. It can be the result of long-term environmental processes, storms, or changes in the waterway. erosion, mining activities, construction, or other events.

B.9.2.2 Contraction Scour. Contraction scour is caused by an increase in water velocity and the resulting shear stress on the mudline due to restriction of the waterway. The movement of large volumes of water from tidal fluctuations, storm surge, floods, or rapid drawdown create accelerated water currents at constricted locations, increasing the erosive forces on the mudline and removing bottom material. The contraction can be caused by a narrdue of the waterway or any hydraulic opening due to the structure's presence in the waterway, adjacent structures, or large deep draft vessels.

B.9.2.3 Local Scour. Local scour is the removal of material from a specific and isolated area of the marine area or riverbed. The main mechanism of local scour is the formation of vortices due to increased current flow. This commonly occurs at obstructions such as piles, bulkheads, seawalls, mooring dolphins, pipelines, anchors, or debris. Irregular waterway sections causing changes in waterflow direction can cause vortices in the water flow causing local scour. Local scour can also be caused by vessel propeller wash or thrusters. Vessel propeller wash or thrusters aimed at a bulkhead or quay wall can not only undermine sheet piles or block elements but also erode backfill through unsealed joints.

B.9.3 Causes of Aggradation and Scour

Causes of aggradation and scour are many. Wind-induced currents, storm surges, or floods result in the rapid movement of large volumes of water and may cause large volumes of sediment to be transported. Aggradation can particularly occur at structure foundations where water currents slow and entrap sediment. Conversely, the same occurrences may cause scour of the mudline in areas where the water current is increased. Debris, ice jams, or any additional obstruction in the waterway will likely increase the effect (Figure B-61).

Changes in nearby structures or features in the waterway such as the addition of new piers or coastal structures or dams upstream or downstream

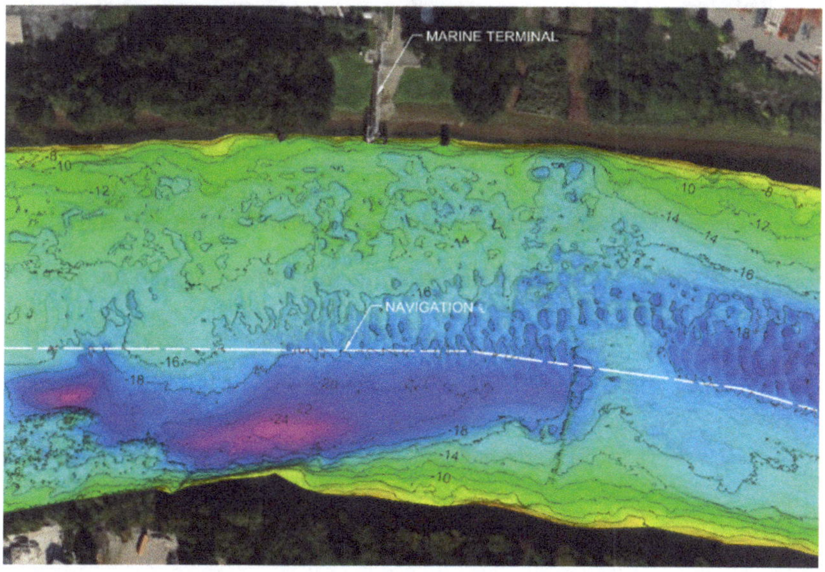

Figure B-61. Hydrographic survey of a river channel shdue scour and aggradation at a marine terminal.
Source: Courtesy of Marine Solutions Inc., reproduced with permission.

of the facility can change the direction and magnitude of water currents. Construction of a jetty for harbor protection may reduce the effect of wave action and tidal currents resulting in aggradation in areas of the harbor and increase scour in other areas. Construction of a dam upstream or downstream of a facility may result in aggradation and inadequate depths. Urbanization of the surrounding lands may increase or decrease sediment in the harbor.

Changes in the facility use can result in changes in the mudline conditions. New applied vessel types may have deeper propeller or jet drive components that cause local scour. Changes in tugboat operations to berth and send-off vessels can also cause local scour. Large deep draft vessels may constrict the waterway causing contraction scour. Berth depths are often increased at a facility to accommodate deeper drafting vessels. The structure should be confirmed or upgraded for the new berth depth requirements. However, this practice is not always performed and even if so, additional scour at the deeper mudline may remove the needed support for the structure (Figure B-62).

Figure B-62. Underwater sonar image shdue large scour depression from ship propulsion along piles supporting a pier.
Source: Courtesy of Marine Solutions Inc., reproduced with permission.

APPENDIX C
OVERVIEW OF SPECIALIZED TECHNIQUES

C.1 INTRODUCTION

Waterfront structures of all types and materials are susceptible to many defects and deterioration mechanisms. Some defects are unknowingly built into a structure, whereas others develop with time because of natural or manmade events. Although some of these defects may be detected by visual or tactile means, others may be hidden within the structure's members or components. In some cases, an indication of a defect may be readily observed, but the true extent or cause of the defect is not evident.

For these cases, the use of specialized inspection techniques often involving various nondestructive testing (NDT) and partially destructive testing (PDT) can be used to thoroughly investigate a structure's condition. Nondestructive testing methods permit the inspection of an element without inflicting damage, whereas PDT typically causes minor localized, repairable damage. Specifically, NDT and PDT are used to investigate the material integrity of the test component and not the function of the component beyond material failure. In addition to NDT and PDT devices, several specialized techniques may benefit waterfront facility owners when data or images are not easily obtainable because of certain conditions.

This appendix presents an overview of various specialized inspection techniques as they apply to waterfront structures and miscellaneous support systems. Although many of these techniques are often performed by specialists, all waterfront facility owners and structure inspectors should be familiar with available advanced techniques so that they can recommend appropriate testing procedures and recognize the limitations of the data.

Many specialized techniques have been developed and are commonly employed in the inspection of waterfront facilities. Several of these techniques are described herein. The inspector should be aware that many other methods are also currently available, and new techniques are constantly being developed.

C.1.1 Qualifications for Specialized Techniques

Previously in this manual, inspector qualifications were discussed. Some specialized inspection techniques require additional qualifications and credentials. Nondestructive testing and PDT methods range from simple timber coring to complex methods such as ultrasonic testing. The person conducting the test and the personnel interpreting the test data must be properly trained in the applied method. Additional qualifications should include both an understanding of the theory behind the test and practical experience.

All inspection methods should be conducted in accordance with an acceptable standard of practice. The American Society for Non-Destructive Testing (ASNT) and ASTM have many well-defined procedures and standards. In addition, other organizations [e.g., the American Concrete Institute (ACI) and the American Welding Society (AWS)] provide guidelines for some specialized techniques. Appropriate training credentials are warranted to ensure accurate data and to ensure that the data are defensible if questioned by other parties such as during legal disputes.

C.1.2 Data Collection and Interpretation

Upon embarking on an NDT or PDT program, a plan should be developed that details the type(s) of testing to be performed, the amount of data needed, test locations, criteria for data interpretation, and follow-up procedures for handling unanticipated test results. Many testing methods produce considerable test data. Data should be collected on the applicable forms, which include the location and test results. When possible, copies of all field data should be retained as part of the structure's official file.

Interpretation of NDT/PDT data should be performed by people knowledgeable both in the test theory and in the analysis or evaluation of the structure being tested. For some tests, such as ultrasonic weld inspection, recognized criteria exist for evaluating the effects of any detected anomalies. However, for many other test methods, the NDT/PDT data must be evaluated on the basis of each individual structure's behavior.

Most nondestructive programs detect and assist in evaluating flaws and discontinuities and determining the strength or serviceability by indirect methods. The tests typically indicate the existence, extent, and location of discontinuities or abnormalities. However, the influence of the discontinuity or abnormality on the strength or serviceability of the structural component is often more difficult to determine.

The validity of an NDT test depends on sound engineering judgment based on experience or fully destructive testing validation results. The information collected by NDT/PDT is typically raw data and must be interpreted to correlate it to a usable parameter. Also, certain techniques may provide false data under certain conditions; therefore, it is important to be familiar with the technique and be able to recognize the false readings. Likewise, the inspector should not recommend testing, or accept testing results, without being familiar with the techniques used.

C.2 INFRARED THERMOGRAPHY

One type of deterioration encountered in waterfront structure decks is delamination within the concrete deck. Delamination is defined as a horizontal fracture plane at or above the top layer of the reinforcing. Overlay surfaces can hide delamination until they are well advanced, and distinguishing between deterioration in the concrete deck slab and debonding of the overlay can be difficult.

Traditional inspection methods involve chain dragging and hammer sounding. These audible methods require inspector judgment and a great deal of effort, traffic control, and possible operational disruptions because of area closures. The presence of an asphalt or concrete overlay reduces the effectiveness of these traditional audible methods.

Infrared thermography is an alternative tool for locating and mapping delamination in structure decks and pavements. A technique using an infrared scanner and control video camera, infrared thermography senses temperature differences between delaminated and nondelaminated areas. A delamination in a concrete deck creates a thermal discontinuity that acts as an insulator. Thermography operates on the principle that when the sun warms the deck, the delaminated area heats up at a faster rate and reaches a higher temperature than the solid areas.

A temperature difference between delaminated and solid areas is normally established only on sunny or partially sunny days. The deck must be dry and winds must be less than 25 mi/h. The temperature difference is primarily related to the amount of sun, not the ambient air temperature, and therefore, inspections can be undertaken under various temperatures.

The procedure involves scanning the concrete deck with an infrared camera and recording the video signal on videotape for detailed analysis in the office. A single pass, with a vehicle speed of approximately 5 mi/h, is typically made for each strip of the concrete deck. At the same time, a real-life control image of the deck surface is recorded. Distance footage is superimposed onto both videotape signals to locate defects.

Field confirmation of the infrared data consists of sounding several suspect deteriorated areas and measuring surface temperatures of both suspect and solid areas. Furthermore, select deck cores are typically taken for confirmation. These proposed core locations are typically marked at the time of the inspection.

Analysis of the infrared data is completed with the aid of a computer digitization program. During the analysis, the recorded temperature variations are interpreted to identify specific delaminated areas. Each delamination is identified and plotted onto plan view drawings of the deck. Square footage and percentage of the delaminated deck are calculated. The real-life control data are examined to make sure that temperature variations were not caused by concrete spalls, discoloration, patches, tar, or debris. In addition, the real-life control data are used to plot existing repair patches and spalls.

The use of infrared thermography for structure decks is covered in ASTM D4788 (ASTM 2022b). In addition, handheld infrared thermography cameras are available, which are more portable than vehicle-mounted devices. These portable infrared thermography cameras may be used in other areas at a waterfront facility. However, manufacturer recommendations should be followed as the surface location or environment may affect the data obtained.

Infrared thermography is most commonly used on concrete decks with or without overlays; however, it can also be used on other concrete components. This method is proven to be accurate and easily repeatable.

Infrared thermography also provides quicker data collection, because the equipment can be vehicle-mounted and driven over the waterfront structure deck. By mounting the equipment to a vehicle, the process typically results in minimal disruption. Infrared thermography can be used in busy areas or where noise levels are high, unlike an audible inspection by a person with a hammer or chain-drag device.

Vehicle-mounted data collection for infrared thermography is completed with the aid of computer-logging software, and the image can be digitally processed for an overall assessment of the deck. Handheld infrared thermography cameras may be useful in other concrete material areas at a waterfront facility. However, environmental conditions (e.g., material moisture content, ambient humidity, temperature, and other factors) may dictate applicable locations at a particular facility (Figures C-1 and C-2).

Frequently, infrared thermography requires a temperature differential of approximately 32°F (0.5 °C) between the delaminated or the debonded areas and solid regions of a concrete surface. This fact typically requires that inspections be done on days with approximately 50% sunshine. Wet areas, shadows, and debris on a concrete surface do not allow a temperature difference to be established, and, therefore, these areas cannot be inspected.

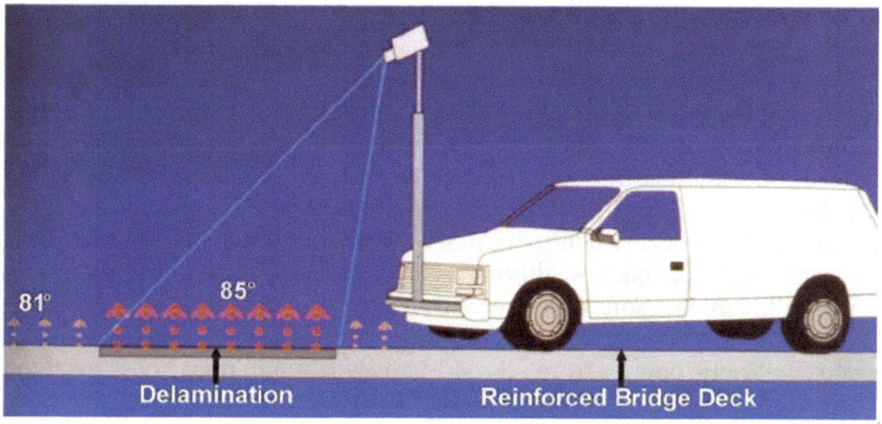

Figure C-1. Infrared thermography equipment illustration.
Source: Courtesy of Collins Engineers Inc., reproduced with permission.

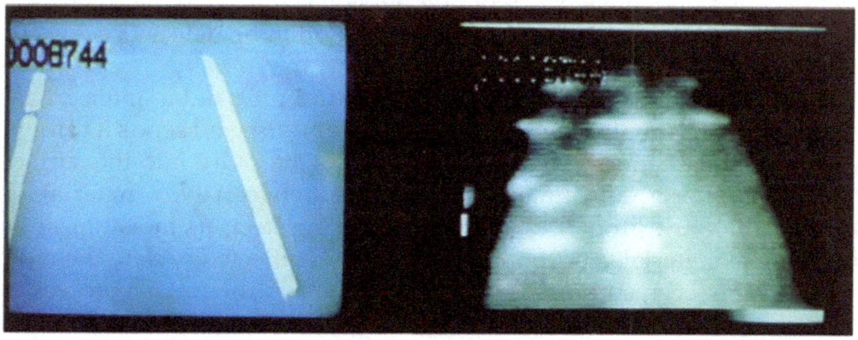

Figure C-2. Infrared thermography images.
Source: Courtesy of Collins Engineers Inc., reproduced with permission.

Infrared thermography locates the delaminated areas in the horizontal plane and does not provide any information on the depth of the delaminated layer where the defect occurs. If confirmation on depth is desired, cores can be taken.

If an inspector plans to use this technology at only a few structures or is considering setting up and operating a data collection vehicle, the cost may be a prohibitive factor, because the scanning equipment and data processing software are expensive. Also, the vehicle must typically be operated by at least two inspectors. However, when compared to manually sounding a large deck or several smaller decks, infrared thermography may be cost-effective. Furthermore, it may require less area disruption or coordination with port operations personnel.

C.3 GROUND-PENETRATING RADAR

Ground-penetrating radar (GPR) is a recognized NDT technique with many applications to structures, including sinkhole detection behind bulkheads, structure deck condition evaluation, overlay thickness determination, location of voiding under structure approach slabs, reinforcing steel location, foundation investigation, and underwater profiling. This section mainly discusses the use of GPR on decks, structural elements, and approaches, although GPR can also be used over the water for mudline evaluations and over ground for subsurface utility location.

Over a body of freshwater, GPR systems radiate short pulses of electromagnetic energy from a broad-bandwidth antenna. These systems typically use a signal of 80, 100, or 300 MHz. Depending on the GPR system used, penetration of up to 40 ft (12 m) into resistive granular material can be attained, and layers as thin as 2 ft (0.6 m) can be detected. However, GPR systems will not work in soils or waters that are highly conductive because of chlorides or pollution. Ground-penetrating radar over seawater does not work. Scour depression geometry, scour depression infill thickness, and riverbed deposition can often be detected using this technique.

A radar system typically consists of a control unit, radar antenna, and display unit. The control unit generates a radar pulse and sends it through a cable to the antenna. The antenna transmits the pulse into the surface. When this energy encounters an interface between two materials of differing dielectric properties, such as reinforcing steel, air, moisture, or the base-course material, a portion of the energy is reflected back to the radar antenna. The received pulse is sent back to the control unit for processing and storage. The display unit (video or chart recorder) presents the data.

The reflected energy is received by the transducer, amplified, and recorded. The electromagnetic pulse is repeated at a rapid rate, and the resultant stream of radar data produces a continuous record of the subsurface. The radar system creates a linear profile of the materials beneath the antenna pass.

There have been advances in GPR technology. The larger units as described previously are still considered useful for most applications, but smaller more portable all-in-one units have been developed. These smaller units are handheld and were designed for concrete applications like identifying reinforcement depth and spacing, locating prestressing and posttension ducts, and looking at cover and overlay thicknesses (Figure C-3).

For GPR over land, two different types of transducers (contact or horn type) can be mounted on a data-collection vehicle or hand towed. The location of the transducers can be varied across the width of the pavement, and, if additional information is required, several passes with the antenna in different locations can be made.

Figure C-3. Handheld GPR unit used to identify (left) reinforcing strand, (right) prestressing strand.
Source: Courtesy of NDT Corporation, reproduced with permission.

For most surveys, the antennae are mounted over the wheel tracks. The data are typically collected at a vehicle speed of 5 mi/h (8 km/h) or less. Faster speeds are attainable, but the longitudinal and vertical resolution of the system is reduced. Horizontal data positioning is accomplished by using a global positioning systems (GPS) or a distance transducer connected to the drive train of the data-collection vehicle (Figure C-4).

An event mark is automatically placed on the data at user-defined intervals, allowing defects to be located accurately. Once the survey is completed, a computer processes the data, and the results of the survey can then be presented in various formats.

Ground-penetrating radar is most commonly used on the ground to detect utilities, behind bulkheads to detect sinkholes, on concrete decks of

(a) (b)

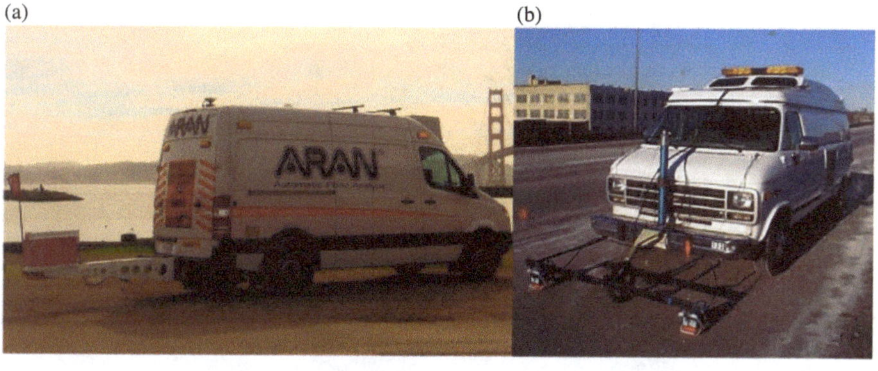

Figure C-4. GPR vehicle equipment.
Source: (a) Courtesy of Fugro Inc., reproduced with permission; (b) courtesy of Collins Engineers Inc., reproduced with permission.

a structure, or over the water to evaluate mudline characteristics. Ground-penetrating radar also allows for an inspection of the concrete deck surface, which might be hidden by an overlay surface. If the concrete deck is not covered, GPR is not often used, because it is not as accurate or rapid as thermography.

On land or on top of a deck, the GPR system provides a means of determining the following items:

- Deck, pavement, and/or overlay thickness;
- Depth or location of reinforcing steel, bulkhead tie-back rods, or utilities; and
- Presence and size of voids beneath a pavement.

Over the water, GPR is typically useful only in shallow [less than 20 ft (6 m)] freshwater with granular bottom and sub-bottom sediments.

Over water or ground, GPR is limited by depth, conductive soils, and chlorides in the water (i.e., seawater or brackish water). With regard to concrete surfaces, GPR identifies areas of the concrete surface with different dielectric properties or conductivities. Some concretes, such as dry low permeability concrete, affect the ability of GPR to detect areas of delamination. GPR is also sensitive to the presence of water and chlorides on the deck and between overlays and the base concrete and the presence of debris on the deck surface. These conditions can significantly influence the accuracy of the data.

GPR should be scanned in both directions to determine the direction of the top reinforcing. Therefore, an inspection of some structures will require the survey to be conducted perpendicular to the flow of traffic. This will require traffic to be restricted or stopped altogether while the survey is being conducted. Frequently, several passes must be made on the deck area, and the cost may be prohibitive (Figure C-5).

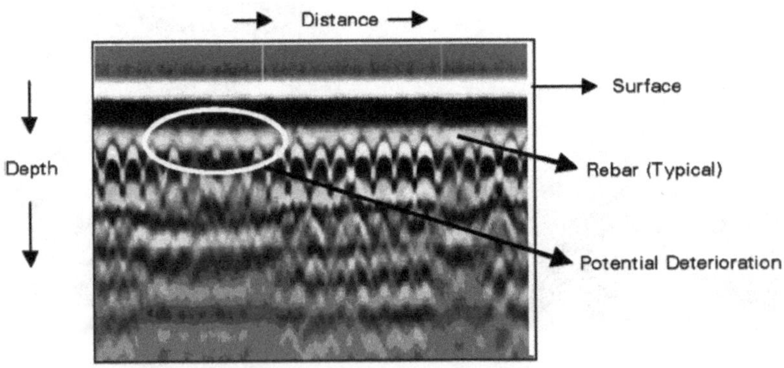

Figure C-5. GPR image of concrete deck delamination.
Source: Courtesy of Collins Engineers Inc., reproduced with permission.

Because of the age and environment of the concrete, the dielectric may vary. It is recommended that a top reinforcing bar be located and a drill hole to the bar be made. Once the bar is reached, a measurement from the concrete surface to the bar should be made and the dielectric changed to match the reinforcing depth.

C.4 PILE INTEGRITY TESTING

Pile integrity testing (PIT) is a nondestructive testing method to evaluate the integrity of various types of piles and caissons that are used as foundation elements for structures such as piers and wharves. This test method uses the pulse-echo type of testing and is based on the reflection of high-frequency sound waves that are transmitted into the pile. The purpose of the PIT method is to detect any defects or anomalies in the pile that could compromise its structural integrity and lead to failure.

The PIT method involves striking the pile with an impact hammer and measuring the resulting vibrations using accelerometers placed against the pile. The accelerometers detect the waves generated by the impact, and the data collected is analyzed to determine the pile's integrity. The analysis can identify the location and extent of any cracks, voids, or other defects in the pile. Figure C-6 shows PIT tests being performed on timber and sheet piles.

The pulse-echo PIT involves placing a small accelerometer or sensor on the surface (top or side) of the pile and striking the pile with a small

Figure C-6. (Left) Timber pile test, (right) sheet pile test using pulse-echo PIT. Source: Courtesy of NDT Corporation, reproduced with permission.

hammer. The hammer's impact generates a high-frequency compressional wave (p-wave) that travels down the length of the pile. When the p-wave encounters a change in internal material, such as a crack or void, it is reflected back to the surface of the pile, where it is detected by the accelerometer (sensor).

The time taken for the wave to travel down the length of the pile and back to the surface is measured, and the data are analyzed to determine the pile's length. A change in the wave's travel time or amplitude may indicate a defect, such as a crack or void, in the pile. Wave reflections can also happen from changes in material external to the pile. A change from organics or clay to a dense till is a good example of this.

The pulse-echo data display the initial impact of the hammer, which is considered zero time (T0) in milliseconds. The signal received is the compressional wave reflecting off the bottom of the pile. This reflection is selected and is considered T1. The compressional wave velocity should be measured before the pulse-echo test. The wave velocity would be halved for two-way travel of the signal (12,000 ft/s would be 6,000 ft/s). Figure C-7 is an annotated example of a timber pile test.

Limiting factors are involved in preparing for a pile test. On poured concrete from auger cast piles, cleaning the surface is essential. Dirt and loose or soft concrete in areas where the test will be performed must be removed. For timber piles, avoid areas with rotted wood. The wave will not propagate through the soft-rotted areas. Steel sheet, H, or pipe piles should be cleaned with a wire brush to ensure good contact with sensors.

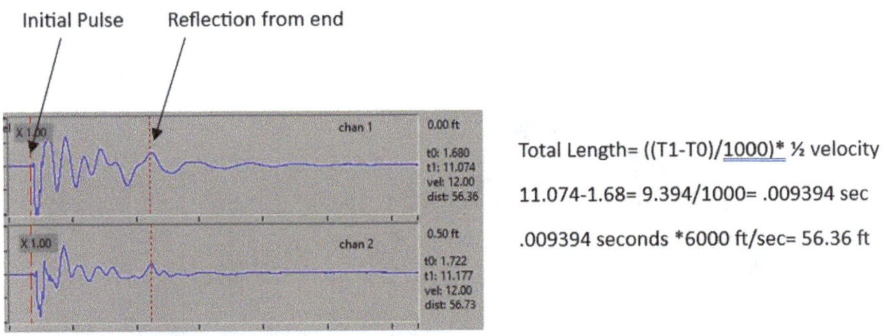

Figure C-7. Timber pile test using pulse echo testing.
Source: Courtesy of NDT Corporation, reproduced with permission.

C.5 SEISMIC REFRACTION TESTING

Seismic refraction (ASTM D5777) (ASTM 2018a) is a method of measuring soil and rock properties, as well as providing data to calculate the soil/rock layer depth profiles. This is a very useful testing method to profile bedrock for unknown foundations of substructures like piers and abutments. Currently, most information is gained by borings at a set spacing apart. Seismic refraction testing provides a comprehensive profile of the soil and bedrock layering across an entire spread of sensors, whether it is 100, 200, or 400 ft long.

Seismic testing methods utilize the natural energy propagating properties of the soils and rocks and are based on the principle that the velocity at which seismic waves travel through the earth is a function of the physical properties (elastic moduli and Poisson's ratio) of the materials. The waves are typically measured on a 12- or 24-channel seismograph using a series or line of sensors positioned at known distances (Figure C-8).

Energy in the form of a wave or vibration is generated at both ends and at the center of the seismic sensor line. The geophone/hydrophone sensors are in direct contact with the earth/water and convert the soil or rock's motion, resulting from the energy-generated wave, into electric signals with a voltage proportional to the particle velocity of the ground motion.

Energy can be input into the ground through the use of a sledgehammer, a weight drop, or small amounts of explosives (black powder shells).

The field technician can collect raw data or add filtering to the seismic signals to minimize background noise. Seismic data are recorded on a seismograph and is archived. The data can be displayed or printed in the field. Interpretations are based on the time required for a seismic wave to travel from an energy source to a series of geophones/hydrophones located at specific intervals along the ground or underwater surface.

The resultant seismic velocities are used for material identification (identifying water table, soils, and dense tills) and bedrock depth and profile determinations.

Figure C-8. (Left) Seismograph, (right) black powder Seisgun energy source. Source: Courtesy of NDT Corporation, reproduced with permission.

The use of Snell's law is a fundamental principle of physics that describes how seismic waves change direction and speed when they cross the boundary between two different materials. In the case of seismic refraction, Snell's law is used to determine the velocity of seismic waves as they pass through different layers of rock or soil.

Seismic waves travel through the Earth at different speeds depending on the density and composition of the material they are passing through. When seismic waves encounter a boundary between two layers of different material, they are partially reflected and partially refracted (bent) at the interface between the two layers as illustrated in Figure C-9. The amount of refraction that occurs depends on the angle of incidence of the incoming wave and the velocities of the two materials, Snell's law.

Snell's law states that the ratio of the sine of the angle of incidence to the sine of the angle of refraction is equal to the ratio of the velocities of the two materials.

This can be written mathematically as

$$\frac{\sin(i)}{\sin(r)} = \frac{v1}{v2}$$

where

i = Angle of incidence,
r = Angle of refraction,
$v1$ = Velocity of the first material, and
$v2$ = Velocity of the second material.

By measuring the angles of incidence and refraction of seismic waves as they pass through different layers of the earth and knowing the velocity of the seismic waves in one of the layers, we can use Snell's law to calculate

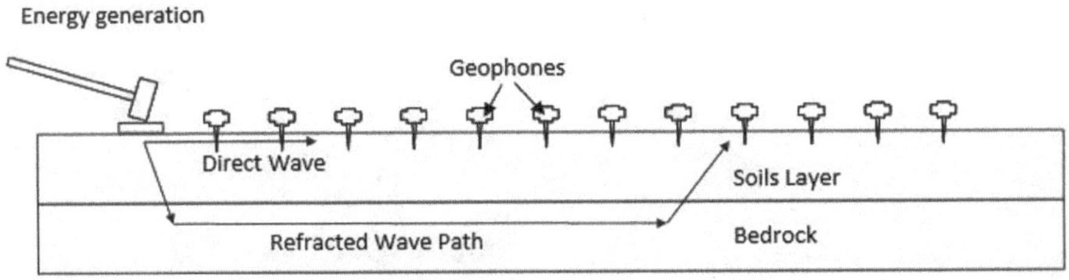

Figure C-9. Seismic wave paths for direct wave and refracted wave illustrating the effects of a boundary between materials with different elastic properties. Source: Courtesy of NDT Corporation, reproduced with permission.

the velocity of the waves in the other layers. This information can be used to create a model of the subsurface structure of the Earth, which is useful in a variety of applications such as oil and gas exploration, engineering, and seismology.

The seismic refraction technique is an accurate and effective method for determining the thickness of subsurface geologic layers. Applications for engineering design, assessment, and remediation, as well as groundwater and hydrogeologic studies, include

- Continuous profiling of subsurface layers, including the bedrock surface;
- Water-table depth determinations;
- Mapping and general identification of significant stratigraphic layers;
- Detection of sinkholes and cavities;
- Detection of bedrock fracture zones; and
- Detection of filled-in areas.

Seismic refraction investigations are particularly useful because seismic velocities can be used for material identification. The Caterpillar D8R Ripper Performance Chart of Caterpillar Tractor Co. Handbook of Ripping (Caterpillar Tractor Co. 2000) equates seismic velocities into three categories of rippable, marginal, and nonrippable to present a guide to material identification on the basis of P wave seismic velocities.

C.6 ACOUSTIC EMISSION TESTING

Noises occurring in nature are accompanied by and are the result of an energy release of some kind. Fatigue cracks, weld discontinuities, and many other failure-causing mechanisms also produce sound energy. Although a portion of the sound produced by materials under stress may exist as audible sound, most are low energy and inaudible. This depends on deformation magnitude and type and flaw growth or failure.

For the purposes of this discussion, an acoustic emission (AE) is defined as inaudible sound energy released within a material undergoing deformation or flaw growth. An AE test is described as a method used to detect this sound energy. One or more listening transducers are attached to the test object to detect AE. Positioning of AE transducers in the path of anticipated sound propagation enables detection. The detected signals are then electronically processed to derive information on the location and severity of growing flaws. Note that guard transducers are also used in conjunction with the listening transducers to differentiate the flaws from normal structure noises.

Acoustic emission testing differs significantly from the other NDT methods discussed in this appendix. Perhaps the most notable differences are the following:

1. Detected signal is produced by the test material itself and not by an external source. The AE transducers need to act only as receivers.
2. Acoustic emission tests detect movement, whereas most other methods typically detect only existing geometrical discontinuities.
3. Applied stress is required to cause flaw growth, and, hence, the acoustic emission. The applied stress can be the result of the component's service and dead loads or an induced load used specifically for the AE test. In many tests, a combination of the two is necessary.

Acoustic emission testing is used to detect cracks, corrosion, weld defects, and material embrittlement. This method can be used on various materials, such as metal, timber, concrete, fiberglass, polymer composites, and ceramics. An entire structure can be monitored with AE testing from a few locations, reducing the amount of access required. Acoustic emission testing can also be conducted while the structure is in service.

Acoustic emission testing is a real-time NDT method. In other words, it monitors the actual condition of the component during the test. The AE test method can also be used to record an accumulation of damage occurring within a structure. The data obtained can be used as history for a structure and possibly to predict failure.

A primary limitation of AE testing on structures is the requirement to differentiate the sound energy released by a growing flaw from background noise. Many background noise generators, such as bolts, joint friction, traffic, port operations, and so forth, can mimic or mask the sound energy released from growing cracks. Some AE test methods avoid this problem by isolating areas known to contain possible background noise generators.

When a global AE inspection is conducted to determine areas where structural problems exist, additional NDT or PDT methods may be required to identify the exact nature of the emission source defect.

C.7 STEEL REINFORCEMENT TESTING

Covermeters or pachometers are electromagnetic devices that detect the reinforcing steel in concrete and measure its size and depth of cover. The device produces a magnetic field and locates the reinforcing steel by measuring the distortion of the magnetic field created by the presence of the steel. The received signal increases with increasing bar size and decreases with increasing cover thickness. Making certain assumptions, the pachometer can be calibrated to convert the signal into a distance, which indicates the depth of cover.

The relationship between cover depth and deterioration mechanisms highlights the importance of measuring the depth of concrete surface cover. Inadequate cover can undermine the concrete's protection from corrosion in the steel reinforcement. Carbonation begins as soon as concrete is exposed to air, where carbon dioxide and moisture mix with products in the concrete to cause chemical changes.

The process of carbonation neutralizes the protective alkaline nature of the concrete. If the cover is too shallow, the carbonation will reach the level of the reinforcing steel, and the alkaline protection will be lost, causing corrosion to begin on the steel. The exposure of the concrete and reinforcing steel to chlorides and moisture will further accelerate the corrosion process once cracks or spalls have breached the cover. If the cover is too deep, the possibility exists of increased crack widths and decreased effective depth, affecting design parameters on a concrete member.

Accurately locating the reinforcing steel in concrete allows the inspector to determine whether the steel is placed outside the carbonation zone. Covermeters, in general, can accurately measure the cover depth within 1/4 in. in the range of 0 to 3 in. in lightly reinforced structural members. Covermeters can also be used to locate reinforcing steel to *tie in* a new structural member to an existing structure. This process typically occurs during rehabilitation and involves drilling into the existing structure and adding reinforcing steel, which spans from the old to the new components.

Several factors may limit the effectiveness of a covermeter as an NDT method. Traditional covermeters only locate the reinforcing steel and may not provide any information about defects or the material's deterioration. Furthermore, it may not distinguish if one or more bars are present at a certain location, and, therefore, the intensity of the signal may be misinterpreted, and the cover depth can be incorrectly noted as shallower than the true depth. This problem is most pronounced in heavily reinforced structures or when large steel objects, such as scaffolds, are near the test area. Also, some reports indicate that the epoxy coatings on reinforcing steel can distort the readings of certain devices. Likewise, the relative material properties of the concrete often must be assumed to utilize conversion charts for the readings (Figure C-10).

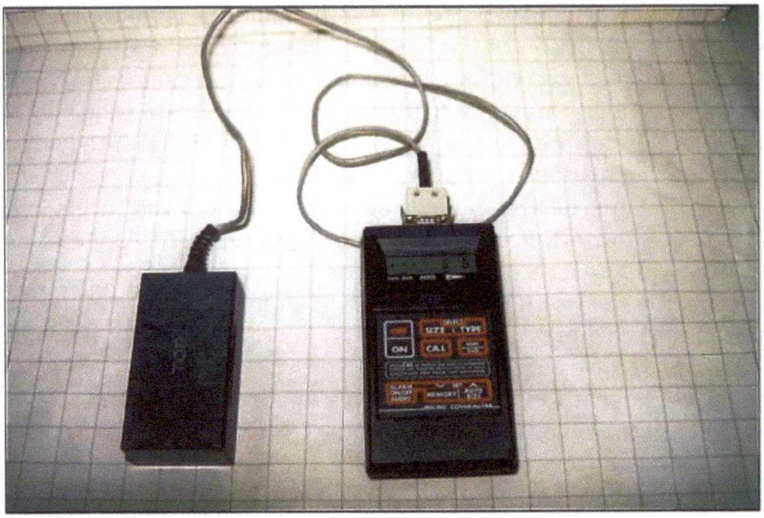

Figure C-10. Pachometer unit.
Source: Courtesy of Collins Engineers Inc., reproduced with permission.

C.8 SCHMIDT HAMMER

A rebound hammer, commonly referred to as a Schmidt hammer, is a mechanical device used to measure the compressive strength of in-place concrete. The device consists of a plunger and a spring-loaded hammer. When triggered, the hammer strikes the free end of the plunger that is in contact with the concrete, which, in turn, causes the plunger to rebound. The extent of the rebound is measured on a linear scale attached to the device. This test is covered in ASTM C805/C805M (ASTM 2018b).

The rebound hammer is used to assess the uniformity of in situ concrete and to delineate zones of poor quality or deteriorated concrete. It is also useful for detecting changes in concrete characteristics over time, such as hydration of cement, to remove forms or shoring. The advantages of the rebound hammer are that it is portable, easy to use, and low cost, and can be used to evaluate large areas quickly.

The rebound hammer is valuable purely as a qualitative tool. It measures the hardness of the concrete at the surface by comparing the amount of hammer rebound on a scale of 1 to 100 with an estimated compressive strength of the concrete. Other tests, such as a compression test, must be used to determine the actual strength of the concrete. Several factors govern the rebound measurement, including the size, age, and finish of the concrete and the aggregate type and moisture content. The Schmidt hammer test should not be used on overexposed aggregate in the concrete because a false reading will be displayed (Figure C-11).

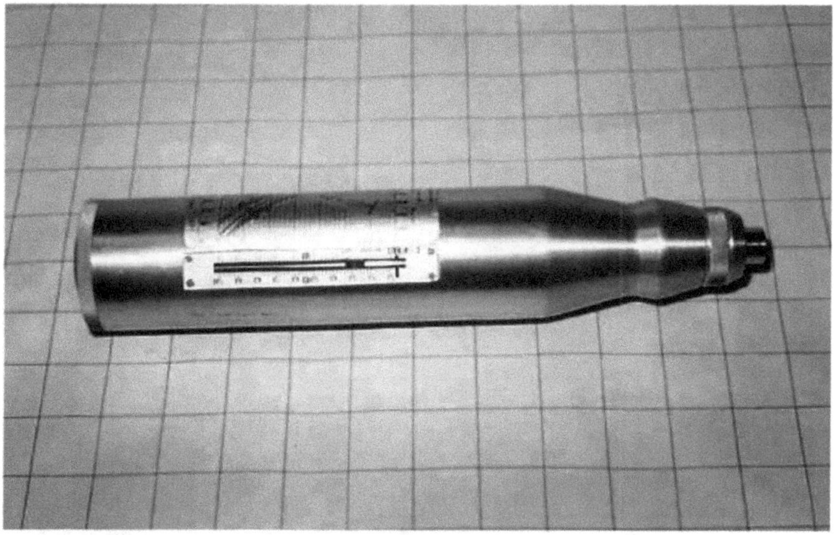

Figure C-11. Standard rebound hammer.
Source: Courtesy of Collins Engineers Inc., reproduced with permission.

C.9 IMPACT-ECHO TESTING

The impact-echo method is an NDT technique used for detecting internal flaws in concrete. It has been used on various members, particularly slab, beam, and wall type members. The impact-echo test method produces a transient stress pulse in a member by means of a point impact. This pulse produces a surface wave and waves that travel into the element. Internal defects and the boundaries of the element reflect these waves.

The testing apparatus consists of a handheld unit that generates an impact that produces a wave and a receiving transducer that receives the reflected waves. A computer-based system is then used to process the data and display the echo waveform data. The operator interprets the data to determine the presence and extent of defects found. Impact-echo testing is covered in ASTM C1383 (ASTM 2022a).

The impact-echo technique utilizes easily transportable equipment and can be performed by a single individual. Testing is fairly rapid, and only minimal surface preparation is needed to ensure proper impact energy transfer to the structure. Tests are often done in a grid pattern, with the size of the grid determined by the suspected damage. Experience has shown that the technique can be used to locate subsurface delamination, honeycomb, cracks, voids, etc. In the hands of a skilled technician, it may also locate voids around reinforcing steel and within grouted prestressing strands and posttensioned tendon conduits (Figure C-12).

Figure C-12. Impact-echo testing to detect flaws in concrete.
Source: Courtesy of Collins Engineers Inc., reproduced with permission.

The impact-echo method requires the interpretation of the waveform output for each test by the field technician. The testing technician must be trained and experienced to properly interpret the output data. Prior to testing, design plans should be carefully reviewed for embedded items or other details that may affect wave behavior and test results. The presence of steel reinforcement bars must also be properly accounted for. The maximum element thickness for this test is approximately 6.5 ft (2 m).

C.10 WINDSOR PROBE TESTING

The penetration method, typically utilizing the Windsor probe test system, consists of a device that drives a probe into the concrete using constant energy. The probe is made of a hardened steel alloy specifically designed to crack the aggregate particles and compress the tested concrete. Once fired, the length of the probe projecting from the concrete is measured. A test typically consists of firing three probes and averaging the projecting lengths. This test is covered in ASTM C803/C803M (ASTM 2023).

Penetration tests assess the uniformity of in situ concrete and delineate zones of poor quality or deteriorated concrete. They are also well suited for estimating the compressive strength of concrete and the relative strength of concrete across the same structure. Penetration tests are commonly used to estimate concrete's early age strength for stripping forms.

Figure C-13. (Left) Windsor probe kit, (right) Windsor probe testing.
Source: (Left) Courtesy of Collins Engineers Inc., reproduced with permission; (right) Courtesy of Jacobs reproduced with permission.

The penetration test method is a qualitative tool, and like the rebound hammer, requires that other tests be conducted to determine the actual strength of the concrete being tested. The penetration test method also damages the concrete at the test location. The probes must be removed, and the holes must be patched (Figure C-13).

C.11 HALF-CELL TESTING CORROSION SURVEY

Half-cell testing can be conducted on coated steel members or embedded steel reinforcement in concrete to determine whether corrosion is occurring. It can be conducted on elements located above or below the water. Steel reinforcement is typically protected from corrosion by the alkaline nature of concrete. If the alkalinity of the concrete is compromised, corrosion on the steel will commence, provided moisture and oxygen are present. The corrosion reaction will promote anodic and cathodic activity along the reinforcing steel. The corrosion of the reinforcement produces a corrosion cell caused by these differences in electrical potential.

The half-cell testing method uses this process to detect whether the reinforcing steel is under active corrosion. This method utilizes a multimeter to measure the potential difference between the steel and a half-cell apparatus. An analysis of the potential difference can indicate whether active corrosion is taking place on the reinforcing steel. Likewise, this method can be used on a coated steel pile to detect corrosion below the coating. Last, half-cell testing can be used to obtain voltage readings to ensure that a cathodic protection system is working properly.

Although commonly used on steel piles or concrete decks, the half-cell test can be performed on any reinforced concrete component, provided a direct electrical connection can be made to the reinforcing steel. Because the test can detect corrosion only directly under the device, a systematic grid of test points should be created to map the potential readings throughout the concrete component. This map can then be analyzed to determine the probable areas of active corrosion.

Half-cell testing requires specialized equipment, including a half-cell reference apparatus and a multimeter. For topside applications, such as investigating the condition of steel reinforcement in a concrete deck, a copper–copper sulfate (Cu-$CUSO_4$) half-cell reference apparatus is typically used. A silver–silver chloride (Ag-AgCl) reference half-cell would be more appropriate for underwater seawater applications.

The potential measurements for Cu-$CUSO_4$ half-cell reference apparatus can be interpreted as follows, but it depends on the type of half-cell:

- To −0.20 V indicates greater than 90% probability of no corrosion;
- −0.20 to −0.35 V indicates that corrosion is uncertain;
- −0.35 V indicates greater than 90% probability that corrosion is occurring; and
- Positive number indicates that the moisture content of the concrete is insufficient, and therefore, the test is not valid.

The potential measurements for an Ag-AgCl half-cell reference apparatus can be interpreted as follows, but it depends on the type of half-cell:

- To −0.15 V indicates greater than 90% probability of no corrosion;
- −0.15 to −0.30 V indicates that corrosion is uncertain;
- −0.30 V indicates greater than 90% probability that corrosion is occurring; and
- Positive number indicates that the moisture content of the concrete is insufficient, and therefore, the test is not valid.

Note that a connection with the reinforcing steel is required; therefore, holes may need to be drilled into the concrete to locate and connect to the steel. This test method only indicates the probability of corrosion present at the time of testing and does not indicate the extent or rate of corrosion (Figure C-14).

Where structure elements are fully submerged and locating a topside multimeter voltage device nearby is impractical, proprietary devices, such as the Buckley bathycorrometer and Polatrak CP gun, can operate underwater in a self-contained waterproof containment unit.

Using this half-cell testing methodology, information can be obtained about members protected by cathodic protection. For example, elements are usually considered adequately protected by cathodic protection if a reading is −0.80 to −1.2 V with a silver-silver chloride (Ag-AgCl) reference half-cell, and varies depending on factors such as the water salinity and type of half-cell. Furthermore, hydrogen embrittlement of welded members underwater may be a concern if readings of −1.3 V are obtained due to an overactive impressed current cathodic protection system.

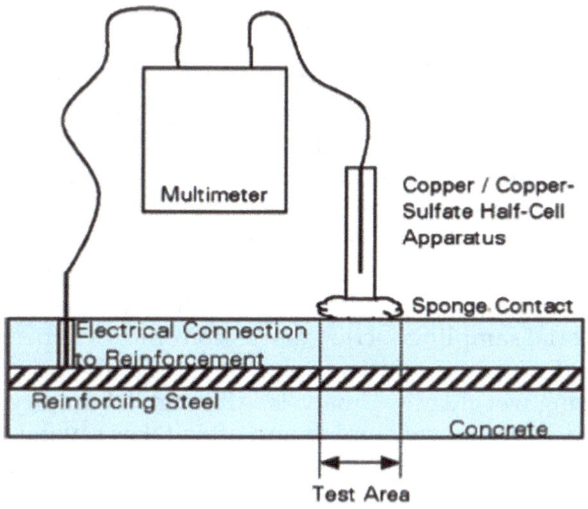

Figure C-14. Half-cell testing illustration.
Source: Courtesy of Collins Engineers Inc., reproduced with permission.

C.12 CHLORIDE ION TESTING

Chloride ions (CIs) are the major cause of reinforcing steel corrosion in concrete. Chloride ions are most often provided from seawater, although they may also be available as contaminants in the original concrete mix. Although present, these CIs are not likely to cause problems unless they exist in unusually high concentrations. Because corrosion of steel reinforcing is usually considered to begin at a chloride ion content of between 0.025% and 0.033% by weight of concrete, a knowledge of chloride content can aid in determining the likelihood of the onset or presence of corrosion and aid in estimating the remaining life of a facility. Note that chloride content can be expressed in several ways. Primarily, these include the percentage of chloride ion in the bulk solid or the chloride content in parts per million (ppm) of a solution containing dissolved material. Both results, percent Cl^- in bulk and ppm Cl^- in solution, have been shown in the past to be accurate. If discussing ppm Cl^-, the threshold is considered typically 500 ppm, although empirically, each site may be unique.

In evaluating chloride content, a chloride profile (chloride concentration percentage versus depth measurement below the concrete surface) should be developed. This profile is important for assessing the future corrosion susceptibility of steel reinforcing and for determining the primary source of chlorides.

The chloride content in concrete is typically determined through laboratory analysis of powdered concrete samples. The powdered samples can be obtained on-site or in the laboratory. Field-collected powdered samples are typically taken by drilling at different depths down to and beyond the level of the reinforcing steel. Extreme care should be exercised to avoid inadvertent contamination of the samples.

Alternatively, cores can be collected and powdered samples can be obtained at different depths by grinding the core sample in the laboratory. The collection of these samples, in essence, destroys a portion of the component, making this test procedure similar to material sampling as described later in this appendix. However, because this test can be performed in the field and results obtained quickly, it has been separated from the material sampling section discussions in this appendix.

The chloride ion content of concrete is usually measured in the laboratory using wet chemical analysis. Although laboratory testing is the most accurate, it is time-consuming and often takes several weeks before results are available. As a result, field test kits have been developed. The use of field test kits allows rapid determination of chloride levels to be made on-site. Although the field kits are not as accurate as the laboratory method, they do provide a good correlation with laboratory tests when a correction factor is used.

Various detailed procedures for chloride testing are published by ASTM International. However, the ASTM International standards typically apply to testing in the laboratory, not in the field. However, AASHTO T 260 (AASHTO 2021), "Sampling and Testing for Chloride Ion in Concrete and Concrete Raw Materials," is a useful reference by the American Association of State Highway and Transportation Officials in Washington, DC.

Chloride ion testing can be performed on any concrete component. Field kits allow inspectors to perform the test on-site and determine chloride levels immediately.

Collecting samples to perform this test requires damaging a portion of the concrete member and potentially compromising the tested area, thus requiring repair. Therefore, several samples cannot be taken from a single location to validate results. This method is also time-consuming and requires access to the member. After conducting this test in an area, the area should be repaired or patched to prevent future deterioration.

C.13 TIMBER RESISTANCE DRILL

A timber resistance drill is a helpful tool for an inspector to ascertain interior damage by detecting voids or areas of decay within the cross section of timber elements. As indicated in Appendix B, "Types and Causes of Defects," timber elements are subject to a variety of deterioration mechanisms, some of which are internal to the element and obscured from view. Marine borers, fungal decay, and rot can diminish timber cross section without always affecting the outer shell material.

Although designs vary among resistance drill manufacturers, the basic premise is similar among models. A calibrated metal probe is advanced into the material and the feed and torsional resistances are logged. This provides the inspector with an assessment of timber quality through the cross section of the probed element. Models are available for use in both above and underwater applications.

A limitation of timber resistance drill testing is the localized nature of the data. That is, the drill can only detect deterioration/low-quality material at the probed location, and achieving a representative set of defect data can require extended inspection time. In addition, the probe does not provide engineering values for the inspected element, such as strength properties, and so interpretation of the probe results requires a careful evaluation of the inspected elements beyond the drill readings. Put another way, the drill readings provide a qualitative and not quantitative assessment of likely element integrity.

Inspection planning to incorporate a timber resistance drill should consider the following:

- Key elements to be probed and quantity of probes;
- Access requirements as well as member orientation (e.g., overhead elements requiring staging or special access);
- Embedded elements such as anchor hardware or conduits; and
- Protective elements, such as wraps, and methods to reseal after drill usage

Figures C-15 and C-16 provide sample resistance drill output data.

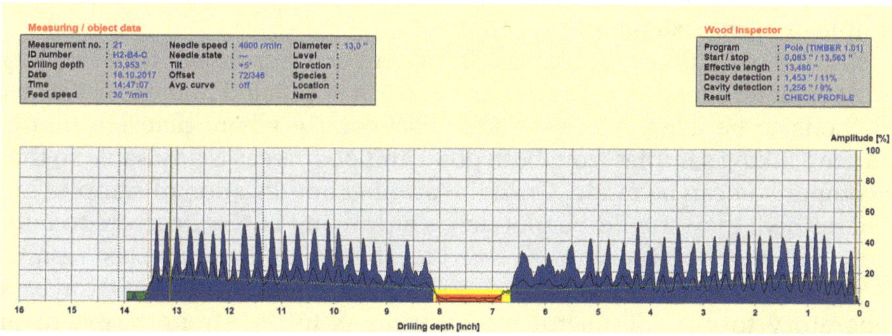

Figure C-15. Sample resistance drill output indicating a void at the center of the pile. Note growth rings.
Source: Courtesy of Appledore Marine Engineering, LLC, reproduced with permission.

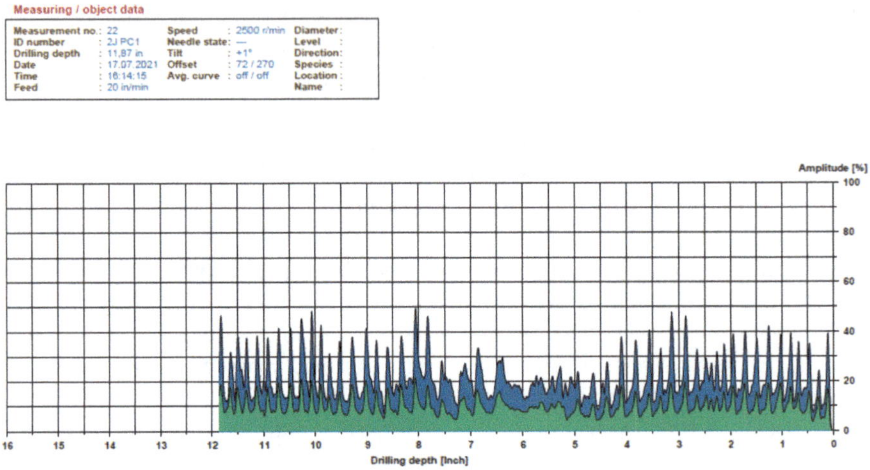

Figure C-16. Sample resistance drill output indicating a sound cross section. Note growth rings.
Source: Courtesy of Appledore Marine Engineering, LLC, reproduced with permission.

C.14 MATERIAL SAMPLING

To fully determine the condition of a structure, extracting material samples from the structure may sometimes be necessary so that laboratory tests can be conducted to better determine the condition of the structure's materials or the states of deterioration or damage. Typical laboratory tests may include compressive tests and petrographic examination of concrete; tension tests, charpy tests, or crack surface investigations of steel; or even simple integrity examination of timber.

Prior to obtaining any samples, the extent and purpose of the sampling must be determined. The sample size is often stipulated in the specific test methods to be used. In most cases, particularly where deterioration is present, taking samples from both good and poor areas is advisable so that a comparison can be made. Once the number and location of samples are determined, they should be plotted on a drawing of the structure to both aid in fieldwork and serve as a record for the evaluation of the test results.

All materials can be sampled and tested either in the field or in a laboratory to provide useful information as to the strength, extent of deterioration, and material characteristics. Specimens should come from representative areas of the structure using a statistically representative sampling number with a minimum of three samples.

All material samples should be collected, and tests conducted in accordance with ASTM and methods for the respective materials. The removal of material from the structure should be conducted only when a specific piece of information is required and the information obtained provides useful information in the evaluation of the structure. Extracting samples will leave holes or voids in the tested component, and, therefore, repairs will be required. Concrete and timber repairs are somewhat easy, but steel repairs may be more complex. Welding requires the use of experienced personnel, and care should be taken to minimize any residual stresses or fatigue-prone details associated with the repair.

C.14.1 Concrete Sampling

Concrete material sampling most often consists of drilled cores, although sections may also be obtained by sawing or breaking off a portion of the component. The core size should be determined by the tests to be run; however, in most cases, a 4 in. diameter core is extracted. Core holes are normally filled with grout; other sample areas should also be repaired with a suitable mortar material. When feasible, steel reinforcement should be avoided, unless sampling specifically requires it to be part of the core (Figure C-17).

Figure C-17. Concrete deck core photographs.
Source: Courtesy of Collins Engineers Inc., reproduced with permission.

Samples should be marked for location and orientation and packaged to prevent damage during transport. As part of the sampling operation, reinforcing steel is typically located and marked to avoid cutting it during the sample extraction. In some instances, including reinforcing steel as part of the sample may be desirable. In these cases, confirming that the cut reinforcing steel will not jeopardize the structure's integrity is necessary.

Some concrete tests that require samples include

- Carbonation,
- Permeability,
- Cement content,
- Percent air content,
- Moisture content,
- Steel reinforcing yield strength,
- Concrete compression strength,
- Modulus of elasticity (static and dynamic), and
- Concrete splitting tensile strength.

C.14.2 Steel Testing

Material coupons for steel members are usually obtained by sawing, coring, or collecting drill shavings. Coupons may be flame cut; however, the heat induced by the cutting operation alters the material's properties in the vicinity of the cut both in the sample and in the remaining base material. These heat-affected areas must then be removed by grinding prior to testing. Repairing the base material is also often required.

For these reasons, flame cutting should typically be avoided. In selecting coupon locations to test material properties, such as yield strength or toughness, the investigator must remember that the properties of steel members vary over the cross section as a result of differing rates of heat loss resulting from fabrication techniques and rolling or production practices. The orientation of the steel samples should be recorded prior to removal (Figure C-18).

Some of the steel tests that require samples are

- Brinell hardness test,
- Charpy impact test,
- Chemical analysis, and
- Tensile strength test.

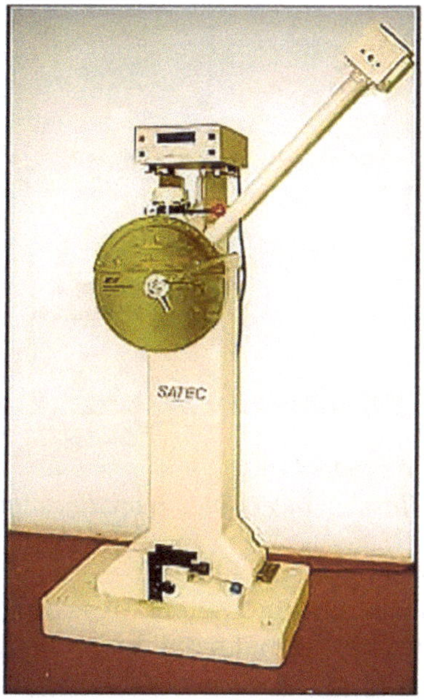

Figure C-18. Charpy impact metal testing machine.
Source: Courtesy of Collins Engineers Inc., reproduced with permission.

C.14.3 Timber Testing

Timber sampling often consists of the use of incremental borer (up to 1/4 in. diameter for void verification or preservative penetration examination) or extraction cores (typically 2 in. diameter or greater for marine borer investigations and strength testing), although sections may also be obtained by sawing off a portion of the component. Boreholes should be plugged with a treated hardwood dowel. Large-diameter core holes are more challenging to plug and should be handled on the basis of the required life span of the repair. Drilling, boring, and probing are most often used to assess the presence of marine borers or voids, the extent of rot, and the depth of preservative penetration (Figure C-19).

Timber sample cores are assessed to determine whether bacterial or fungal decay is present and the extent of interior rot and to determine the species of timber, if required. These methods typically do not produce a global sample specimen; however, several local specimens from random locations can be effective. Any holes made should be plugged with a treated hardwood dowel.

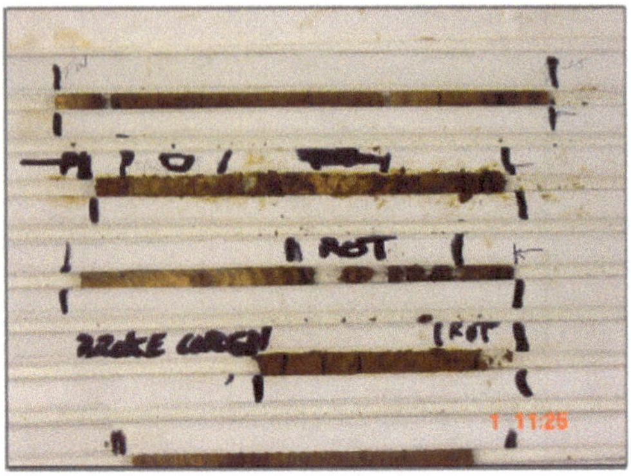

Figure C-19. Timber cores placed in tray.
Source: Courtesy of Collins Engineers Inc., reproduced with permission.

Moisture content and rot can also be assessed on specimens using electrical devices such as the Shigometer. These devices require electrodes to be driven into the timber or that small holes be drilled to insert probes into the timber. These detect the presence of timber rot; however, drilling or coring should be conducted to determine the extent of the rot.

C.15 ULTRASONIC TESTING

Ultrasonic testing (UT) is used to evaluate various materials and examine the internal (volumetric) condition of materials. Specifically, it is used to confirm suspected discontinuities or cracks and check questionable material thicknesses or lengths in steel, concrete, or timber. Typical discontinuities, which are detectable by use of UT, include delamination, surface cracks, and many surface and subsurface weld-related discontinuities (lack of fusion, porosity, etc.). Although ultrasonics may be used in timber, there are mixed opinions within the industry as to the effectiveness of the nonhomogeneous material in a marine environment. Therefore, the user should be familiar with the applications and limitations of UT on timber in the marine environment.

A transducer is a device that is capable of converting energy from one form to another. In the case of UT, electrical energy is changed to mechanical energy and vice versa. UT transducers convert electrical energy into mechanical vibrations, which, in turn, produce high-frequency sound waves. They also convert high-frequency sound waves back into electrical energy upon receiving the return echoes.

The most common UT technique currently in use in the United States is called *pulse echo*. The pulse-echo method employs short bursts, or pulses, of waves, which are transmitted into the specimen by the transducer, which must be in integral contact with the specimen. Any returning unexpected echo from these pulses is evaluated to determine the reflector location and size.

The signal height, or amplitude, relates to the amount of reflected sound energy. Large reflectors, causing total reflection of sound, produce signal responses of higher amplitude than smaller reflectors, which reflect only a portion of sound energy. Larger return echo amplitudes suggest larger-sized flaws. Echo indications are normally retested from another position to confirm flaw size and position. Basic ultrasonic pulse-echo systems utilize a power supply, pulser, receiver/amplifier, oscilloscope (cathode ray tube or CRT), timer (clock), and a transducer (Figures C-20 through C-22).

Power for the testing equipment is supplied by portable battery packs or by an external alternating current (AC) source. The pulser, also called the *pulse generator*, produces the short-duration voltage burst, which is

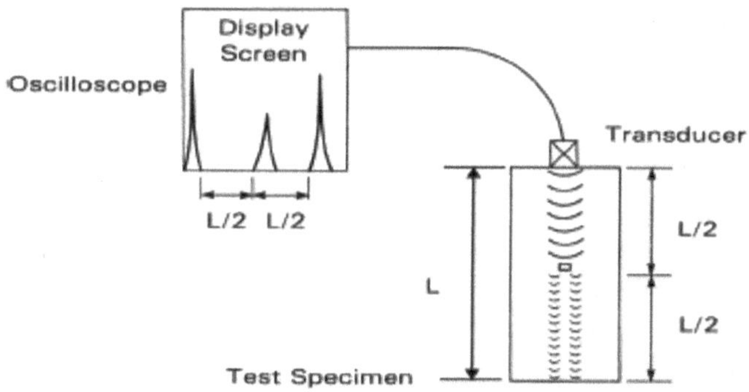

Figure C-20. Pulse-echo UT illustration.
Source: Courtesy of Collins Engineers Inc., reproduced with permission.

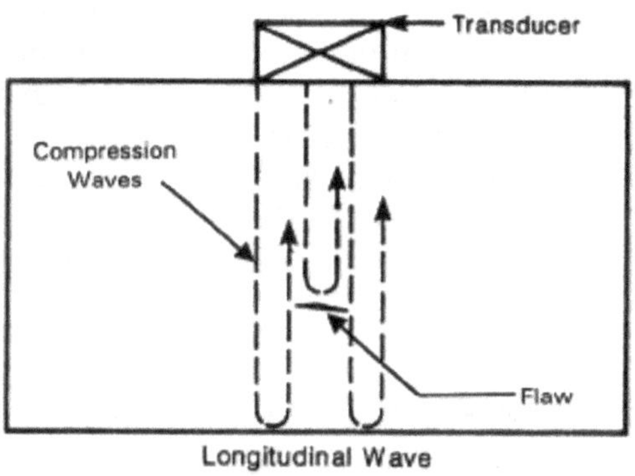

Figure C-21. Pulse-echo UT compression wave schematic.
Source: Courtesy of Collins Engineers Inc., reproduced with permission.

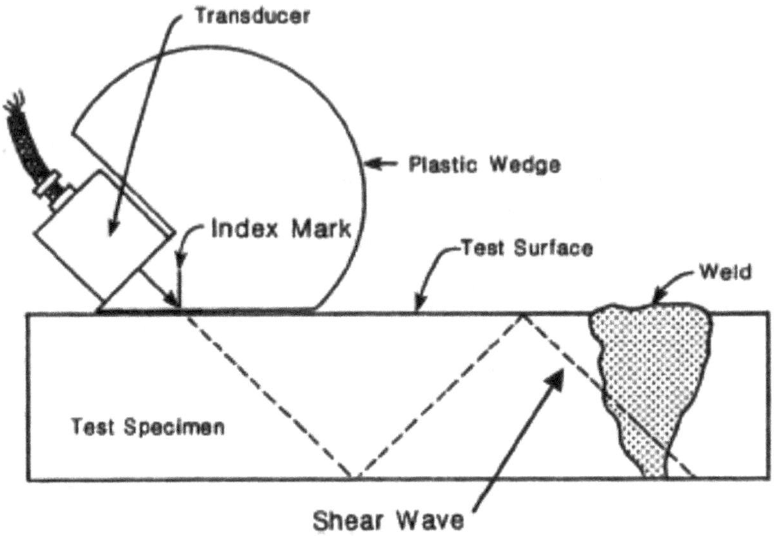

Figure C-22. Pulse-echo UT shear wave schematic.
Source: Courtesy of Collins Engineers Inc., reproduced with permission.

applied to the transducer. A clock or timer controls the rate of these voltage bursts. Sound echoes returning to the transducer are relayed to the receiver, amplified, filtered, and sent to the CRT for display on the screen. Pulse-echo methods include compression, shear, and surface wave modes.

Phased array is a unique type of UT and has several advantages over traditional UT devices. However, phased array units are more complex and require additional training.

Ultrasonic testing should not be performed on rough surfaces, on parts with complicated geometries, on highly attenuative materials, or where the discontinuity size is expected to be smaller than one-half of the wavelength. Rough surfaces may require grinding in surface preparation. Other factors that limit the successful application of UT are a lack of properly trained personnel, overestimation of the accuracy of flaw locating and sizing, and poorly written testing procedures. Typically, a certified Level II ASNT specialist should conduct all testing, and a Level III ASNT specialist should develop written testing procedures for uncommon applications.

C.16 LIQUID PENETRANT TESTING

Liquid-penetrant testing is used to confirm the presence of a crack or flaw. Liquid-penetrant tests can be conducted on many nonporous materials, including metallic and nonmetallic, magnetic and nonmagnetic, and conductive and nonconductive materials. This method is highly sensitive to small surface discontinuities and produces indications directly on the surface of the component, providing a visual representation of the flaw.

Liquid-penetrant testing relies on the capacity of a liquid to enter into a discontinuity; therefore, it can only find discontinuities that are open to the surface of the material. It can be used with any material provided the material is nonporous and is not adversely affected by the penetrant material.

The basic procedure requires that the material be prepared by removing all surface contaminants and applying a liquid (penetrating oil) to the surface being tested. The penetrant will seek out and enter small surface openings. Penetrant is then removed from the test surface by wiping or water rinsing. A drying developer is next applied. The penetrant remaining in the discontinuity bleeds out, forming a highly visible, contrasting indication on the test surface (Figures C-23 and C-24).

A good penetrant is defined by the ability of the fluid to be drawn into small openings, even against gravity. Many variables affect this penetrating ability, including surface tension of the liquid, wetting ability, surface condition, surface contamination, and temperature.

Two major types of penetrants are used: (1) visible dye penetrants and (2) fluorescent penetrants.

Visible dye penetrants are normally red, providing contrast with the applied white developer under visible light. Fluorescent penetrants contain dyes that fluoresce brilliantly when viewed under a black light in a darkened

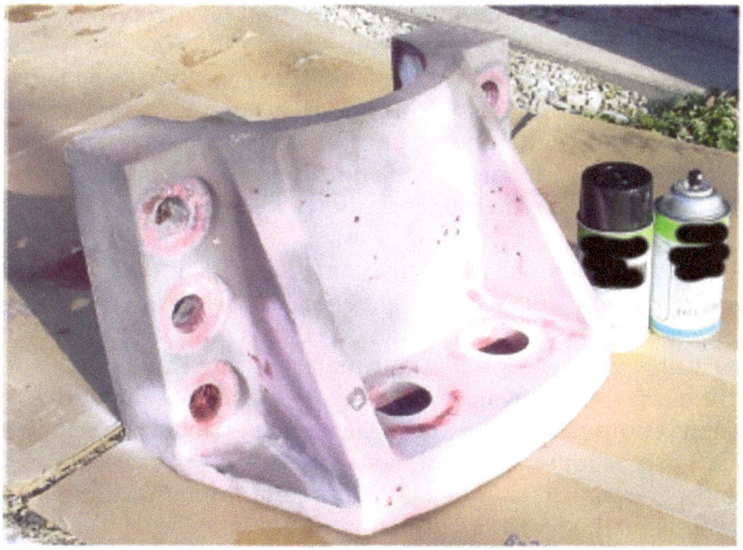

Figure C-23. Liquid penetrant and developer applied to a casting.
Source: Courtesy of Collins Engineers Inc., reproduced with permission.

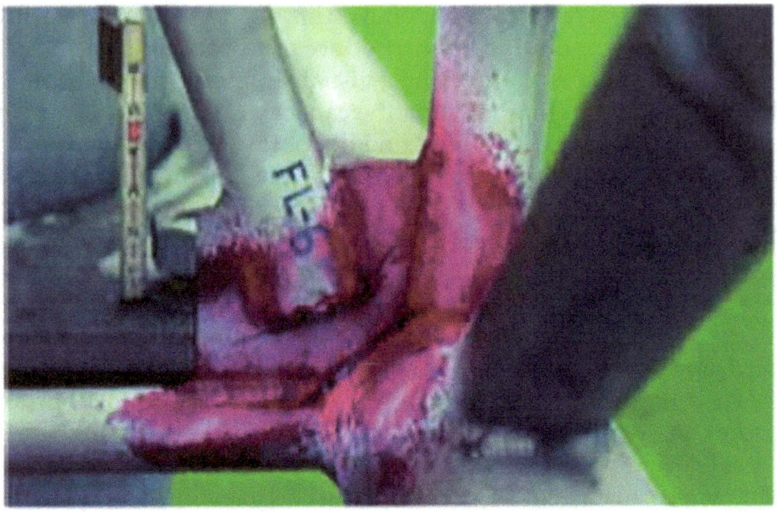

Figure C-24. Gusset plate with an applied dye penetrant.
Source: Courtesy of Collins Engineers Inc., reproduced with permission.

area. The ability to see penetrant indications on the test surface relates to the contrast provided between the penetrant and the test surface. Fluorescent penetrants provide better contrast than visible dye penetrants. For this reason, fluorescent penetrants are more accurate than visible dye penetrants.

Interpretation of the liquid penetrant indication involves determining what condition is present to have caused the indication, evaluating the condition as to its effect and seriousness from the standpoint of usability of the part, and reporting inspection results accurately and clearly. Proper interpretation and evaluation of liquid penetrant indications requires a knowledge of the types, causes, and appearance of indications, knowledge of the test method and material fabrication process, adequate illumination, good eyesight, and experience.

The penetrant materials typically come in aerosol form, making them highly portable and well-adapted to field use. This also allows large areas of a component to be tested rapidly even if the component has a complex geometric shape. Powder-penetrant materials are also available but are typically cumbersome in the field. Finally, penetrant materials and the associated equipment are somewhat inexpensive, especially when compared to most other NDT methods.

Liquid-penetrant testing does have several limitations, including the fact that this method works only on nonporous materials, surface finish and roughness can affect the sensitivity of the test, and it can only detect discontinuities that are open to the surface. Discontinuities filled with contaminants, paint, rust, oxidation, or corrosion products may not be detected. Therefore, surface preparation is critical. The process also requires multiple time-consuming steps, including preparing the surface, applying the dye or fluorescent, cleaning off the dye or fluorescent, applying the developer, and cleaning off the developer after the test is completed. This effort requires the safe handling of chemicals and the proper disposal of saturated cleaning rags and empty aerosol cans.

Finally, the test sensitivity is lowered at reduced temperatures because crack widths are typically reduced, and the test medium is less fluid.

C.17 MAGNETIC PARTICLE TESTING

The magnetic-particle test (MT) is used for testing ferromagnetic materials (steel, wrought iron, cast iron, etc.). The magnetic-particle test is used to confirm suspected cracks or test suspect details. The primary advantage of MT is high sensitivity in the detection of tight surface cracks and other small discontinuities. Typical detectable discontinuities include cracks, lack of fusion, and other weld-related surface discontinuities. Base metal discontinuities such as seams, laps, and stringers are also easily detected (Figures C-25 and C-26).

As defined by the American National Standards Institute (ANSI), the MT method utilizes the principle that magnetic lines of force, when present in a ferromagnetic material, will be distorted by a change in material

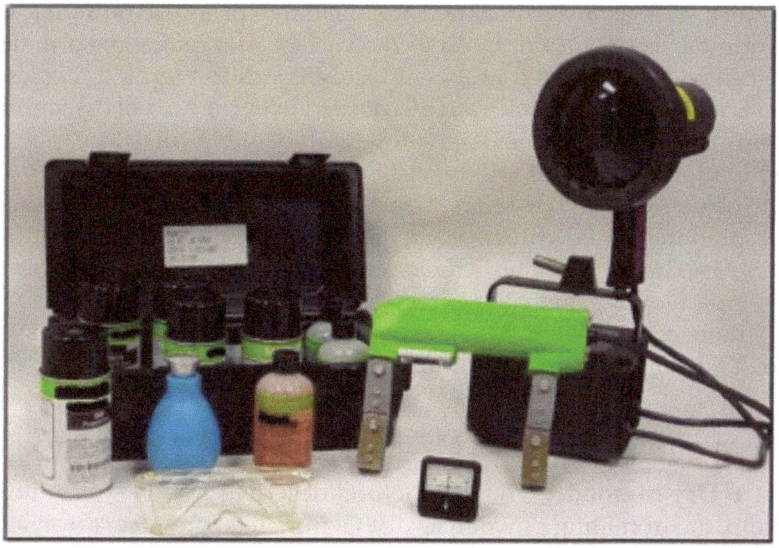

Figure C-25. Magnetic particle testing kit.
Source: Courtesy of Collins Engineers Inc., reproduced with permission.

Figure C-26. Crack identified by a magnetic particle.
Source: Courtesy of Collins Engineers Inc., reproduced with permission.

continuity such as a sharp dimensional change or a discontinuity. If the discontinuity is open to or close to the surface of a magnetized material, flux lines will be distorted at the surface; a condition termed *flux leakage*. When fine magnetic particles are distributed over the area of discontinuity while the flux leakage exists, they will be held in place, and the accumulation of particles will be visible.

The objective of MT is to create a magnetic field of sufficient strength and a predetermined direction to identify a leak if discontinuities are present. The inspector detects these leaks by sprinkling the test area with iron filings, blowing away the excess, and then looking for areas where the filings have accumulated. These areas of accumulation indicate a surface or possible subsurface discontinuity.

C.17.1 Magnetic-Particle Test Methods and Implementation Procedures

C.17.1.1 Dry Method. The dry method describes the type of indicating medium, the iron filing powder, as dry. Commercial powders are available in various colors, including red, black, gray, or yellow. The color selection should be based on the maximum color contrast with the material to be tested. Dry fluorescent particles are also available for use with a black light.

Dry particles are finely divided ferromagnetic materials with high permeability and low retentivity. The powder consists of a mixture of particle sizes, smaller ones being attracted by weak leakage fields and larger ones for detecting larger discontinuities.

C.17.1.2 Wet Method. If powders or particles are suspended in oil or water, the method is considered wet. Wet suspensions are also available in various colors and are fluorescent. They can be sprayed onto the part, or the part can be bathed in a suspension. Wet fluorescent particles provide maximum sensitivity (superior visibility) if used with the proper current, lighting, and surface preparation. Wet particles are mixed with the suspension in predetermined concentrations and particle sizes. The concentration of the particles within the wet suspensions will affect the test sensitivity. Light concentrations will produce faint indications, and heavy concentrations may provide too much coverage. Wet particles are usually smaller in size and lower in permeability than dry particles.

C.17.1.3 Continuous Procedure. This term is used if magnetizing force is applied prior to the application of the particles and terminated after excess powder has been blown away.

C.17.1.4 Residual Procedure. The term *residual* is used where the particles are applied after the part has been magnetized and the magnetizing current has been terminated.

C.17.2 Field Testing of Structures

Field-testing of structures and related structures uses portable units, which include the small portable prod or yoke units with AC/half-wave direct current capability. Portable prod equipment is commonly available in amperages up to 1,500. However, 115 V single-phase AC can also power the equipment. Some magnetic units do not require electricity, but they are less common.

Yokes are lightweight portable units easily carried to the job site. On some yokes, the legs are fixed at a set distance; on others, the legs are adjustable for various pole spacings. Yokes operate with 115 V AC.

The magnetic-particle test is a sensitive means of locating small and shallow surface cracks and can locate near-surface discontinuities with direct current. Unlike liquid penetrants, cracks filled with foreign material can be detected and no elaborate surface preparation is necessary. This test is also effective on painted surfaces.

This method is reasonably fast and inexpensive, especially compared to some other NDT methods, and the equipment is highly portable. The method also presents a few limitations on the size or shape of the part being inspected. The magnetic-particle test does have some limitations. This test will work only on ferromagnetic material, and the magnetic field must be in a direction perpendicular to the principal plane of the discontinuity for best detection. The magnetic-particle test will not disclose fine porosity. The deeper the discontinuity lies below the test surface, the larger the discontinuity must be to provide a readable indication.

C.18 MICROBIAL-INDUCED CORROSION TESTING

For marine structures where accelerated corrosion is observed and microbial-induced corrosion (MIC) is suspected, confirmation screening for MIC is the logical next step. If initial screenings indicate the presence of MIC, more detailed sampling and testing may be warranted. It is also essential to consider the site conditions and potential contributing factors that can lead to accelerated low-water corrosion and MIC. Water quality testing can be conducted to provide a better picture of what is occurring at the site. Handheld sondes can be employed to examine salinity, pH, temperature, dissolved oxygen, and electrical conductivity of the water at the site. Samples of the water can be taken for lab analysis of composition and understanding of what nutrients might be present that could contribute to the on-site MIC.

Biological Activity Reaction Testing (BART) can be used to confirm the presence of different types of bacteria as well as estimate approximate population size and aggressivity. Biological Activity Reaction Testing is a self-contained testing that requires no special laboratory equipment. Samples taken from the field are added to a preprepared BART tube. Nutrients are already at the base of the tube, prefilled by the testing manufacturer. The test is conducted at room temperature and is viewed daily.

C.18.1 Sulfate-Reducing Bacteria Testing

Sulfate-reducing bacteria (SRB) activity in the BART test produces the black sulfides in the base of the tube, in the anaerobic zone of the tube (as black precipitates), or around the floating ball in the aerobic zone of the tube (as an irregular black ring). If the black sulfides form around the ball, they may form as part of an aerobic community of bacteria, forming around and on the floating ball. When this happens, the black sulfides may appear as discrete, granular pieces of black sulfides suspended within a nonblack slime ring around the floating ball.

The SRB BART test uses short-chain fatty acids as the nutrient compound for the growth of the SRB. In some instances, other types of bacteria may grow, potentially heterotrophic anaerobic bacteria in the BART tube.

On occasion, these types of bacteria can potentially grow faster than the SRB, and when this occurs, the test tube liquid will likely go cloudy and appear as a gel-like clouding in the anaerobic portion of the tube (in the bottom third of the tube).

In rare occurrences, the SRB BART test may immediately blacken (within 30 min of starting the test). In this instance, H_2S already contained within the water sample is likely reacting immediately with the ferrous iron. If this occurs, another sample should be prepared, and the sample should be aerated through agitation to remove the H_2S from the sample water by shaking the sample in the outer vial for 10 s prior to pouring the sample into the test bottle (Figure C-27).

Once the test is prepared, it should remain upright, and with the lid sealed in a room temperature area, outside of direct sunlight. When choosing a room in which to store the sample, the temperature of the environment from which the sample was taken should be considered. If the room in which the sample is stored is colder (e.g., an air-conditioned office) or warmer (e.g., a storage unit or detached garage) than the native environment, it can affect the results of the testing. Daily observations of the test tube should be conducted, looking for the appearance of any reaction.

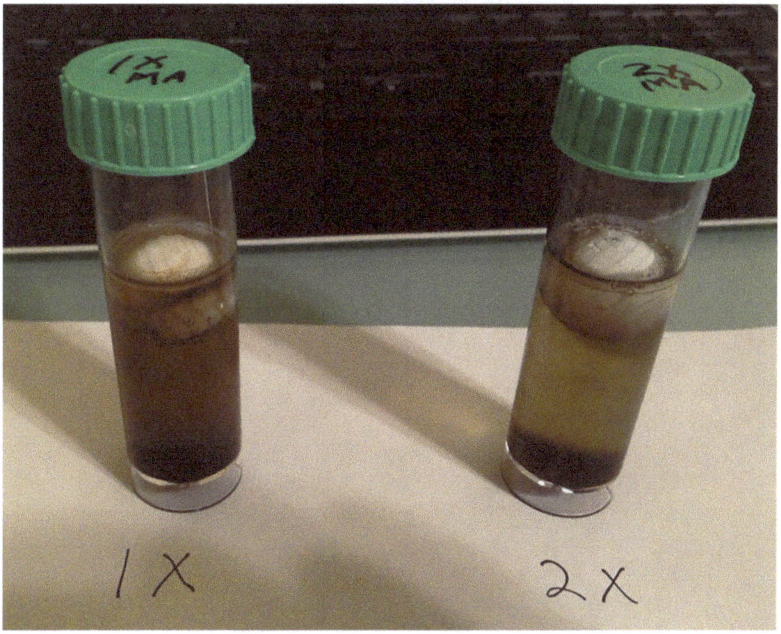

Figure C-27. Bart testing sulfate-reducing bacteria.
Source: Courtesy of Jacobs, reproduced with permission.

The duration of time elapsed before a reaction is observed (i.e., the time lag) is indicative of how much bacteria and how aggressive the bacteria is in the sample. The observer should carefully monitor the test tube daily and note and photo-document reactions occurring in the sample over time. The approximate sample bacteria population density, in terms of colony-forming units per milliliter (cfu/mL) and the aggressiveness of the bacteria, can be estimated by recording the time until the first reaction is observed. Reactions observed within a day or two after starting the test, indicate a high level of bacteria aggressivity, with an approximate bacteria population of 700,000 cfu/mL or more. The test should be run for at least eight days; however, reactions can occur after day eight. Observing reactions four to seven days after starting the test indicates moderate aggressivity with approximate colony size of 18,000 cfu/mL to 500 cfu/mL, respectively. Droycon Bioconcepts Inc. provides detailed guidance on BART testing in the Biological Activity Reaction Test User's Manual (Droycon Bioconcepts, Inc. 2024a) and Sulfate Reducing Bacteria, SRB-BART™ Quality Control (Droycon Bioconcepts, Inc. 2024b) (Figure C-28).

Figure C-28. Bart testing sulfate-reducing bacteria.
Source: Courtesy of Jacobs, reproduced with permission.

C.18.2 Iron-Reducing Bacteria and Acid-Producing Bacteria Testing

Depending on the MIC at the site, additional BART testing may be warranted to confirm the presence of other bacteria. Biological Activity Reaction Testing can also be performed to confirm the presence of iron-reducing bacteria (IRB) and acid-producing bacteria (APB). Similar to the BART testing for SRB, the self-contained test contains a test tube that has a floating intercedent device to create anaerobic and aerobic zones. At the base of the tube, a nutrient culture is designed to promote the growth of the bacteria that the test is targeting.

This test is conducted by filling the sample tube and watching and noting reactions over time. The order and type of reaction offers indications of the type of bacteria, the approximate population, and its aggressivity. It is worth noting that a negative result does not necessarily indicate that there is no MIC on-site; the bacteria live within the biofilms and are present at differing populations within the different strata of the biofilm. It is possible that you can observe the effects of MIC on-site but take test samples from the site that are not representative of the full microbial picture on-site.

C.18.3 Additional Microbial-Induced Corrosion Testing Methods

Outside of BART testing, there are commercially available dip-slide preprepared culture tube tests that confirm the presence of a type of bacteria within the sample (Figure C-29). Prefilled vials contain a gel or liquid culture medium. In the field, samples are gathered and placed within the tube. The tubes are observed over time and reactions are noted

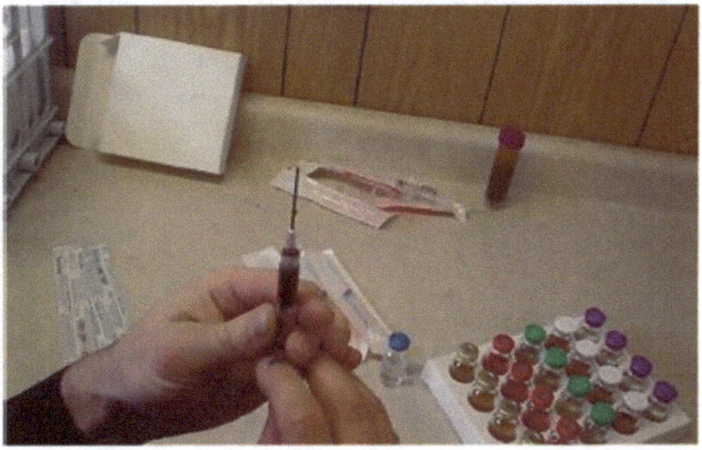

Figure C-29. Bacteria testing of samples obtained at a steel sheeting bulkhead with MIC.
Source: Courtesy of Collins Engineers Inc., reproduced with permission.

by the observer. When testing for APB, for example, the tube culture medium within the tube turns from red to yellow when APB is present. Similar to BART testing, the time it takes for the reaction to occur is then used to estimate the population of the bacteria and its aggressivity.

Once screening level testing has confirmed the presence of MIC, further laboratory analysis may be warranted to understand more precisely the type of bacteria present and the exact colony size and aggressivity in laboratory cultured testing. Outside of the bacteria testing, a more intensive inspection of the assets, including more frequent ultrasonic thickness measurements, may be warranted to better understand the degradation of the asset at the site and the extent of MIC that has occurred to date.

C.19 STRUCTURE-MONITORING SYSTEMS

In contrast to other NDT testing methods where the inspector conducts an inspection over a finite time interval, monitoring systems provide continuous data over an indefinite time interval. Monitoring devices are typically mounted to the structure, connected to a data collection device, and left by the inspector to monitor the structure. These systems can be set to monitor a specific component or section of the structure or be designed to encompass the entire structure. The scope of monitoring depends on the desired data, the potential problem areas, and/or the potential areas of structure movement (Figure C-30).

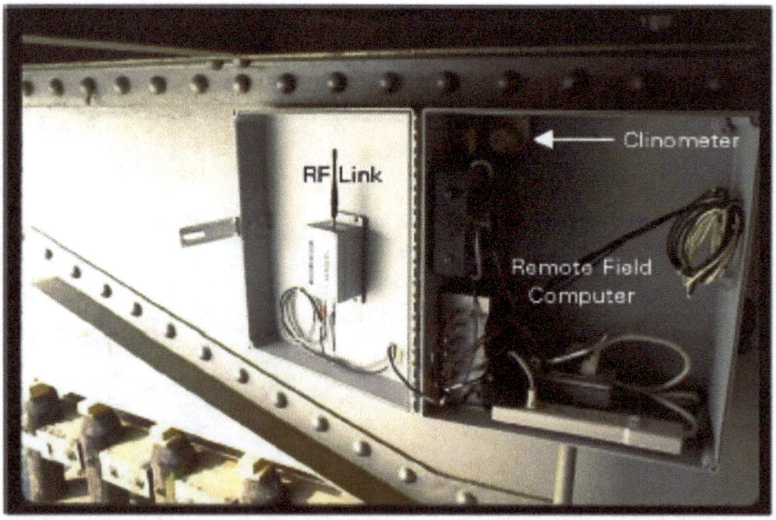

Figure C-30. Remote field computer, clinometers, and radio frequency link mounted in protective enclosure.
Source: Courtesy of Collins Engineers Inc., reproduced with permission.

C.19.1 Monitoring Systems

A monitoring system can comprise various sensors; a data collection device (computer); and, in the case of a remote system, a communication device, which transmits the data to the monitoring station for analysis. Sensor types include strain gauges, clinometers (tilt meters), accelerometers, thermocouples, and various other devices.

- Strain gauges: The term *strain gauge* typically covers a wide range of devices that are used, as their name implies, to measure the strains of structural members under load.
- Clinometers: Measures the inclination or tilt of a structure or structural component.
- Accelerometers: Measures structure dynamics under conditions such as high winds, seismic activity, and/or vehicular traffic.
- Thermocouple: Measures the temperature of a structure or its environment.

C.19.2 Strain Gauges

Strain gauges are the predominant sensors in use today. Several strain gauges are available, and selection should be made on the basis of location of their use and the specific data type and quantity to be collected. A strain gauge will have an established initial set length that is used as a datum, and the gauge will electronically measure an elongation. The resulting elongation divided by the gauge length yields the strain at that point. This calculation is typically performed automatically by the data acquisition system.

Strain gauges can also be used to measure rotational strain. Strain gauges can be placed either external to or, in certain circumstances, within the object being examined. Groups of gauges are usually installed in patterns determined by the type of data desired. Strain gauges are typically small and flat and do not interfere with the use of the structure. The strain gauges are then connected to a data acquisition system that records the strain data. Under real-time loading situations, the acquisition system can automatically collect data at a given time interval, perhaps every 10 min, to record a strain-time history.

In the area of waterfront structures, strain gauges may be installed at carefully selected locations on a structure to measure strains under live load conditions. These strains may be caused by daily crane loads, vessel impact forces, wind, temperature, or specifically applied test loads. These strain data are then evaluated directly, or more often, converted to stresses that can be compared to calculated design stresses. Thus, strain data allow the real performance of a structure to be compared to the theoretical design and enable the development of an analytical model to more accurately predict actual performance.

Strain gauges are also often employed to study the performance of a local area or detail for which theoretical analysis may be difficult. In field situations, strain gauges may also be used as monitoring systems at locations of great concern to detect either movement or changes in stresses. This work may be required to verify safety in areas of uncertain stability or strength. In a sufficiently sophisticated system, movement beyond a certain range may cause alarms to sound at the structure or at an off-site monitoring station.

Strain gauges may be used at locations where replacing an existing structure of uncertain strength is difficult or prohibitively expensive, while still allowing safe use of the facility (Figures C-31 through C-33).

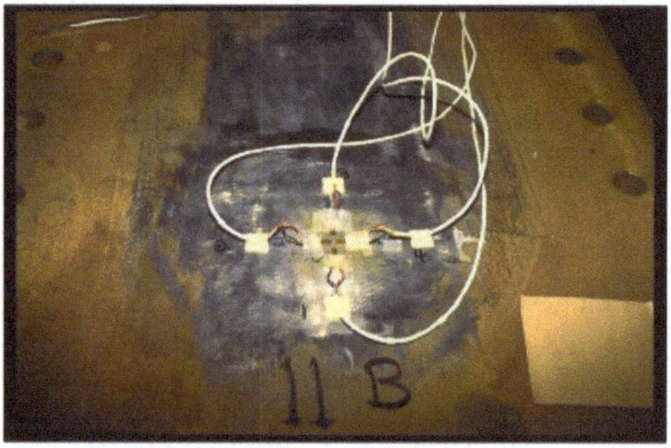

Figure C-31. Strain gauges attached to a steel member.
Source: Courtesy of Collins Engineers Inc., reproduced with permission.

Figure C-32. Submerged strain gauges attached to a sheet pile bulkhead.
Source: Courtesy of Collins Engineers Inc., reproduced with permission.

Figure C-33. Submerged seafloor displacement gauge attached to a sheet pile bulkhead.
Source: Courtesy of Collins Engineers Inc., reproduced with permission.

C.19.3 Applications of Monitoring Systems

Monitoring systems have many applications in the inspection of structure components. The sensors are highly versatile and can be applied to many materials. They are typically small and can be attached in tight-fitting areas. Furthermore, many of the sensors have a high level of accuracy and can be applied in both static and dynamic situations. Once installed, the sensors can provide data for an indefinite period of time. The ability to continuously monitor structures or specific components allows the owner to record and observe performance and detect deterioration. These systems work well from a preventative maintenance stance. Monitoring systems, however, do have some limitations. The sensors, although typically inexpensive, are often one-time-use devices. Once mounted to a particular structure, they cannot be removed and reused for another application. The sensors also typically require a high level of expertise to install.

Although the sensors are inexpensive, the data collection and transmission devices can be expensive and require specialized individuals to maintain and process the data. The gathered data must also be analyzed and manipulated to provide usable information.

C.20 UNKNOWN FOUNDATION INVESTIGATION

Many older waterfront structures do not have any design or as-built plans on file to document the type, depth, geometry, or materials incorporated into their foundations. Structures with unknown foundations pose a potential problem from a scour safety perspective. In addition to scour concerns, unknown foundations also pose a concern when a structure is considered for operational changes such as increased vertical or lateral loading or berth-area dredging for deeper draft vessels.

The evaluation of unknown foundations can be conducted either by conventional methods, such as physically disruptive excavation, coring, or boring methods, as well as less-invasive NDT methods. Conventional methods can be more disruptive and expensive. Therefore, research emphasis has recently been placed on NDT methods that can reliably determine the foundation property parameters with less cost and disruption to the site.

The important parameters in the evaluation of unknown foundations are

- Foundation depth: the bottom elevation of the footing, piles, and so on.
- Foundation type: shallow (footings), deep (piles or shafts), and so on.
- Foundation geometry: buried substructure dimensions, pile locations, and so on.
- Foundation materials: steel, timber, concrete, masonry, and so on.
- Foundation integrity: condition of foundation materials.

The foundation depth and type, if unknown, are considered to be the most critical pieces of information in a scour evaluation. The foundation geometry, materials, and integrity are frequently desired when improvements are being considered near to a structure (Figure C-34).

In determining which NDT methods might be useful, the method's capacity to detect and delineate the foundation components from the surrounding environment is often the deciding factor. The subsurface environment typically consists of a mixture of air, water, riprap materials, soil, and/or rock. Thus, the method must be chosen considering the wide range of substructure, geological, and hydrological conditions at a particular site.

The NDT methods used for unknown foundation investigation can be categorized as surface methods or borehole methods. Surface methods are usually less invasive because they do not require soil disruption like borehole methods. Although the following list provides a brief sample of applicable methods, the inspector should know that other methods are also being researched and implemented.

Figure C-34. Excavation of a tie rod to measure the remaining tie rod cross section. Source: Courtesy of Collins Engineers Inc., reproduced with permission.

C.20.1 Surface Methods

C.20.1.1 Sonic Echo/Impulse Response Test. This test involves placing a source and receiver on the top and/or sides of the exposed pile or columnar-shaped substructure. The depth of the reflector is calculated using the identified echo time(s) for sonic echo tests or resonant peaks for impulse response tests.

C.20.1.2 Bending Wave Test. This test involves mounting two horizontal receivers a few feet apart on one side of an exposed pile, and then the pile is impacted horizontally on the opposite side a few feet above the topmost receiver. This method is based on the dispersion characteristics and echoes of bending waves traveling along very slender members like piles.

C.20.1.3 Ultraseismic Test. This test involves impacting an exposed substructure with an impulse hammer to generate and record the travel of compression or flexural waves down and up the substructure at multiple receiver locations.

C.20.1.4 Spectral Analysis of Surface Waves Test. This test involves determining the variation of surface wave velocity versus depth in layered systems. The bottom depths of exposed substructures or footings are indicated by slower velocities of surface wave travel in underlying soils.

C.20.1.5 Ground-Penetrating Radar. This method uses a radio frequency signal that is transmitted into the subsurface and records the reflection echoes from the concrete/soil interface to determine the unknown depth.

C.20.2 Borehole Methods

C.20.2.1 Parallel Seismic Test. This test involves impacting an exposed foundation substructure, either vertically or horizontally, with an impulse hammer to generate compression or flexural waves that travel down the foundation and are reflected by the surrounding soil. The reflected compression wave arrival is tracked at regular intervals by either a hydrophone receiver or a geophone receiver.

C.20.2.2 Borehole Radar Test. This test involves using a transmitter/receiver radar antenna to measure the reflection of radar echoes from the side of the substructure foundation.

C.20.2.3 Induction Field. A magnetic field is induced around the steel of the pile or reinforced concrete foundation. The field strength will decrease significantly below the bottom of the foundation.

C.21 UNDERWATER ACOUSTIC IMAGING AND MUDLINE SOUNDINGS

Waterfront inspections require understanding the surrounding mudline and the condition of submerged structural elements. Sonar is a useful tool to determine and document underwater conditions. However, numerous different types of sonar technologies are available depending on the inspector's objectives. The inspector may wish to use sonar technology to evaluate and document the adjacent mudline, scour depressions, submerged debris, and submerged portions of the structure.

Sonar is now commonly used to provide engineers and clients with high-resolution images of elements that can be used at various stages of the work: planning the dive, identification of potential hazards for the diver, identification of areas of concern that merit close attention during the dive, and postconstruction images that can form part of the site's baseline report. Sonar has become increasingly specialized, based on the required task.

The needs of engineers are different from the needs of dredgers or surveyors collecting soundings to produce maritime charts. This section introduces the use of sonar for underwater acoustic imaging and mudline sounding surveys that are fit for structural engineering purposes. It does not prescribe dredge-quality hydrographic surveys

C.21.1 Mudline Sonar Data

Basic mudline soundings are often known as water depth soundings. This process is used to obtain underwater surface elevation data for evaluating the mudline surrounding a structure and the waterway in general. Water depth soundings, as part of a waterfront structure inspection, can utilize simplified methods or more specialized equipment depending on the required level of detail.

Mudline depth soundings may be conducted by simple single-beam sonar (fathometers), lead lines, or sounding poles. These techniques result in profiles under the vessel or spot elevations at each observation point. Other situations may require more detailed multibeam bathymetric surveys, for example, assessing slopes, revetments, breakwaters, and other long linear features. Locations that are subject to sediment movement, especially sand waves, frequently require multibeam surveys, because spot elevations and profiles do not represent volumes and sand waves appropriately. Multibeam bathymetry is also more reliable in identifying pits, mounds, and localized scouring and accretion around foundation or fender features.

C.21.1.1 Sub-Bottom Profiling. Another method of employing sonar techniques is in sub-bottom profiling of the seafloor to investigate the subsurface geology. Sub-bottom profiling is a geophysical technique that requires calibration against physical inspection data, such as boring cores. Sub-bottom profiling data can be a valuable asset for identifying the limits of mobile sediment (erosion potential), investigating soil heterogeneity between borings, and considering soil rippability/pile drivability during dredging or construction. Sub-bottom profilers are also valuable in identifying potential hazards such as uncharted buried cables or pipelines.

C.21.1.2 Hydrographic Surveys. The level of accuracy in hydrographic surveying varies greatly on the basis of equipment and methods used. The surveys must be properly referenced to the appropriate geodetic or local control and referenced to the correct datum—both horizontal and vertical. Mistaking NAVD88 vertical control for local MLLW can be a critical but easily made error. In addition, data (GPS-derived vertical data in particular) are subject to revision over time. Thus, for all methods of marine survey, the inspector needs to consider when the best interests of the project are served by having a certified hydrographer complete and validate the survey—especially as the conditions observed could identify deficiencies requiring remediation.

Hydrographic surveys may involve using a small survey boat equipped with single-beam, multibeam, or side-scan sonar systems. These systems can be used independently or together. They can also be coupled with above-water laser scanners to develop a three-dimensional image or model. These systems rely heavily on accurate GPS coordinates and are only as accurate as the horizontal and vertical controls. New options of unmanned remote-controlled vessels with sophisticated sonars, motion reference units, and accurate GPS are now readily available at decreasing costs. For a more in-depth discussion of these systems, see USACE EM 110-2-1003 (USACE 2013) and IHO S-44 Edition 6.1.0 (IHO 2022).

C.21.1.3 Side-Scan Sonar. Side-scan sonar is used for acoustic imaging of the seafloor, foreign objects, and the submerged portion of the structure, especially in low- or no-visibility water conditions. Side-scan sonar is usually deployed on a towfish that is towed behind a vessel or can be mounted on the hull of the vessel. Where water depths make towfish contact with the bottom risky, engineers should consider multibeam sonars.

C.21.1.4 Multibeam Sonar. Multibeam sonars are used successfully for structural inspection, producing high-resolution two-dimensional images as well as three-dimensional point cloud data similar to LiDAR data.

C.21.1.5 Utilizing Survey Results. The results of all marine surveys should be compared to any previous survey data available and retained for future comparisons to monitor changes in sea-floor conditions affecting the structure. New sonars designed to be permanently mounted are increasingly being specified to take continuous soundings. These sonars are deployed to monitor and report change over time, including localized extreme scour and/or shoaling.

C.21.1.6 Sonar Survey Applications. Applications for sonar surveys primarily include vessel operations and maintenance dredging activities. Besides sonar surveys, other methods of recording water depth soundings include a lead line or a sounding pole. Although these methods are of very low cost, the soundings are correspondingly coarse and may miss key variations in bottom conditions.

C.21.1.7 Other Methods

C.21.1.7.1 Lead Line. A lead line is a simple device typically consisting of standard surveyor's tape with a weight attached to the end. The inspector simply lowers the lead line until the weight comes to rest on the mudline. The inspector then pulls the line taut and records the reading from the mudline to the waterline or top of the deck. An inspector working from the top of the deck most often obtains lead line readings along the fascia of a structure at predetermined intervals (e.g., measurements obtained at a certain bent numbering or stationing along a bulkhead).

C.21.1.7.2 Sounding Pole. A sounding pole is another simple device that typically consists of an extendable, graduated rod. An inspector on the deck or in a boat typically places the pole vertically on the mudline and records the measurement at the waterline. The inspector then records the distance from the waterline elevation to a known elevation on the structure.

C.21.1.7.3 Single-Beam Sonar Fathometer. One commonly used electronic sounding device is a single-beam sonar fathometer. This device uses a transducer just below the waterline and repeatedly transmits sound energy through the water column. The time interval between the transmission of the sound pulse and the returning echo from the mudline is used to automatically calculate a depth measurement that is recorded onto the device. Determining water sound velocity with adjustment and the use of a bar check provides quality control measures for sonar soundings.

Depending on the complexity of the sonar device, single-beam technology is limited by its inability to detect refilled scour holes, false readings from heavy drift or heavy turbulence, distorted scale on the readout due to varying boat speed, and inability to provide information about the geophysical sub-bottom. Soundings also vary because of differences in the speed of sound through changing water conditions—the effect of temperature differences at the surface through the water column, and where freshwater mixes with salt water. Care must be taken to account for these changes in tidal zones and between mornings and afternoons. Furthermore, the technology must compensate for any water surface elevation changes during tidal swings while taking mudline to waterline measurements.

C.21.1.7.4 Limitations. All these methods are limited by the softness of the mudline. The swiftness of the current can also affect lead line and sounding pole measurements because the water velocity can introduce horizontal drift into the line or cause a lightly weighted tape to drift downstream. This method also may be more prone to inaccuracies resulting from the experience of the inspector.

C.21.2 Underwater Acoustic Imaging Data

Some sonar devices produce high-resolution images to enable documentation and evaluation of underwater conditions such as structure material surface condition, mudline elevation location, and the presence of debris or underwater objects that represent a berthing concern. These sonar devices can obtain photograph-like quality images of bulkheads, piles, and armored slopes under wharves (Figures C-35 and C-36).

Underwater acoustic imaging provides photograph-like documentation of a structure element's surface, but it is limited in its ability to penetrate beyond the line of sight. Also, the images may not accurately represent the structural element material if heavy marine growth is present. Sonar imaging can also be used for diver tracking in real time, which can be a valuable safety consideration. Depending on the type of sonar device used, acoustic images may be displayed in 2D exhibits or 3D point-cloud data sets or models.

Underwater acoustic images vary in quality, resolution, and dimensional perspective (2D or 3D), depending on the particular sonar device. Sonar

Figure C-35. Sector scanning sonar image of a steel bulkhead with typical defects circled.
Source: Courtesy of Collins Engineers Inc., reproduced with permission.

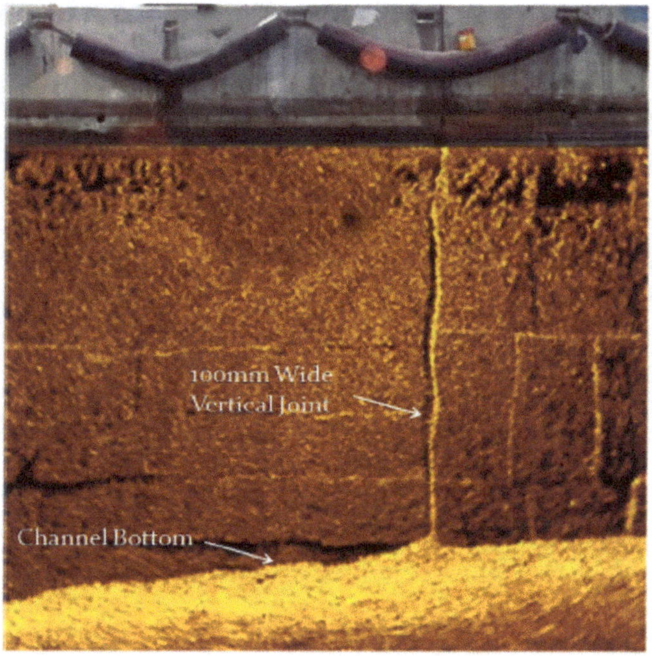

Figure C-36. Sector scanning sonar image of a quay wall with offset blocks and noted vertical joint.
Source: Courtesy of Collins Engineers Inc., reproduced with permission.

technologies can be classified into two broad categories on the basis of the type of data—2D or 3D—that they produce. Two-dimensional sonar systems can be modeled into a 3D space or just plotted in a 2D view. For 3D sonar data collection, multibeam sonar systems, also referred to as swath echosounders, can collect mudline data and submerged vertical surfaces.

Another form of multibeam sonar is 3D mechanical scanning sonar, which is essentially a multibeam sonar unit fitted with a mechanical stepping motor set on a tripod resting on the mudline. Three-dimensional images can also be produced using a Dual Axis Sonar, a single-beam transducer mounted on a precise dual-axis actuator. This sonar can be used to characterize complex shapes and is designed to be deployed to produce continuous soundings in situ. Because of a rugged form factor, it is suitable for installation in regions of severe scour or shoaling and long-term deployment (Figures C-37 through C-44).

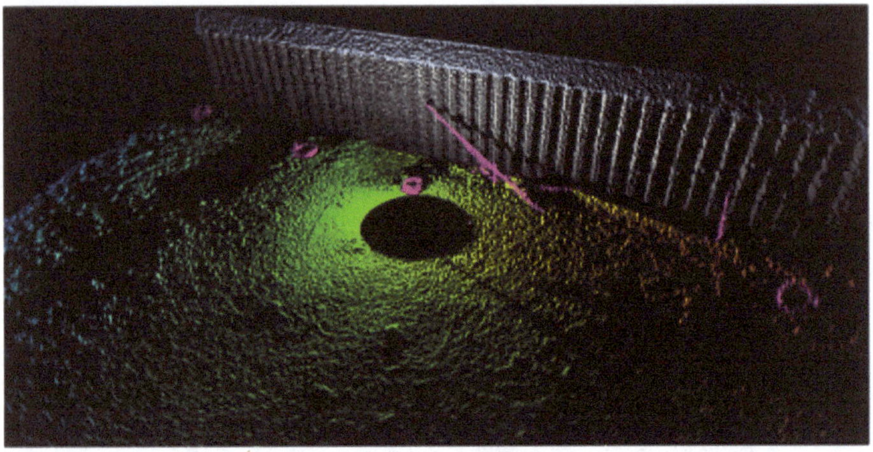

Figure C-37. Multibeam sonar image of steel sheeting bulkhead with debris on the mudline.
Source: Courtesy of Collins Engineers Inc., reproduced with permission.

Figure C-38. Multibeam sonar data integrated with topside mobile laser scanner data.
Source: Courtesy of Fugro Inc., reproduced with permission.

Figure C-39. Side-scan sonar image of a sunken vessel on the mudline.
Source: Courtesy of Fugro Inc., reproduced with permission.

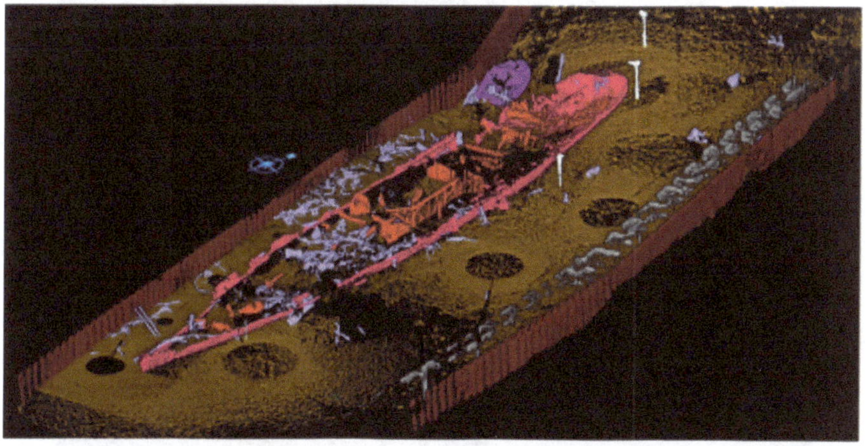

Figure C-40. Multibeam image of a sunken vessel in a navigational channel between steel sheet pile bulkheads.
Source: Courtesy of Collins Engineers Inc., reproduced with permission.

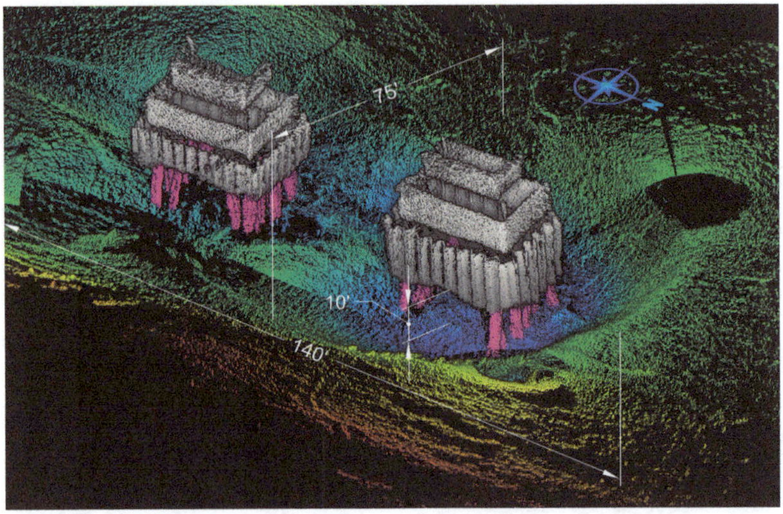

Figure C-41. Multibeam sonar image of undermined foundation.
Source: Courtesy of Collins Engineers Inc., reproduced with permission.

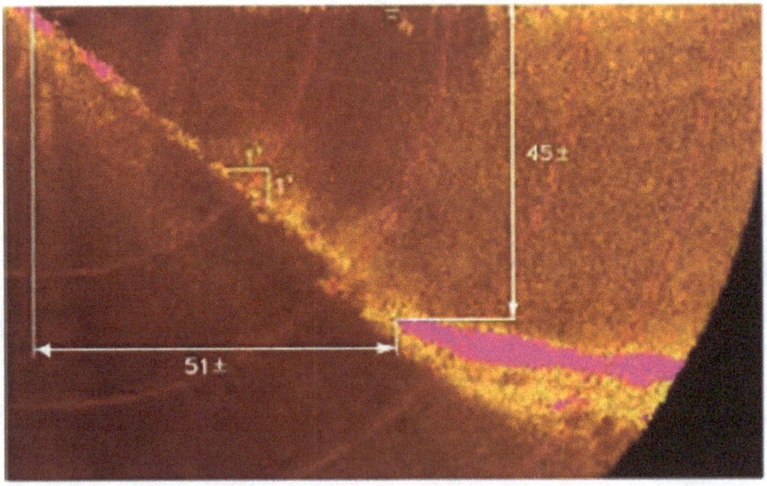

Figure C-42. Sector-scanning sonar image of embankment slope erosion.
Source: Courtesy of Collins Engineers Inc., reproduced with permission.

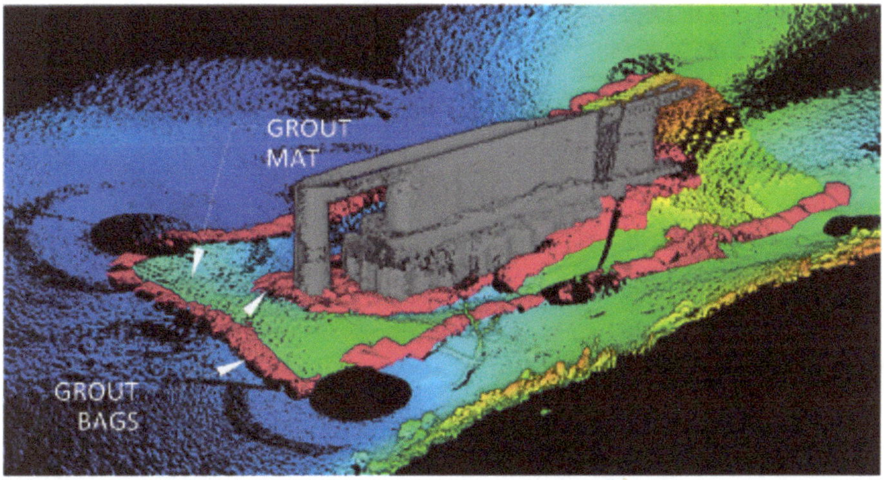

Figure C-43. Multibeam sonar image of repaired undermining with grout bags.
Source: Courtesy of Collins Engineers Inc., reproduced with permission.

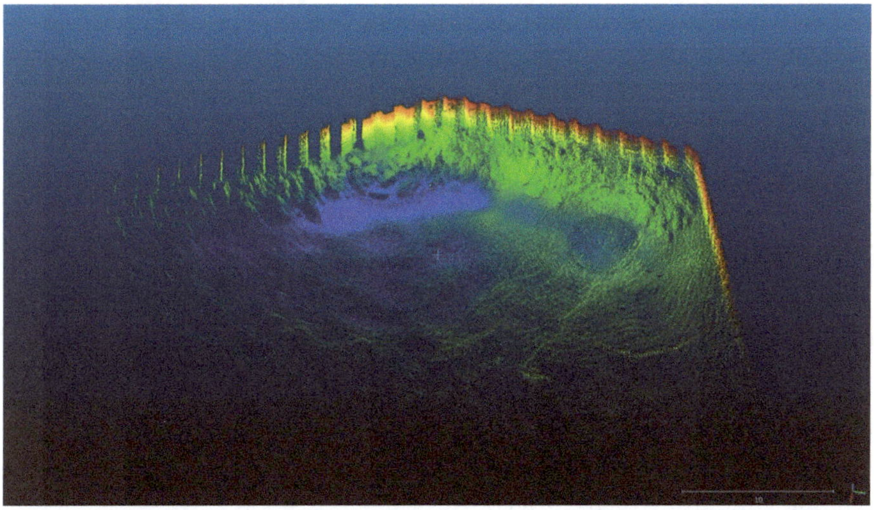

Figure C-44. Dual-axis sonar point cloud data of a submerged steel sheet pile bulkhead.
Source: Courtesy of Kongsberg Discovery Canada Ltd., reproduced with permission.

C.22 LIGHT DETECTION AND RANGING

Light detection and ranging (LiDAR) is commonly utilized as a remote sensor for both manned aircraft and unmanned aerial systems, as well as terrestrial laser scanners and handheld scanners.

Understanding the equipment and its limitations is important as it can be a determining factor in whether it is an applicable method of collection for the analysis and or data extraction taking place. Another important thing to understand is how the data are processed to get usable data. Whereas a manned aircraft sensor may be integrated with high-quality onboard inertial measurement units and GPS, a terrestrial laser scanner may be processed to survey control, scanning targets, or onto itself in a process known as *cloud-to-cloud* registration. Most handheld scanners utilize algorithms called simultaneous localization and mapping computations, where the software attempts to correct itself unto itself in addition to being able to tie it to a survey control network.

Acceptable aerial mapping standards collected by LiDAR are published by the American Society for Photogrammetry and Remote Sensing. These standards can be utilized to get a sense of the quality control of the data and the reporting procedures to determine the level of confidence of the data. Terrestrial and handheld scanners have limited to nonexistent published standards and usually fall back on individual client standards (DOT, USACE, GSA, etc.). Considerations should be made on determining the methods for reporting the level of confidence of the data.

Light detection and ranging data are delivered as "point clouds" and can be very large in size. Point clouds are categorized into two formats, structured and unstructured. E57 files, for example, are structured files and are typically from terrestrial scanners. These files contain information such as X Y Z of the point, its R G B color value, the intensity of the return value back, and the scanner position. Aerial mapping collected data are typically delivered in LAS/LAZ format, which is an unstructured data file. LAS is uncompressed and the latter is. These files record the same information as the E57 but can also contain data such as point classification (ground versus nonground), GPS time at which the point was collected, and the scanning angle at which it was retrieved.

When reviewing LiDAR data, a couple of methods to review these data are to spot-check elevations against other known points that were not utilized in the postprocessing solution. These can be additional survey control points. Another method is to do cross-section slices through the point clouds. Cross sections should be taken over flat surfaces to be able to validate the data and should be done on both horizontal and vertical planes. Depending on the scale of the project, for mapping, it is acceptable to do 30 cm sections (Figures C-45 and C-46).

Figure C-45. Unmanned aerial vehicle equipped with a LiDAR sensor.
Source: Courtesy of Jacobs, reproduced with permission.

Figure C-46. Three-dimensional point-cloud model of a pier structure generated with a LiDAR sensor.
Source: Courtesy of Jacobs, reproduced with permission.

C.23 ROBOTIC INSPECTION AIDS

Recent advancements in robotics and unmanned vehicles provide cost-effective solutions to enhance and streamline waterfront inspections. Robotic tools continue to grow smaller, less expensive, and more autonomous. Although robotic inspection aids can augment in-person above and underwater inspection efforts, they are not a replacement for an engineer diver or an in-person inspector.

C.23.1 Unmanned Aerial Systems

Unmanned aerial systems (UASs), more commonly called drones, can provide additional perspectives to waterfront inspectors. Most drones used for waterfront inspection are under 55 lb and are classified as small unmanned aerial systems (sUAS). Small unmanned aerial system may be fixed-wing or rotary-wing aircraft, although rotary-wing drones are much more common. Typically equipped with high-definition cameras, some sUAS can also carry payloads, including thermal imaging cameras and LiDAR.

Drones provide inspectors with visual access to hard-to-reach portions of waterfront structures, allowing inspectors to conduct initial investigations. The inspection team leader can utilize this information to focus hands-on inspection efforts. It is important to note that drones lack the capacity for tactile inspection and cannot fulfill the requirement for a hands-on inspection as recommended in Chapter 3. Many modern drones are equipped with obstacle avoidance, autonomous route planning, and photogrammetry software. See Chapter 5, Section 5.4, and Section C.24 for more information on the use of photogrammetry to create digital twins.

Drones should be operated only by qualified personnel. Federal, state, and local regulations must be considered when planning a drone inspection. The planning process should also consider any airspace restrictions, particularly when working around airports and military installations. Temporary airspace restrictions are also common around mass gatherings such as sporting events and parades. Drones should never be operated over people, and special care should be taken when operating at night. In the United States, an FAA Part 107 remote pilot license is required to operate a drone for commercial purposes, including waterfront inspections (Figure C-47).

Figure C-47. (Left) sUAS, (right) drone operator investigating a high-level platform. Source: Courtesy of Jacobs, reproduced with permission.

C.23.2 Unmanned Underwater Vehicles

Unmanned underwater vehicles (UUVs) are submersible vehicles that operate without a human pilot on board. Unmanned underwater vehicles can be divided into two categories: remote operated vehicles (ROVs) and autonomous underwater vehicles (AUVs). Remote operated vehicles are operated by a human pilot on the surface, who sends commands through a tether or data cable. By following preprogrammed survey routes, AUVs operate without continuous human inputs and without the need for a tether.

Remote operated vehicles can be further subdivided into two classes: working class and survey class (Figure C-48). Working-class ROVs are not typically suitable for most waterfront inspection tasks. Although they are highly versatile, they typically weigh well over 1 ton, requiring a large crew to operate and a crane or davit to launch and recover. Survey-class ROVs typically weigh between 10 and 100 pounds (4.5 and 45 kg) and can be deployed, operated, and recovered by one or two people.

Most useful as a platform for survey equipment, survey-class ROVs commonly carry lights, high-resolution cameras, SONAR, material samplers, water samplers, and manipulators (Figure C-49). ROVs are also a valuable tool for mitigating risk to diving inspectors. Remote operated vehicles can augment diving inspections in hazardous environments, reducing the exposure of divers to these dangers. For example, an ROV can conduct an initial investigation of a potentially unstable structure, to determine whether it is safe for a diving inspector. Other hazards that can be mitigated with an ROV include diving in narrow discharge pipes, contaminated waters, and excessive water depths.

Figure C-48. (Left) working-class ROV being deployed, (right) working-class ROV control station.
Source: Courtesy of Appledore Marine Engineering, LLC., reproduced with permission.

Figure C-49. (Left) Survey-class ROV investigating a raw water intake structure, (right) survey-class ROV and control unit.
Source: Courtesy of Jacobs, reproduced with permission.

Survey-class ROVs cannot remove marine growth and lack the capacity for tactile inspections. Like drones, ROVs do not replace the need for diving inspectors as specified in Chapter 3. When planning an ROV inspection, it is important to consider the length of the tether, current, and other metocean conditions, visibility, entanglement risks, and vessel traffic.

Commonly equipped with various SONAR packages or ADCPs, AUVs can collect hydrographic, bathymetric, and metocean data in support of a structural repair/upgrade design inspection or special purpose inspection. Because AUVs do not require a tether, they can operate more easily over larger areas and at depths deeper than ROVs. Autonomous underwater vehicles are most often utilized to survey very large areas of the seafloor or mudline.

C.23.3 Other Robotic Inspection Aids

Unmanned surface vessels (USVs) are small remote-controlled, programable, or autonomous floating devices (Figure C-50). Like AUVs, USVs are not commonly utilized for structural inspections. However, USVs can be equipped with various survey equipment to collect data over large areas of relatively shallow water to inform rehabilitation, design of new construction, and near-shore dredging programs.

Propeled by either wheels or treads, crawler robots can be utilized to inspect confined spaces or other hazardous environments, both above and below water. These small remote-controlled vehicles carry cameras and facilitate visual inspections inside spaces too small to be entered by an inspector. For example, pipe crawlers can be utilized to inspect dry or partially filled outfalls penetrating seawalls or quays.

Figure C-50. Autonomous surface vessel.
Source: Courtesy of Kongsberg Discovery Canada Ltd., reproduced with permission.

C.24 DIGITAL TWINS

Multiple definitions are there for a digital twin. Depending on the application, the organization, or the level of complexity, all can be true. A digital twin is a virtual representation of real-world entities and related processes, relationships, behaviors, and data (Figure C-51). A digital twin can take multiple forms depending on the integration platform and supporting infrastructure, local or geospatial context, the types of data and information associated with the digital asset, how the data and information are synchronized with the physical object, and the purpose for which the models are intended to drive planning, analysis, visualization, and intelligence.

A digital twin is an effective tool for the visualization, interaction, and management of facilities, natural environments, emergency and disaster response, climate impacts and scenarios, mitigation, operations, and utility networks. Defining the objective for the development of the digital twin is critical in structuring the data management framework and is key to developing an operational data and information model.

Numerous challenges exist to implementing digital twins effectively. The terminology used to define the digital twin between industries and professionals is not synonymous and often requires further descriptions of the application to effectively communicate the intent and purpose between entities. Data and information stovepipes are commonplace within many industries. Specialized fields of study create data and information for specific purposes without the foresight or standardization to allow for

Figure C-51. Digital twin of a large port facility.
Source: Courtesy of VERSAR, reproduced with permission.

incorporation further along the workflow pipeline, thus inhibiting collaboration and interoperability.

The objective of a digital twin is to assimilate and synthesize multiple types of data into a common operating environment and associate the data and information with real property and objects. Multiple platforms, software, licensing, and enterprises present their own inherent difficulties for integration and interoperability as well as unique data formats, schemas, ontology, semantics, quality, and integrity. The ability to navigate the different systems and data characteristics and develop a framework to access, integrate, store, and associate is imperative.

Requirements, procedures, protocols, and documentation must be standardized to enable interoperability or crosswalking to facilitate integration. Data drive the system, and adhering to a FAIR (Findable, Accessible, Interoperable, and Reusable) approach, a scalable platform can be constructed to support growth and expand capabilities if the data meet these criteria.

C.25 MOORING HARDWARE TESTING

Mooring hardware testing is performed to verify the capability of a mooring point to hold its designed load. Failure of mooring points can be the result of line loads and other incidents such as vehicular impact or other accidental interactions. Damage to the fitting identified prior to testing should be noted and carefully considered as catastrophic or sudden failure during testing can be hazardous to personnel.

Failure of mooring hardware can occur within the casting or fitting itself, the fasteners that secure the fitting, or the foundation upon which the hardware is secured. Failure of mooring points being tested can be identified by visual observation of distress or movement, permanent yielding of the fasteners, and cracking or failure of the foundation surrounding or supporting a fitting. The Unified Facilities Criteria (UFC) provide detailed guidance on testing of mooring hardware in Appendix A of the UFC 4-150-08 (NAVFAC 2020) Mooring Hardware Inspection. This document identifies three levels of testing:

- Level 1—Bolt Pull Test (Tension): Fastening hardware is tested individually to verify the tensile strength of the bolt and its anchorage.
- Level 2—Indirect Line Loading: Mooring fittings are pull-tested with a line force but not in the actual direction of typical use. This level of testing will confirm the strength of the system including the casting, fasteners, and foundation. The as-built design should be considered when selecting this level of testing as the design capacity may be specific for the direction of typical loading.
- Level 3—Load applied in the actual direction of typical mooring forces. This will confirm the working load of the entire mooring system.

C.25.1 Bolt Testing

Bolt testing is typically part of new mooring hardware installation but can also be performed if the mooring fitting is in good condition, but the fasteners show signs of corrosion. Fasteners that secure fittings to steel or timber take all loads and, if tested, should be tested in tension and shear. Fasteners that secure fittings to timber and steel are also typically accessible and could be removed for inspection, thus eliminating the need for in-situ load testing. Fasteners that secure fittings set in concrete need be tested only in tension as the concrete or grout bed provides the shear resistance. Bolts are loaded to 110% of their working load in accordance with ASTM E488/E488M (ASTM 2022c), "Standards Test Methods for Strength of Anchors in Concrete and Masonry Elements."

C.25.2 Pull Testing

Pull testing can be performed with a test rig customized for the specific location such as shown in Figure C-52, a land-based crane or winch, a water-based crane or winch, or by testing between two similar mooring fittings loading one against the other. Load testing of the mooring hardware should be to 110% of its rated capacity. Loading should occur in 10% increments and be held at each increment for 2 min while monitoring for movement or other failure modes. Loading of the fitting at 110% should be held for 10 min.

Figure C-52. Mooring hardware pull test rig.
Source: Courtesy of Jacobs, reproduced with permission.

REFERENCES

AASHTO (American Association of State Highway and Transporation Officials). 2021. *Standard method of test for sampling and testing for chloride ion in concrete and concrete*, AASHTO T 260. Washington, DC: AASHTO.

ASTM International. 2018a. *Standard guide for using the seismic-refraction method for subsurface investigation*, ASTM D5777. West Conshohocken, PA: ASTM.

ASTM. 2018b. *Standard test method for rebound number of hardened concrete*, ASTM C805/C805M. West Conshohocken, PA: ASTM.

ASTM. 2022a. *Standard test method for measuring the P-wave speed and the thickness of concrete plates using the impact-echo method*, ASTM C1383. West Conshohocken, PA: ASTM.

ASTM. 2022b. *Standard test method for detecting delaminations in bridge decks using infrared thermography*, ASTM D4788. West Conshohocken, PA: ASTM.

ASTM. 2022c. *Standard test methods for strength of anchors in concrete elements*, ASTM E488/E488M. West Conshohocken, PA: ASTM.

ASTM. 2023. *Standard test method for penetration resistance of hardened concrete*, ASTM C803/C803M. West Conshohocken, PA: ASTM.

Caterpillar Tractor Co. 2000. Handbook of Ripping, Caterpillar Inc. 12th ed. AEDK0752-02. Irving, TX: Caterpillar Tractor Co.

Droycon Bioconcepts Inc. 2024a. Biological Activity Reaction Test Users Manual. Saskatchewan, Canada. Droycon Bioconcepts Inc.

Droycon Bioconcepts Inc. 2024b. Sulfate Reducing Bacteria, SRB-BART™ Quality Control. Saskatchewan, Canada. Droycon Bioconcepts Inc.

IHO (International Hydrographic Organization). 2022. *Standards for hydrographic surveys*, S-44 Edition 6.1.0. Monaco: IHO.

NAVFAC (Naval Facilities Engineering Command). 2020. *Unified facilities criteria mooring hardware inspection*, UFC 4-150-08. Washington, DC: NAVFAC.

USACE (US Army Corps of Engineers). 2013. *Hydrographic surveying*, EM 110-2-1003. Washington, DC: USACE.

APPENDIX D
INSPECTION NOMENCLATURE

Helping to ensure the consistency of waterfront inspection and assessment information between in-field data collection and reporting efforts is a standard system of shorthand and references/orientations to capture defect and configuration information. The exemplary nomenclature included herein is not exhaustive and is intended as a guide, whereas the actual nomenclature will be determined by the client or engineer in charge of the inspection and assessment effort.

D.1 DATA COLLECTION NOMENCLATURE

The following includes exemplary defect and configuration recording nomenclature that is frequently used:

- Gridlines should be established to standardize defects and asset locations at pier and wharf facilities (Figures D-1 and D-2).
    - Bents: On piers, bents should be numbered beginning at the shore side extent of the facility and increasing as you move outboard. On wharves, bents should be numbered west to east or south to north depending on the orientation of the wharf.
    - Rows: On piers, rows should be labeled alphabetically beginning with A on the left (facing outboard). On wharves, rows should be labeled beginning with the outboard row.

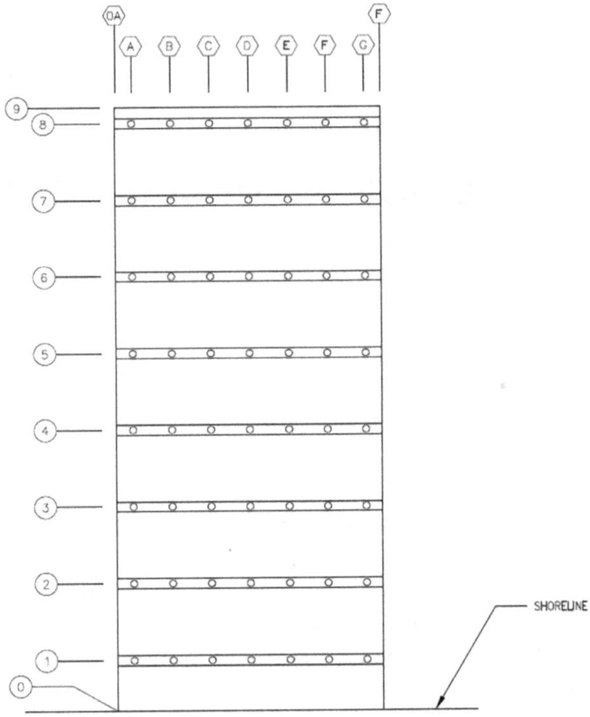

Figure D-1. Bent and row gridline example for a pier structure.
Source: Courtesy of Appledore Marine Engineering, LLC, reproduced with permission.

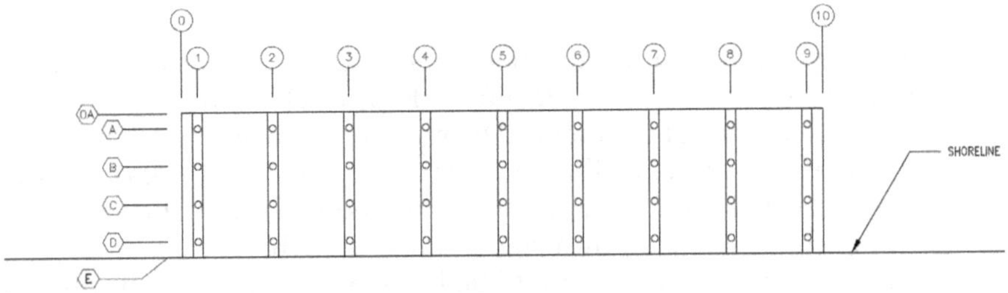

Figure D-2. Bent and row gridline example for a wharf structure.
Source: Courtesy of Appledore Marine Engineering, LLC, reproduced with permission.

- "0" defines the edge of any deck structure adjacent to Gridline 1 (or the lowest number). The next higher number defines the edge of any deck structure adjacent to the last Gridline number.
- "0A" defines the edge of any deck structure adjacent to Gridline A (or the lowest letter). The next higher letter defines the edge of any deck structure adjacent to the last gridline letter.
- Locating defects between bents and rows should utilize a decimal system to the nearest 0.1.

• Stationing should be established to standardize defects and asset locations at the bulkhead and seawall facilities (Figure D-3).
  - Station baseline is typically established along the face of the bulkhead or seawall.
  - Station (STA) 0+00 should begin at the westernmost or southernmost extent of the facility.
  - Use of offset measurements from the station baseline is recommended for locating defects perpendicular to the baseline.

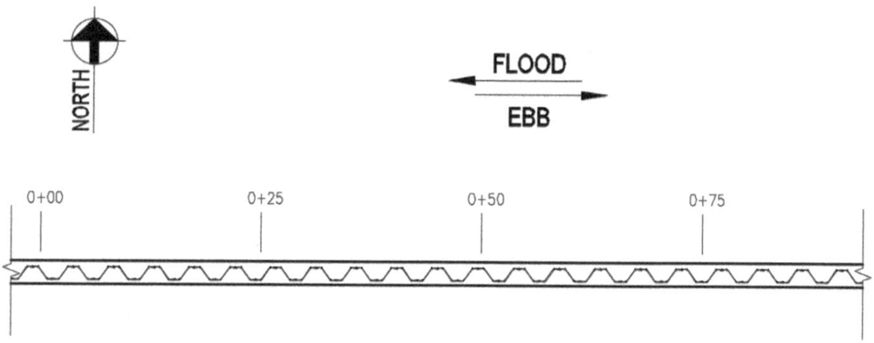

Figure D-3. Stationing example for a bulkhead structure.
Source: Courtesy of Appledore Marine Engineering, LLC, reproduced with permission.

• Data should be collected in accordance with Table D-1 and Figure D-4. All numerical data should be rounded up to the minimum tolerances noted.

Table D-1. Typical Field Data Tolerances.

Element	Item	Notation	Unit	Tolerance
Pile (Figure D-4)	Face	F	ft	0.5
	Corner	C	ft	0.5
	Quadrant	Q	ft	0.5
	Vertical position	P	ft	0.5
Pile cap	Horizontal	H	ft	0.5
	Vertical	V	ft	0.5
Pile cap/deck	Area	A	SF	1
Deck	Parallel to bent	X	ft	0.5
	Perpendicular to bent	Y	ft	0.5
	Depth	D	in.	1
Various	Crack width	W	in.	1/16
	Section loss	—	%	5

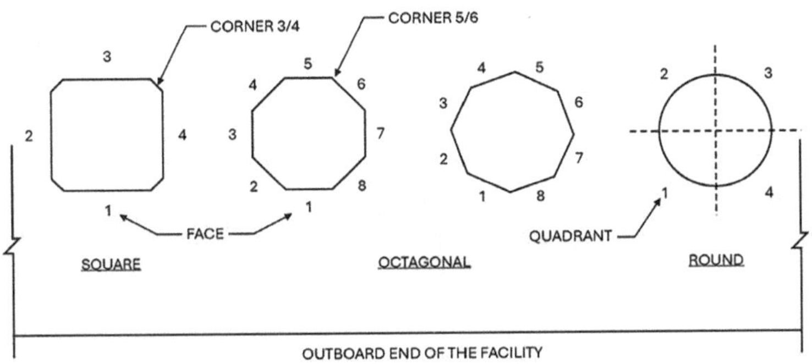

Figure D-4. Pile orientation key.
Source: Courtesy of Appledore Marine Engineering, LLC., reproduced with permission.

Table D-2 presents a list of typical defects commonly observed on waterfront facilities. This list is organized primarily by material type and construction. Common abbreviations are listed to facilitate field notes. The list is not comprehensive, and specific waterfront facilities may require additional defects not listed subsequently.

Table D-2. Defect Types.

Defect	Abbreviation	Description
General Defects		
Abrasion	ABS	Physical wear and gradual degradation of materials, typically caused by friction or repeated contact with sea debris, material, and so on
Broken	BRK	When a structure is present but no longer has any capacity
General crack	CRK	Use this if nothing else fits for cracks
Through crack	CKT	Crack through the entire element
General deterioration	DET	Usually deteriorated element
Displaced element	DIS	Displaced element
Loose element	LOS	A loose element with broken or missing fasteners
Missing	MIS	If the element is missing and would need to be replaced
Construction defect	CDF	General damage caused during construction or improperly installed repair
Undermining	UND	Scour at the toe of a structure caused by currents or waves
Exposed top of pile	EXP	The top of the driven pile is exposed and not properly contained in a cap or extension
Gap	GAP	Separation between two objects
Nonbearing	NBR	The pile is no longer supporting the cap

(*Continued*)

Table D-2. (*Continued*).

Defect	Abbreviation	Description
Settlement	SET	Settling of a wall or structure
Sinkholes	SNK	Topside holes caused by loss of fill; typically found behind seawalls or bulkheads
Voids	VOD	A wall section where the fill material has been washed out
Concrete Defects		
Horizontal crack	CRH	Horizontal cracking
Vertical crack on multiple faces	CRM	Vertical cracking on multiple faces of a concrete element
Vertical crack	CRV	Vertical cracking on a single face of a concrete element
Pattern crack	CRP	A network of cracks within an area that appear to resemble a road map; often found on concrete decks due to the shrinkage of concrete during the curing process
Closed corrosion spall	CCS	Delaminated concrete usually due to rebar expansion
Chemical disintegration	DSC	Term for general deterioration that causes softening of the concrete exterior and is typically found in the tidal zone but can be found underwater depending on the type of chemical attack
Delayed ettringite formation	DEF	A chemical attack resulting in volumetric expansion of the concrete; can be found above water in moist humid conditions but is typically found underwater

(*Continued*)

Table D-2. (*Continued*).

Defect	Abbreviation	Description
Scaling	SCL	The loss of surface mortar that exposes the aggregates; typically found in the tidal zone
Freeze–thaw disintegration	DSF	Material deterioration caused by the repeated freezing and thawing of water within the element's pores or cracks; typically found in areas exposed to saltwater/freshwater spray
Fracture spall	FRS	An impact causing a section of the structure to fail
Honeycombing	HNC	Voids/hallows in concrete due to poor vibration of concrete
Open corrosion spall	OCS	When DEL or CCS deteriorates further and exposes rebar
Open mechanical spall	OMS	Loss of concrete cover, with rebar exposed, caused by impact or other mechanical means
Voids	VOD	Voids/hallows in concrete due to poor vibration of concrete
Timber Defects		
Marine borer deterioration	MBD	Damage caused by aquatic organisms; can initially present as 1/4 in. deep scrapes along the timber but can also be difficult to identify whether the damage is restricted to the interior of the timber element
Fungal decay	FDY	Rot caused by the wetting and drying of the timber

(*Continued*)

Table D-2. (*Continued*).

Defect	Abbreviation	Description
Split	SPL	Longitudinal crack or separation of the wood fibers that occurs along the grain of the timber
Crushed	CSH	The timber element is crushed due to overload and usually occurs at supports; deformed grain
Steel Defects		
Coating loss	CTL	Section exhibiting no coating from initial COR or defects/damage to the coating system
Pitting	PIT	Deep narrow penetrations from corrosion; often lead to holes
Corrosion	COR	An area usually covered in orange-black by-product
Knife edge	KNE	Typically observed on H piles with flange edges that have corroded to a point
Corrosion hole	COH	Advanced COR or PIT
Weld crack	CRW	Narrow cracks along the edge of a weld
Anode loss	ANL	Percent of an anode that has been depleted
Buckling	BUC	Deformation of an element usually due to overload
Impact damage	IMP	Deformation caused by an impact may become COH

(*Continued*)

Table D-2. (*Continued*).

Defect	Abbreviation	Description
Chain wear	CHW	The wear caused by chain-to-chain contact
Masonry Defects		
Mortar degradation	MDE	Deteriorated, loose, or missing mortar between blocks
Marine borer deterioration	MBD	Pear-shaped burrows caused by pholads
Separation	SEP	Gaps between blocks
Other Defects		
Overgrown vegetation	OVG	Overgrown vegetation in an area
Potential safety hazard	HAZ	Potential safety hazard such as missing vault cover
Torn wrap	TRN	Pile wrap that has been torn or broken
Loose wrap	LSW	Pile wrap that allows water to flush in and out of it
Open seam	OPN	Pile wrap where the seam is ineffective
Failed connection	FCN	A connection that has failed
Failed joint material	FJM	Typically found at expansion joints where the joint filler has failed
Missing nut pocket sealant	MNP	Mooring hardware with missing nut pocket sealant
Leak	LEK	Typically found at a utility connection
Listing	LIS	Typically a floating dock that is tilting or leaning to one side

D.2 REPORTING NOMENCLATURE

Figure D-5 includes typical symbols used to graphically depict and communicate both inspected elements and observed defects.

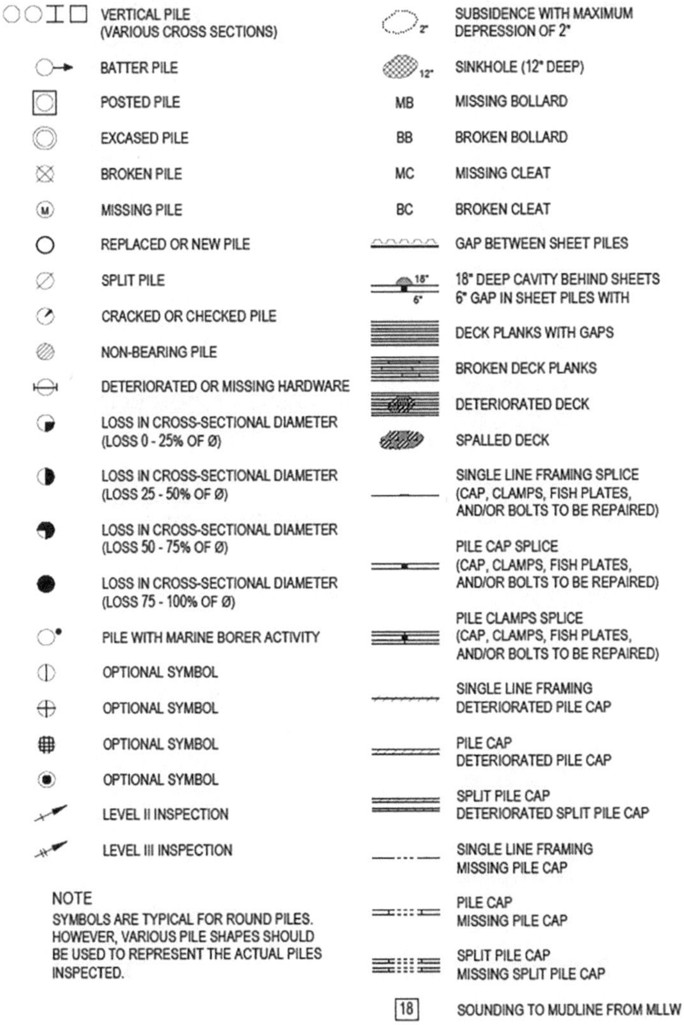

Figure D-5. Typical symbols used to graphically depict and communicate both inspected elements and observed defects.
Source: Courtesy of Appledore Marine Engineering, LLC., reproduced with permission.

GLOSSARY

AASHTO (American Association of State Highway and Transportation Officials): A standards setting body that publishes specifications, test protocols, and guidelines that are used in highway design and construction throughout the United States. Despite its name, the association represents not only state highways but air, rail, water, and public transportation as well.

AAR (Alkali–Aggregate Reaction): Two most common forms are alkali–silica reaction and alkali–carbonate reaction. Both are caused by a reaction between chemical ions in the alkaline cement solution in concrete and reactive forms of aggregate.

Abrasion: Wearing away of surfaces by friction.

Abutment: Foundation/retaining structure at the approach and departure ends of a bridge.

ACA (Ammoniacal Copper Arsenate): A wood preservative used for timber treatment.

ACC (Acid Copper Chromate): Wood preservative used for timber treatment.

ACR (Alkali–Carbonate Reaction): A reaction that occurs between some of the dolomitic limestone aggregates and alkalis in the cement; the reaction forms a gel, which swells when sufficient moisture is present, causing cracking around the reacting aggregates.

ACZA (Ammoniacal Copper Zinc Arsenate): A wood preservative used for timber treatment.

AHJ (Authority Having Jurisdiction): Acronym for governmental or regulatory agency having jurisdiction over any new development or maintenance/repair to an existing structure.

Aluminum: A structural lightweight metal used in ships and docks and a metal used in sheathing; may be used as a sacrificial anode in cathodic protection against corrosion.

Amphipod: A smaller division (order) of the larger group (class) of invertebrates known as Crustacea. Chelura is an example.

Amphoteric: A metal that is susceptible to corrosion in both acid and alkaline environments. Aluminum is an example of an amphoteric metal.

Analysis: Involves the application of scientific/mathematical analytic principles and processes to determine the properties and condition of a member or structure.

Anode: Positive electrode of an electrolytic cell.

ASR (Alkali–Silica Reaction): A potentially expansive chemical reaction between siliceous aggregate and the hydroxide ions associated with the ions of sodium and potassium in solution in the paste. The siliceous rocks involved in this reaction are those with an imperfect crystal structure or those that are not crystalline, which can be highly deleterious if the forces generated by the expanding silica gels exceed the cohesive forces of the placement.

Assessment: Evaluation of the inspection results to determine the significance of observed damage and deterioration on the design capacity of the structure.

ASTM (American Society for Testing and Materials): A globally recognized leader in the development and delivery of international voluntary consensus standards.

Atmospheric Pressure: Normal pressure of air at sea level, 14.7 pounds per square in. (101.4 kPA).

AWPA (American Wood Preservers Association): A nonprofit organization that is responsible for promulgating voluntary wood preservation standards.

Bankia: Genus of molluscan marine borers.

Bark: Outside layer of a tree, composed of living, inner bark called phloem and an outer bark of dead tissue.

Barnacle: Encrusting fouling organism belonging to the large general group (class) Crustacea.

Batter Pile: Pile driven at an angle such that the pile can develop both axial and lateral load resistance.

Beams and Stringers: Lumber of a rectangular cross section, 5 in. (127 mm) or more thick and 8 in. (203 mm) or more wide, graded with respect to its strength when loaded on the narrow face.

Bearing Piles: Those piles in a structure that support the load.

Bench Capping: Method of replacing damaged piles at higher elevations when more than one pile in line is to be repaired.

Bent: Transverse row of piles fastened together by a pile cap.

Biological Deterioration: Deterioration or damage caused by living organisms.

Biofouling or Biological Fouling: The accumulation of microorganisms, plants, algae, or animals on wetted surfaces.

Bitt: Vertical post, usually one of a pair, used to secure ropes or cables.

Bitumastic: Coating made from higher boiling point materials found in tar.

Bleeding: Exudation of liquid preservative from treated wood, the exudate may evaporate, remain liquid, or harden into a semisolid or solid state.

Bollard: Cast steel cylindrical capped head extending up from a base plate for fastening ships to piers.

Borers, Marine: Marine organisms that attack wood in the submerged portions of structures placed in salt or brackish waters. Two general groups of borers are recognized, the Crustacea and the Mollusca.

Boring: Sample taken from wood for detection of deterioration or preservative penetration; the movement of certain organisms through wood.

Bracing: Wood or supports supplying additional strength to a structure.

Brackish Water: Water that is partly salt and partly fresh.

Branding: Permanent marking on a treated wood product to identify the supplier and date of treatment; other information may be included in a brand when so specified.

Breakwater: Structure used to eliminate or reduce the magnitude of wave impinging on the shoreline or vessels behind it.

Breasting Dolphin: Free-standing independent structure that a vessel will bear against when current, wind, or berthing motion moves the ship into the pier or wharf; breasting dolphins are typically equipped with energy-absorbing fender systems and are pile-supported or solid-filled structures.

Brinell Hardness Test: Oldest of the hardness test methods in common use today, the Brinell test is frequently used to determine the hardness of forgings and castings that have a grain structure too course for Rockwell or Vickers testing. Therefore, Brinell tests are frequently done on large parts. By varying the test force and ball size, nearly all metals can be tested using a Brinell test. Brinell values are considered test-force-independent as long as the ball size/test force relationship is the same.

Brown Rot: Deterioration caused by a group of fungi producing a brown residue or powder.

Bulkhead: Structure used in waterfront construction to retain earth fill.

Bullrail: Wide low curb along the outboard edge of the pier or wharf, it may be cast-in-place concrete, steel or timber; fixed or removable; mooring hardware is often mounted on top of the bull rail.

Burrow: Tunnel or excavation made by marine borers.

Caliper: Compass or divider with curved legs for measuring the diameter of pipes, rods, or piles.

Camel: Floating device acting as a fender and used to separate a moored vessel from a pier, wharf, quay, or other vessel, camels are used with ships that have hull configurations that do not match well with typical pier or wharf fender systems, such as submarines or where vessels require an offset from the pier or wharf because of deck or superstructure overhangs, such as an aircraft carrier.

Cap Log: Timber member connecting and protecting the heads of piles; usually not a structural member.

Capstan: Vertical axled rotating machine used to apply force to ropes, cables, and hawsers. The principle is similar to that of the windlass, which has a horizontal axle.

Carbonation: Occurs when the calcium in concrete is attacked by carbon dioxide of the air and converted to calcium carbonate.

Catenary: Curve that an idealized hanging chain or cable assumes under its own weight when supported only at its ends.

Cathode: Negative electrode of an electrolytic cell.

CCA (Chromated Copper Arsenate): Wood preservative used for timber treatment since the mid-1930s.

Chafing: Abrasion caused by material rubbing against a structure.

Charpy Impact Test: Also known as the Charpy V-notch test, a standardized high strain-rate test that determines the amount of energy absorbed by a material during fracture. This absorbed energy is a measure of a given material's notch toughness and acts as a tool to study temperature-dependent ductile-brittle transition.

Check: Separation along the grain of the wood, the separation occurring across the annual rings.

Chelura: Genus of Crustacea borers.

Chemical Damage: Deterioration of structural members caused by the effect of chemical reactions.

Chock: Piece of wood fitted between two piles to prevent the piles from rolling upon impact.

Clamped: Method of fastening, where one member is sandwiched between two other members.

Clear Water Box: Transparent container filled with clean water to provide a transparent path for viewing or photographing objects submerged in dark or dirty water.

Cleat: Wood or metal fitting usually with two projecting horns around which a rope may be made fast; a piece fastened to or projecting from something and serving as a support or check.

Coal Tar Derivative: Preservative obtained from the distillation of coal tar.

Coastal Waters: Seawaters bordering the continents subject to tidal flow.

Coatings: Protective covers to prevent corrosion.

Cofferdam: Temporary enclosure built within, or in pairs across, a body of water and constructed to allow the enclosed area to be pumped out, creating a dry work environment for the major work to proceed.

Cold Iron: Describes the condition of a ship when all shipboard boilers, engines, and generators are inoperative during repairs or due to intentional shutdown and can furnish none of the ship's required services.

Composite: Structural member or members made up of disparate materials.

Concrete Forms: Usually temporary wood or steel structures constructed to retain wet concrete until the concrete sets.

Conditioning: Heating or removal of moisture from unseasoned or partially seasoned wood as a preliminary to preservative treatment and as a means of improving the penetrability and absorptive properties of the wood.

Connectors, Timber: Devices, such as metal rings and plate and wood discs that, when embedded in each member, increase the efficiency of a timber joint.

Coping: Top course of stone or concrete to tie a structure together or to distribute the pressure from exterior loading.

Copper–Copper Sulfate: Reference cell for electrolyte potential measurements in seawater.

Copper Naphthenate: Toxic chemical preservative particularly effective against insects and destructive fungi.

Core: Cylinder of wood, removed by means of an increment borer, from which may be determined, by linear measurement, sapwood thickness and preservative penetration, and, by assay, preservative retention and distribution.

Corrosion: Destruction of a metal by a chemical or electrochemical reaction with its environment.

CP (Cathodic Protection): Using anodes or direct current systems to minimize or stop the corrosion process by establishing the steel as a cathode of an electrochemical cell.

Crack: Split or separation of material.

Creosote, Coal Tar: Distillate derived from coal tar. As used in the wood-preserving industry, creosote denotes a distillate of coal tar produced by the high-temperature carbonization of bituminous coal. Creosote consists principally of liquid and solid aromatic hydrocarbons and contains some tar acids and tar bases; it is heavier than water and has a continuous boiling range beginning at about 200°C.

Creosote–Coal Tar Solution: Solution of coal tar and creosote in selected proportions; usually contains 20% to 50% coal tar.

Creosote, Marine Grade: Coal tar creosote meeting special requirements as specified for the treatment of materials for marine use.

Crevice Corrosion: Corrosion of a metal at an area where contact is made with a nonmetallic material.

Crib: Cellular framework of timber, concrete, or steel filled with ballast.

Crustacea: Large group (class) of invertebrate animals. Barnacles, Limnoria, Sphaeroma, and Chelura are examples.

Cyclopean Masonry: Type of stonework built with huge boulders, roughly fitted together with minimal clearance between adjacent stones and no use of mortar.

DEF (Delayed Ettringite Formation): Type of internal sulfate attack in concrete, which is common in many precast concrete elements that have been heat-treated beyond a certain temperature and have suppressed the normal ettringite formation, or, in a concrete that is made using a high-sulfate portland cement. Instead of normal formation of ettringite by cement hydration in the plastic state, in these concretes, ettringite forms after the hardening of the concrete. In the continued presence of moisture, components of ettringite (i.e., Ca, Al, S) slowly dissolve out and form ettringite in the confined spaces in hardened paste and thereby cause expansion, subsequent separations around the aggregate particles, stresses in the paste due to restrained expansion, and the resultant eventual cracking of concrete.

Deadman: Buried plate, wall, or block attached at some distance from and forming an anchorage for a retaining wall.

Decay: Disintegration of wood substance because of the action of wood-destroying fungi.

Depassivate: Remove the ability of steel to resist corrosion.

Design Life: Theoretical life span estimated during the planning and design stages of a structure.

DGPS (Differential Global Positioning System): Enhancement to the global positioning system that provides improved location accuracy, from the 15 m nominal GPS accuracy to about 10 cm in the best case implementations.

Dimension Stock: Square of flat wood, usually in pieces smaller than the minimum sizes admitted by standard lumber grades, which is rough, dressed, green, or dry, and cut to the approximate dimension required for the various products of woodworking factories.

Dip Treatment: Total submergence of a structural member in the preservative.

Disbonding: Coating separating from the protected structure.

Discharge Structure: Outfalls at the ends of pipes, where the material flowing in the pipes enters the ocean or lake. Usually concrete or masonry but can be steel or wood.

DO (Dissolved Oxygen Testing): Testing procedure to determine the oxygen content of a body of water in general, or within the interstitial space between a pile and the protective membrane installed as a preventative maintenance technique.

Dolphin: Free-standing, pile-supported or solid-filled structure used for mooring and berthing vessels, protection of the end of piers or wharves, turning ships, or protection of the bridge structure.

Dote: "Dote," "doze," and "rot" are synonymous with decay.

Douglas Fir, Interior: Douglas fir growing east of the summit of the Cascade Mountains; sometimes referred to as *intermountain Douglas fir*. Interior Douglas fir growing in Oregon, Washington, Idaho, Wyoming, and Montana is designated *Douglas fir interior north*.

Douglas Fir, Pacific Coast: Douglas fir growing between the Pacific Ocean and the summit of the Cascade Mountains; sometimes referred to as *coastal Douglas fir*.

Drydock: Specialized facility used for repair of ships, where the vessel is removed from the water or placed within a lock and the water is removed, leaving the ship in the dry to facilitate repairs.

Dry Rot: Term loosely applied to many types of decay but especially to that which permits the wood to be crushed easily to a dry powder when in an advanced stage. The term is actually a misnomer for any decay, because all fungi require considerable moisture for growth, and the wood must have been moist at the time the rot occurred.

DTA (Differential Thermal Analysis): Thermoanalytic technique, similar to differential scanning calorimetry. In DTA, the material under study and an inert reference are made to undergo identical thermal cycles, while recording any temperature difference between sample and reference. This differential temperature is then plotted against time or against temperature (DTA curve or thermogram). Changes in the sample, either exothermic or endothermic, can be detected relative to the inert reference.

Durability: As applied to wood, its lasting qualities or permanence in service with reference to its resistance to decay and other forms of deterioration. Decay resistance is a somewhat more specific term indicating resistance to attack by wood-destroying fungi under conditions favorable to their growth.

Ebb: To recede from the flood, falling tide.

Efflorescence: Surface encrustation caused by the evaporation of solutions seeping out onto the surface of rock or concrete.

Electrochemical: Phenomenon where chemical change occurs through the indirect exchange of electrons.

Electrolyte: Chemical substance or mixture, usually liquid, containing ions that migrate in an electric field.

Embed: Place or fix firmly in surrounding matter; also imbed.

Embrittlement: Causing a loss of ductility, such as hydrogen embrittlement, where the introduction of hydrogen into metal reduces its ductility.

Empty Cell Process: Method of pressure-treating wood without using a preliminary vacuum.

Estuary: Area connecting a harbor or open sea and a freshwater river.

Eurocode: Set of harmonized technical rules developed by the European Committee for Standardization for the structural design of construction works in the European Union.

Evaluation: Statement of value; a spoken or written statement of the value, quality, importance, extent, or condition of something.

Fenders: Energy-absorbing devices used on the face of a pier, wharf, or dolphin to protect the ship and shore facility from damage caused by contact between the two during berthing and mooring.

Fir: Species of wood used in waterfront structures. *See* Douglas fir.

Fishplate: Timber or steel member used to stiffen and tie posts to caps and bench cap or pile.

Fleet Mooring: Offshore anchoring system placed in a fixed location.

Flood: Rising tide.

Flow: Direction of movement.

Fluke: Barbed shaped part of an anchor that digs into the bottom.

Fouling: Organisms growing on the surfaces of submerged materials.

FRP (Fiber-Reinforced Polymer): Composite materials made of a polymer matrix reinforced with fibers. The fibers are usually fiberglass, carbon, or aramid, whereas the polymer is usually an epoxy, vinylester, or polyester thermosetting plastic. FRPs are commonly used in the aerospace, automotive, marine, and construction industries.

Freeboard: Height of a structure or deck above the water level.

Full Cell Process: Method of pressure-treating wood using initial vacuum.

Fungus: Primitive plant in the group that includes molds, mushrooms, and others.

Gabions: Wire container filled with stones used for retaining structures.

Galvanic: Pertaining to the electrochemical interaction between two metals such as in galvanic couple or galvanic series.

Galvanic Anode: Metal that, because of its relative position in the galvanic series, provides sacrificial protection to metal or metals that are more noble in the series, when coupled in an electrolyte. Anodes, such as zinc, magnesium, or aluminum, are the current source in one type of cathodic protection.

Galvanic Corrosion: Electrochemical process in which one metal corrodes preferentially when in electrical contact with a different type of metal and both metals are immersed in an electrolyte.

Girder: Large-sized beam used as a main structural member, normally for the support of other beams.

Global Structural Stability Evaluation: Term that emphasizes the need to include the consideration of a number of factors when developing design or repair design options for a waterfront structure, including but not limited to the stability of the structure and the stability of the soils.

GPR (Ground-Penetrating Radar): Geophysical method that uses radar pulses to image the subsurface. This nondestructive method uses electromagnetic radiation in the microwave band (UHF/VHF frequencies) of the radio spectrum and detects the reflected signals from subsurface structures. GPR can be used in several media, including rock, soil, ice, freshwater, pavements, and structures. It can detect objects, changes in material, and voids and cracks.

Grade: Any of the quality classes into which lumber products are segregated.

Grade Mark: Identification of lumber with symbols or lettering to certify its quality of grade, which is based on the presence or absence of defects such as knots, checks, or decay.

Grain: Direction, size arrangement, appearance, or quality of the fibers in wood.

Grain, Close: Wood with narrow and inconspicuous annual rings. The term is sometimes used to designate wood having small and closely spaced pores, but in this sense, the term *fine texture* is more often used.

Grain, Coarse: Wood with wide and conspicuous annual rings in which there is considerable difference between springwood and summerwood. The term is also used to designate wood with large pores, but in this sense, the term *coarse textured* is more often used.

Grain, Cross: Wood in which the cells or fibers do not run parallel with the axis or sides of a piece.

Grain, Diagonal: Wood in which the annual rings are at an angle with the axis of a piece as a result of sawing at an angle to the axis of the tree.

Grain, Edge: Wood in which the rings (so-called grain) form an angle of 45 degrees or more with the surface of the piece; also called *vertical grain* and *quarter sawn*.

Grain, Flat: Wood in which the rings form an angle of less than 45 degrees with the surface of the piece; also *plain sawn*.

Grain, Interlocking: Wood in which the fibers are inclined in one direction in a number of rings of annual growth, then gradually reverse and are inclined in an opposite direction in succeeding growth rings, then reverse again.

Grain, Open: Common classification of painters for wood with large pores; also called *coarse textured*.

Grain, Plain-Sawn: Another term for flat grain, used usually in hardwoods.

Grain, Quarter-Sawn: Another term for edge grain, used usually in hardwoods.

Grain, Spiral: Type of growth in which the fibers take a spiral course about the bole of a tree instead of the normal vertical course. The spiral may extend right-handed or left-handed around the trunk.

Grain, Vertical: Another term for edge grain.

Grain, Wavy: Wood in which the fibers collectively take the form of waves or undulations.

Green: Unseasoned, wet.

Greenheart: Tropical wood used in marine construction, resistant to marine borer attack.

Greensalt: Chromated copper arsenate (CCA).

Gribble: Common name for the crustacean borer, Limnoria.

Groin: Structure consisting of large rocks, precast concrete units, reinforced or prestressed concrete piles, steel sheet piles, or timber cribbing filled with rock, projecting outward perpendicular to the shoreline designed to control the rate of shifting sand.

Grout: Cement–sand mortar of plastic consistency that can easily be poured.

Hardwoods: Botanical group of trees that are broadleaved. The term has no reference to the actual hardness of the wood.

Heartwood: Inner core of a woody stem, extending from the center to sapwood and usually of darker color.

Heavy Timber Construction: Construction composed of planks or laminated floors supported by beams or girders.

Helical Anchors: Screwed-in steel anchoring systems used for columns, tie backs, marina anchoring systems, and pipeline tie downs.

Honeycombs: Voids or hollows in the concrete.

Hotel Services: Dockside utilities provided for a ship at a berth (also called ship's services, utility services, and cold iron services).

Hydrology: Study of the movement, distribution, and quality of water throughout the Earth.

Imbed: *See* Embed.

Impressed Current: Cathodic protection system that uses an external electrical alternating current to provide corrosion protection to a steel structure.

IMU (Inertial Measurement Unit): Electronic device that measures and reports on a craft's velocity, orientation, and gravitational forces, using a combination of accelerometers and gyroscopes, sometimes also magnetometers.

Increment Borer: Auger-like instrument with a hollow bit used to extract cores of wood from piling.

Infilling: Somewhat slow replacement of scoured-out material with softer, finer silt.

Inland: That part of the land above the waterline (shoreline).

Inspection (of Construction Materials): Scrutiny and supervision of the purchasing, acquiring, manufacturing, treating, and handling of material for compliance with specification requirements.

Inspection (of Structures): Method by which structures are examined for determination of the presence and extent of deterioration.

Intake Structure: At the opposite end of the conduit from the discharge end; can be of the same materials as the discharge.

Interlock: Connection or joint between two adjacent steel sheet piles.

Intertidal: Area between mean low water and mean high water.

Investigation: To inspect carefully, systematically, and thoroughly a complex or hidden structure and to evaluate or set the value of the structure and/or repairs needed.

Ion: Electrically charged atom or group of atoms.

Isopod: Smaller division (order) of the larger group (class) of invertebrates known as Crustacea, which includes Limnoria and Sphaeroma.

Jetty: Dock or breakwater that projects into the water to prevent the formation of sandbars, normally located near harbor entrances and river estuaries.

Jewelry: Hardware fittings on a buoy to which the mooring lines/chains are attached.

Joists and Planks: Lumber of a rectangular cross section, from 2 in. up to but not including 5 in. thick, and 4 in. or more wide, graded with respect to its strength in bending when loaded either on the narrow face (joist) or on the wide face (plank). If 5 in. or more thick, the lumber is known as beams and stringers.

Kenter Joining Link: Constructed in three parts: two half links and a stud. The stud slides in place and locks the whole link. The stud is secured by hammering a tapered pin into the hole drilled diagonally through all three parts of the joining link.

Key: Wedge in rock used as an anchor.

Kiln-Dried: Lumber or other materials that have been dried in drying kilns to a moisture content, usually below that obtained in air-drying, by the application of artificially supplied controlled heat, humidity, and air circulation.

Laitance: Soft, punky weak layer of cement and aggregate fines on a concrete surface that is usually caused by an overwet mixture, overworking the mixture, improper excessive finishing, combination thereof, or on concrete during tremie pours. Often found in layers at the face of concrete piers/abutments.

Laminate: Single layer of wood or plastic in an assembly of layers.

Leaching: Process of removing a soluble substance from a heterogeneous material by means of a solvent (usually water).

Lead Wool: Lead that is spun to form a wool-like material and pounded into masonry joints that are still found in older structures.

LiDAR: Detection system that works on the principle of radar but uses light from a laser. Light detecting and ranging is a method for determining ranges or distances by measuring the time of travel from a light-emitting laser to bounce back from a target surface.

Lighterage: Small craft designed to transport cargo or personnel from ship to shore; includes amphibians, landing craft, discharge lighters, causeways, and barges.

Limnoria: Common "gribble," a genus of Crustacea borers causing serious destruction to marine structures.

Locks: Enclosed chamber in a canal, dam, etc. with gates at each end, for raising or lowering vessels from one level to another by admitting or releasing water.

Magnesium: Metal that may be used as a sacrificial anode in cathodic protection against corrosion.

Marine Borers: *See* Borers.

Marine Organisms: Living entities normally found in natural waters containing measurable salinity.

Masonry: Structure built with stones, bricks, and other materials, usually held together with mortar.

Mediterranean Mooring: A mooring in which a ship's bow is moored to buoys and the stern is tied to the pier.

MHW (Mean High Water): Tidal datum: the average of all the high water heights observed over the National Tidal Datum Epoch.

MHHW (Mean Higher High Water): Tidal datum: the average of the higher high water height of each tidal day observed over the National Tidal Datum Epoch.

MIC (Microbial-Induced Corrosion): Caused by the presence of microbes whose metabolism produces acids and sulfides; microbes and their byproducts in the presence of metals can produce a film conducive to accelerated local corrosion.

Mill Scale: Heavy oxide layer resulting from hot fabrication or heat treatment of metals.

MLW (Mean Low Water): Average of all the low water heights observed over the National Tidal Datum Epoch.

MLLW (Mean Lower Low Water): Tidal datum: the average of the lower low water height of each tidal day observed over the National Tidal Datum Epoch.

Moisture Content: As related to wood, the weight of water contained in wood, usually expressed as a percentage of the oven-dry weight of the wood.

Moisture, Free: Moisture that is held inside the cell cavities of wood in contrast to that within the cell walls.

Mollusca: One of the 11 main divisions (phylla) used in animal classification that includes several of the destructive marine borers.

Monopile Dolphin: Single pile dolphin usually consisting of a large-diameter concrete or steel pipe pile filled with concrete, used as a mooring or breasting dolphin.

Mooring Dolphin: Freestanding pile-supported or solid-filled structure used for mooring vessels, usually placed at the bow or stern of a moored ship to provide mooring points to attach breasting lines, bow lines, and stern lines.

MPI (Magnetic Particle Inspection): Nondestructive testing procedure that magnetizes a component to produce lines of magnetic flux within the component. The process is used to detect cracks or other defects in steel components.

MSL (Mean Sea Level): Arithmetic mean of hourly heights observed over the National Tidal Datum Epoch.

Multibeam: Device typically used by hydrographic surveyors to determine the depth of water and the nature of the sea bed. Most modern systems work by transmitting a broad acoustic fan-shaped pulse from a specially designed transducer across the full swath acrosstrack with a narrow alongtrack, then forming multiple receive beams (beamforming) that are much narrower in the acrosstrack (around 1 degree depending on the system). From this narrow beam, a two-way travel time of the acoustic pulse is then established utilizing a bottom detection algorithm. If the speed of sound in water is known for the full water column profile, the depth and position of the return signal can be determined from the receive angle and the two-way travel time.

Mudline: Point of intersection of the seawater and the bottom soil.

Mushroomed: Head of a pile that has been subjected to excessive axial load and/or deterioration such that the timber fibers have separated, causing the pile head to flatten, resulting in reduced load capacity.

Mussels: Molluscan fouling organisms.

Nacerda: Beetle that can cause damage to the superstructure of wharves.

Nail, Dating: Nail with a date or symbol on its head that is driven into wood to indicate the year of treatment or the date of installation.

NDT (Nondestructive Testing): Testing such as ultrasonic thickness measurement of steel.

Nearshore: Area of the water next to the shoreline.

Neat Cement: Cement mortar without added sand.

NGVD (National Geodetic Vertical Datum): Reference mean sea level elevation measured in 1929.

Nominal Dimension: Dimension of lumber corresponding approximately to the size before dressing to actual size and used for convenience in defining size and in computing quantities.

Nonbearing Pile: Pile that is not connected to a pile cap such that the pile is not carrying any axial load.

NTDE (National Tidal Datum Epoch): The specific 19-year period adopted by the National Ocean Service as the official time segment over which tide observations are taken and reduced to obtain mean values (e.g., Â mean lower low water, etc.) for tidal data. It is necessary for standardization because of periodic and apparent secular trends in sea level. The present NTDE is 1983 through 2001 and is actively considered for revision every 20 to 25 years. Tidal data in certain regions with anomolous sea level changes (Alaska, Gulf of Mexico) are calculated on a Modified 5-Year Epoch.

Oak: Species of hardwood used in waterfront structures.

Ogee Washer: Special cast washer used in waterfront construction to distribute the load from a bolt and nut to a timber face to prevent the wood from being crushed as the nut is tightened.

Oil Borne: Chemical capable of dissolving in an oil solvent.

Opposite Hand: Mirror image.

Pachometer: Device designed to specifically locate reinforcing steel in concrete and to assist in the determination of the size of the hidden reinforcing steel.

Passivation: In physical chemistry and engineering, refers to a material becoming "passive," that is, being less affected by environmental factors such as air or water. It involves a shielding outer layer of corrosion that can be applied as a microcoating or found occurring spontaneously in nature.

PCBs (Polychlorinated Biphenyls): Carcinogenic chemical compound formerly used as insulating and cooling liquids in electrical equipment often found in sediments where waste products from industry have been released into the water. A PCB count is now required, prior to all dredging operations.

PCF (Pounds per Cubic Foot): Way of expressing a material's density, weight divided by volume.

PDT (Partially Destructive Testing): Tests carried out to the specimen's failure to understand a specimen's structural performance or material behavior under different loads. These tests are usually much easier to carry out, yield more information, and are easier to interpret than nondestructive testing. Examples of specimens may be concrete cores or steel coupons taken from existing structures.

Penetration: Depth to which preservative enters the wood.

Penetration Resistance Testing: Penetration resistance test is conducted on concrete structures using the Windsor Probe test machine. In this test method, a steel probe is fired on the concrete surface by a sudden explosion. The penetration is inversely proportional to the strength of concrete. The result of the test is influenced by the aggregate strength and nature of formed surfaces of concrete.

Penstock: Enclosed pipe that delivers water to hydraulic turbines or sewerage systems.

Penta Preservative: Wood-preserving solution made of pentachlorophenol, C6CL5OH, dissolved in hydrocarbon solvent. Often, an auxiliary solvent is added to increase the solubility of the penta.

Pentachlorophenol: Toxic chemical preservative particularly effective against destructive fungi.

Petrographic Analysis: For concrete, this analysis is used to determine the quality of the aggregate, concrete paste, bonding, and so on.

Pfisteria: Skin and nervous system disease that may be passed to human beings by infected fish. Symptoms in fish are massive fish kills and ulcerated lesions in the fish.

Pholad: Molluscan marine borer.

Photogrammetry: Science of making measurements from photographs.

Pier: Structure that projects from the shore, oriented perpendicular, or at an angle to the shore.

Pile Adrift: Old or structurally inadequate pile that has been abandoned for load-carrying capacity and is usually not connected to other members in the structure.

Pile Bent: *See* Bent.

Pile Cap: Beam member connecting pile heads and through which deck loads are transmitted to the piles.

Pile, Marine: Pile that is partly embedded in bottom soil and partly exposed to salt seawater.

Pile Top Cap: Any cover fastened over the cut surface of a pile to prevent exposure to the atmosphere.

Pitting: Form of extremely localized corrosion that leads to the creation of small pits and/or holes in the metal.

Plug, Tie: Wood plug used for filling an old spike hole. The plug is usually treated with preservative.

Plumbness: Quality or state of being plumb or vertical.

POL (Petroleum, Oil, Lubricant): Broad term used to describe all petroleum-containing products.

Pointing: Filling joints or defects in the face of a masonry structure.

Polyethylene: Plastic material used for sheathing.

Polymer Composite Materials: a material composed of a polymer matrix reinforced with fibers, particles, or other fillers to enhance its mechanical, thermal, and chemical properties. The polymer matrix can be either a thermoset (e.g., epoxy, polyester) or a thermoplastic (e.g., polypropylene, polyvinyl chloride), while common reinforcements include glass fibers, carbon fibers, Kevlar, or natural fibers. Polymer composites can be used as primary structural elements or for repairs of structures, such as for jackets around damaged piles.

Pop-Outs: Shallow, cone-shaped holes in the surface of the concrete.

Posting: Method of replacing damaged sections of piles at higher elevations.

Posts and Timbers: Lumber of square or approximately square cross section, 5 × 5 and larger, graded primarily for use as posts or columns carrying axial load.

Preservative: Chemical compound that creates a protective mechanism against destructive organisms.

Preservative, Oil-Borne: Wood preservative that is introduced into wood in the form of a solution in oil.

Preservative, Oil Type: Preservatives such as creosote, creosote-coal tar solutions, creosote-petroleum solutions, and oil-borne preservatives, or other preservatives strictly of an oily nature that are usually insoluble in water.

Preservative, Waterborne: Wood preservative that is introduced into wood in the form of a solution in water.

Preservative, Water Repellent: Solution of one or more chemicals and water-repellent materials that preserves the wood and retards changes in moisture content and the accompanying changes in dimensions.

Preservative, Wood: Term *preservative* is intended to include such chemicals or combinations thereof that will protect wood against deterioration from any one or combination of the following: decay, insects, marine borers, fire, weathering, absorption of water, and chemical action.

Pressure Treatment: Impregnation of wood with a preservative applied under pressure.

Pretreatment Seasoning: Removal of water from the wood before treatment to make the entrance and retention of the preservative possible.

Probing: Penetrating through the surface with a probe to detect deterioration.

PSI (Pounds per Square Inch): Unit of pressure.

PSIA (Pounds per Square Inch Absolute): Pressure measured relative to a vacuum.

PSIG (Pounds per Square Inch Gauge): Unit of pressure relative to the surrounding atmosphere.

Pulse Velocity/Pulse Echo Testing: Use of ultrasonic signals for measuring distances/thickness.

Punky: Area of decay in a timber member exhibiting a soft spongy appearance; structural integrity is lost in these areas.

Punt: Small flat-bottomed boat with square ends.

PVC (Polyvinyl chloride): Plastic material used for sheathing.

Quaywall: Retaining structure used to contain earth or stone behind a wharf.

Rebar: Reinforcing steel used in concrete such that the concrete and reinforcement act together in resisting forces.

Rectifier: Component of an impressed current cathodic protection system that converts the external alternating current to direct current to minimize or stop the corrosion process.

Relieved Edges: Exposed edges of a timber plank cut on an angle to reduce edge splintering (i.e., chamfering).

Repair: Restoring or replacing to a sound or good condition after damage.

Resilient: Capable of withstanding shock without permanent damage.

Resistograph Bit: Small-diameter needle bit that bores into the timber and measures and records a quantifiable resistance value, which is plotted either digitally or on an analog display showing how much resistance the wood is providing versus the distance traveled.

Retaining Wall: Wall for sustaining the pressure of earth or fill deposited behind it.

Retention: Per unit volume the quantity of preservative in the wood.

Revetment: Sloping structures placed on banks or cliffs in such a way as to absorb the energy of incoming water.

Riprap: Rough stone of various sizes placed compactly or irregularly to prevent scour by water.

Rubble: Field stone or rough stone as it comes from the quarry. When it is of a large or massive size, it is termed *block rubble.*

Salinity: Amount of total salt content in proportion to a unit volume of water.

Sapwood: Outer light-colored wood of the tree stem.

Scab: Wood member used in posting to provide a positive connection between the post and the pile and/or the pile cap.

Scaling: Gradual and constant loss of surface mortar and aggregates from an area of concrete.

Schmidt Hammer: Mechanical device utilizing a standard hammer for testing the condition of a concrete surface, also known as Rebound Hammer Testing.

Scour: Condition where bottom material has been washed away from a pile or structure that penetrates the bottom.

Seasoned, Air-Dried, or Air-Seasoned: Dried by exposure to the atmosphere, usually in a yard, without artificial heat.

Seasoned, Kiln-Dried: Dried in a kiln with the use of artificial heat.

Seasoning: Evaporation or extraction of moisture from green or partially dried wood.

Seawall: Coastal structure built along the shoreline to protect coastal areas caused by waves and flooding by heavy waves.

Service Life: Actual life span of the structure.

Shake: Separation along the grain of the wood; the separation usually occurring between the annual rings.

Shank: Long, straight part of an anchor, to which the anchor line attaches at one end and the fluke(s) at the other end.

Sheathing: Exposed face material used in bulkhead construction.

Sheeting: Lining of planks or boards for supporting an embankment, usually placed vertically and supported by walers, braces, and piles.

Shim: Small piece of wood placed between two members of a structure to bring them to a desired relative position.

Shiplapped Lumber: Lumber that is edge-dressed to make a lapped joint.

Side-Scan Sonar: Category of a sonar system that is used to efficiently create an image of large areas of the sea floor. It may be used to conduct surveys for maritime archaeology; in conjunction with seafloor samples, it can provide an understanding of the differences in material and texture type of the sea bed. Side-scan sonar imagery is also a commonly used tool to detect debris items and other obstructions on the sea floor that may be hazardous to shipping or to sea floor installations by the oil and gas industry.

Silver–Silver Chloride: Reference cell for electrolyte potential measurements in seawater.

Skiff: Small boat.

Slip: Space between two approximately parallel piers or the space formed by a cut into the land that provides two approximately parallel mooring faces.

Snooper: Articulating man lift used by inspectors to access under deck or high overhead locations.

Soft Rot: Deterioration of wood components—often without visual distortion or apparent damage to the wood—by certain molds and other fungi that are outside of the common wood-destroying group. The affected wood is likely to be extremely brittle and break without splinters.

Softwoods: Botanical group of trees with needle-like or scale-like leaves, often referred to as *conifers*. The term *softwood* has no reference to the softness of the wood.

Soil Resistivity: Measure of how much the soil resists the flow of electricity.

Sonar, Multibeam: Sonar transmitted in a 180-degree pattern to record bottom topography, scour, or depth over a wide swath in a single pass.

Sonar and Side-Scan Sonar: Ultra-high-frequency sound wave–generating device for measuring distances by reference to time intervals between sending and receiving any pulse. Usual sonar transmits signals vertically, and side-scan transmits signals at an angle less than 90 degrees.

Sounding: Method used to determine interior deterioration in wood and concrete; a method used to determine the depth of water.

Spalling: Chipping or fragmentation of a concrete surface or surface coating.

Specific Gravity: As applied to preservatives, the ratio of weight of a given volume of a preservative to the weight of an equal volume of water.

Splicing: Replacement of the damaged portion of a pile.

Split: Lengthwise separation of the wood extending completely through the piece from one surface to another.

Spud Piles: Piles driven at an angle to develop horizontal resistance to loading.

Staining: Discoloration of wood indicating the presence of fungus activity.

Steel Sheet Pile Wall: Bulkhead composed of driven vertical or near vertical steel sheet sections interlocked to form a continuous wall, sometimes tied back to anchors.

Stringer: Horizontal timber member spanning between pile caps used to support decking.

Structural Deterioration: Failure or damage to a structure due to biological, chemical, or mechanical means.

Structural Lumber: Lumber that is 2 in. or more thick and 4 in. or more wide, intended for use where working stresses are required. The grading of structural lumber is based on both the strength of the piece and the use of the entire piece.

Sub-Decking: Area beneath the surface decking.

Sulfate Attack: Exposure of concrete to sulfates; the most common sulfates are sodium, potassium, magnesium, and calcium sulfates, which react with various hydration products in the presence of moisture to form ettringite, which leads to softening of the concrete.

Surface Treatment: Applying of a preservative to the surface by means of a brush, swab, or spray gun.

Synthetic Resin: Chemical sometimes used for impregnating piles.

Tar: Generic term applied to nonaqueous liquids obtained as residue in the destructive distillation of organic materials such as coal, lignite, petroleum, wood, and others.

Tar, Coal: Nonaqueous portion of the liquid distillate obtained during the carbonization of bituminous coal.

Tender: Individual responsible for the diver's welfare during an inspection; also a small boat.

Teredo: Genus of molluscan marine borers, commonly called the *shipworm*.

Teredo Tube: Tubular residue left by teredo borers.

Tidal: Of, relating to, caused by, or having tides.

Tidal Datum: Vertical reference based on a specific stage of tide that serves as a baseline elevation to which sounding depths or topographic heights are referenced.

Tide: Periodic rising and falling of the surface of the ocean and of water bodies.

Tie Back, Tie Rod: Usually, a tension rod with anchorage used to restrain a wall from movement or displacement.

Timber: Broad term including standing trees and certain products cut from them, including lumber 5 in. or larger in least nominal dimension.

Topography: Configuration of the physical features of a place or region.

Treatment, Dual: Treatment of wood to be used under severe conditions of exposure with two dissimilar synergistic preservatives in two separate treating cycles (e.g., a treatment of marine piles and wood for areas of extreme borer hazard). Usually, the first treatment is with a waterborne salt preservative and the second one with creosote or a creosote-coal tar solution.

Treatment, Empty-Cell: Treatment in which air imprisoned in the wood is employed to force out part of the preservative when treating pressure is released and a final vacuum is applied.

Treatment, Fire-Retardant: Treatment of wood under pressure with chemical to reduce its flame spread, fuel contribution, and smoke development.

Treatment, Full-Cell: Treatment involving a preliminary vacuum followed by pressure impregnation such that the cell cavities in the treated portion of the wood remain partially or completely filled with preservative.

Tremie: Method of placing concrete underwater by gravity or by pumping in which the concrete does not fall directly through the water but is placed below water level through a pipe, the lower end of which is kept immersed in fresh concrete so that the rising concrete from the bottom displaces the water without washing out the cement content.

Tsunami: Solitary wave caused by an underwater earthquake.

Tunicate: Sea grape: a semitransparent organism with the size of a grape that often exists in polluted waters. Some cause rashes on divers.

UFC (Unified Facilities Criteria): Documents that provide planning, design, construction, sustainment, restoration, and modernization criteria and apply to military departments, defense agencies, and Department of Defense field activities.

Ultrasonic Thickness Measurement: Measurement made from one side of a material using ultrasonic wave transmission and return to determine thickness.

USGS (United States Geological Survey): Science organization that provides impartial information on the health of ecosystems and the environment, natural hazards, natural resources, the impacts of climate and land-use change, and the core science systems that provide timely, relevant, and usable information.

Wale or Waler: Horizontal member, usually of wood, used for bracing the sheeting or trench, cofferdam, bulkhead, or similar structures.

Wane: Bark on the edge or corner of a piece, or the absence of wood in a piece from any cause.

Waterborne: Preservative soluble in water.

Wharf: Structure oriented approximately parallel to shore, where ships can be moored at the offshore face.

White Rot: Deterioration caused by a group of fungi that cause "bleaching" of the wood.

Windsor Probe: Device used to determine the strength of concrete by shooting a standardized probe into the concrete and measuring the depth of embedment.

Wolman Salts: Fluor Chrome Arsenate Phenol Type A.

Wood: Broad term including standing trees and certain products cut from them, including lumber 5 in. or larger in least nominal dimension.

Wood Preservation: Art of protecting wood against the action of destructive agents; usually refers to the treatment of wood with chemical substances (preservatives), which reduce its susceptibility to deterioration by fungi, insects, or marine borers.

Xylophaga: Genus of wood-boring pholads.

Zinc: Metal as sacrificial anodes in cathodic protection against corrosion.

INDEX

Page numbers followed by f refer to figures; page numbers followed by t refer to tables.

FSC
www.fsc.org
MIX
Papier | Fördert
gute Waldnutzung
FSC® C083411

Kolibri 360 GmbH
Ferdinand-Jühlke-Straße 7
99095 Erfurt, Deutschland
produktsicherheit@kolibri360.de